T0263840

Game Theory and Learning for Wireless Networks

Fundamentals and Applications

Game Theory and Learning for Wireless Networks

Fundamentals and Applications

Samson Lasaulce

Hamidou Tembine

AMSTERDAM • BOSTON • HEIDELBERG • LONDON
NEW YORK • OXFORD • PARIS • SAN DIEGO
SAN FRANCISCO • SINGAPORE • SYDNEY • TOKYO

Academic Press is an imprint of Elsevier

Academic Press is an imprint of Elsevier
The Boulevard, Langford Lane, Kidlington, Oxford, OX5 1GB
225 Wyman Street, Waltham, MA 02451, USA

First published 2011

Notices
Knowledge and best practice in this field are constantly changing. As new research and experience
broaden our understanding, changes in research methods, professional practices, or medical treatment
may become necessary.

Practitioners and researchers must always rely on their own experience and knowledge in evaluating and
using any information, methods, compounds, or experiments described herein. In using such information
or methods they should be mindful of their own safety and the safety of others, including parties for
whom they have a professional responsibility.

To the fullest extent of the law, neither the Publisher nor the authors, contributors, or editors, assume any
liability for any injury and/or damage to persons or property as a matter of products liability, negligence
or otherwise, or from any use or operation of any methods, products, instructions, or ideas contained in
the material herein.

British Library Cataloguing in Publication Data
A catalogue record for this book is available from the British Library

Library of Congress Number: 2011930747

ISBN: 978-0-12-384698-3

For information on all Academic Press publications
visit our website at *www.elsevierdirect.com*

Typeset by: diacriTech, India

Printed and bound by CPI Group (UK) Ltd, Croydon, CR0 4YY

Transferred to Digital Print 2011

Contents

PART IV   APPENDICES, REFERENCES, AND INDEX

Preface

GAME THEORY AND WIRELESS

In everyday life, it is common to be in situations where the outcome of a situation depends not only on what we do, but also on what other people do. This is clearly the case when participating in an auction on the Internet, voting at a presidential election, negotiating a price with a seller, trying to find a seat in a bus or in the metro, etc. All these situations are referred to as a "game" in game theory. Indeed, they do have something in common with what is usually known as a game (e.g., chess, football, monopoly, video games): several decision makers are involved, and there is interaction between them in the sense that their decisions are interdependent. This broad definition of a game explains why game theory, which is the formal study of interactions between several decision makers who can have conflicting or common interests, has become more and more important in many areas. Among these we find for example, economics, politics, biology, and computer science. Game theory has recently been applied to telecommunications. Indeed, it is now well known that game theory can be used to analyze interactions between entities such as telecoms regulators, operators, manufacturers, and customers; for instance, game theoreticians have been involved in designing radio spectrum auctions in the US and in Europe. More specifically, the spectrum for the third generation mobile system (3G) in Europe has been auctioned in the UK and Germany. Carriers across the continent have paid 110 billion euro for the rights to the radio spectrum reserved for 3G, showing the strategic importance of the radio spectrum in the telecommunications sector. In such a case, the players of the game are typically the governments and the carriers. Although this book is about applying game theory to wireless communications, the players in the games under study are not economic actors such as regulators or carriers, at least not explicitly. Rather, the players of the games under consideration are "machines" such as base stations, laptops, mobile phones, routers, servers, and so on. Obviously, the engineers who design these machines are directed by the economic policy of their company or institution, which indicates that identifying the players is not always obvious in wireless communications. Nonetheless, there is a difference between considering a telecoms operator or a mobile phone to be a player – the nature of the action or decision to be taken is different. For an operator involved in a spectrum auction, the action to be taken may be a price, whereas the possible actions available to a mobile phone may be to decide which access points or base stations to be connected to. Indeed, interaction between wireless terminals like mobile phones is naturally present in wireless networks, since interference often exists or/and common resources must be shared. This book therefore focuses on technical aspects of communication problems and not their economic aspects.

Game theory and communication problems seem to have been connected for the first time by Benoît Mandelbrot in his thesis (in 1952), but it is only during the past decade that game theory has experienced a strong surge of interest in the area of wireless communications. Wireless networks have evolved enormously during this time and new communication paradigms have appeared, making game theory especially relevant in their analysis and design. To emphasize this, let us consider a few examples.

- *Ad hoc networks.* In fixed communication networks, the paradigm of peer-to-peer communications has undergone a powerful surge of interest during the two past decades, in applications such as the Internet. Remarkably, this paradigm has also been found to be very useful for wireless networks, especially ad hoc networks which do not rely on a pre-existing infrastructure, such as routers in wired networks or access points in managed wireless networks. Typically, in such networks, a wireless node or terminal may have to decide whether or not to forward data or packets; that is, it must choose a routing policy.
- *Cognitive radio.* Spectrum congestion has become a more and more critical issue, so the Federal Communications Commission released an important report in 2002 (FCC, 2002), providing a legal framework for deploying networks with cognitive terminals. Roughly, the fundamental idea of cognitive radio is that a terminal is able to sense its environment (mainly radio resources) and react to it accordingly. In the context of cognitive radio, the decision of a terminal (a transmitter or a receiver) is essentially how to use the spectrum (e.g., transmit or not transmit on a given band, listen or not listen to users' activity in a given band).
- *Sensor networks.* Wireless sensor networks comprise several (possibly many) spatially distributed autonomous sensors. A typical objective of these networks is to monitor physical conditions, e.g., estimating a field of temperature or pressure in a distributed manner. For this purpose, communications between sensors may be used to reach a consensus on the estimated quantities or an efficient solution. Sensors have therefore to take decisions, in terms of estimates and what data to send to neighboring nodes.
- *Unlicensed bands.* In the scenario of unlicensed bands, the spectrum available to wireless terminals is not owned by an operator and is free to use, provided some emission levels are respected. Typically, there can be several point-to-point communications operating on the same band, at the same time, and in the same geographical area. Even if the terminals are not cognitive, the engineers who design terminals operating in unlicensed bands must deal with interfering signals especially if good performance is sought. In WiFi systems, channel selection can influence the quality perceived by the user.

To these examples, one might add more classical examples, such as medium access control and power control in multiuser channels. In all these examples, terminals are (quasi) autonomous in their decision making, and may share common resources like energy, power, routes, space, spectrum, and time, which implies potential interaction between them. Modeling the network as a set of independent point-to-point communications may be strongly suboptimal and even inappropriate. Game theory becomes especially relevant when one wants to assume that the environment of the wireless terminals is structured, otherwise classical optimization tools may suffice. The problem in interactive situations, which include several decision makers having a certain knowledge about the situation, is that the optimal decision or strategy is generally unclear, since no one player completely controls the final outcome. This means that the problem needs to be defined before it can be solved. Game theory is a means of proposing, designing interaction models, studying the conditions under which some outcomes can be reached, and designing good strategies. Game-theoretic approaches may lead to more efficient network states than the worst-case approach, where structural elements are ignored or not accounted for. Remarkably, considering a terminal as an intelligent entity which is capable of observing a structured environment and reacting sufficiently rapidly, has become an increasingly realistic assumption with the significant progress in signal processing (e.g., spectrum sensing algorithms and dramatic increase of admissible computational complexity). For a long time, game theory was used quite marginally and more like an analysis tool in communication problems. With recent technological progress and the arrival of new wireless paradigms, the era of using game theory for design has come. Multiuser information theory has inspired researchers and engineers, and led them to invent important concepts such as successive interference cancellation, cooperative diversity, and multiuser diversity. Although applied almost ten years ago to design algorithms such as the iterative water-filling algorithm, much work remains to design algorithms implementable in a real network and compatible with required information assumptions, behavior assumptions, convergence rate, quality of service, or admissible complexity. The right dynamic game models to study distributed wireless networks need to designed.

ABOUT THE CONTENTS OF THE BOOK

This book is dedicated to students, engineers, teachers, and researchers who know about wireless communications. In this respect, almost all the chosen examples and treated case studies concern wireless communication problems. The book comprises of 10 chapters and is divided in three parts: Part I, Part II, and Part III. Part I tries to provide, in a concise manner, a certain number of fundamental notions of game theory, such as mathematical models for a game and game-theoretic solution concepts. Part II is dedicated to games with incomplete information; that is, those in which players do not know all the relevant data regarding the game they are playing. This part is strongly based on the notions discussed in Part I. In particular, iterative algorithms are considered for learning good strategies, and the convergence analysis of these algorithms relies on game-theoretic notions (e.g., potential games and Nash equilibria). Part III comprises three detailed case studies, and shows how game theory has been applied in the wireless literature. It is based strongly on the ideas discussed in Parts I and II.

Moving to a higher resolution level of the description of the book contents, Part I comprises three chapters. In particular, the first chapter contains some motivating examples for using game theory, the main mathematical models for a game (called representations), and comments on various solution concepts for a game. The Nash equilibrium is one of the considered solution concepts. As the concept of Nash equilibrium is a building block of game theory, a complete chapter (Chapter 2) is dedicated to equilibrium analysis (existence, uniqueness, etc). Indeed, throughout the book, it should become clear to the reader that Nash equilibrium is essential to game theory since it both inspires new solution concepts and corresponds to the solution of many interactive situations; in particular, Nash equilibria may result from evolution or learning. In Chapter 3, some important models of dynamic, non-cooperative games are presented. By "non-cooperative" it is meant that the individual objectives of the players are clearly identifiable and are assumed to be selfish. Roughly, dynamic games correspond to general game models, including one-shot games, where the players can observe past moves or data and react to them.

The purpose of Part II is to show some ways of relaxing information and behavior assumptions for the players, which is important in wireless communications (typically only local observations are available, quantities are estimated, etc.). Chapter 4 is a very short chapter presenting Bayesian games, which considers how to take into account the fact that only partial knowledge is available to the players of a game, and shows how the classical machinery of equilibrium analysis in games with complete information can be adapted. Chapters 5 and 6 are devoted to learning; that is, how players learn their strategies. In such a framework, the game is assumed to be played several/many times, so that repeated interaction between players who are partially informed may converge to some steady points: players learn their strategies over time. Chapter 5 focuses on some learning techniques which are partially distributed; that is, a player has some knowledge about the others. Chapter 6 focuses on fully distributed learning techniques including reinforcement learning techniques. In this setting, players are modeled as automata who are therefore not necessarily rational, do not assume that rationality is common knowledge, and only observe a feedback on their own performance metric.

Finally, Part III comprises three case studies. Chapter 7 describes typical power allocation games, where transmission rates have to be optimized. Chapter 8 focuses on energy-efficient power control, and Chapter 9 is devoted to medium access control. The purpose of these chapters is show how the important notions and models, presented in Parts I and II (one-shot games, dynamic games, learning algorithms), can be applied to typical wireless problems. Finally, Part IV provides the appendix, references, and index. The purpose of the appendix is to review some notions (mainly in mathematics, stochastic approximation, and information theory) which can make the reading easier.

ACKNOWLEDGMENTS

The authors would like to thank all the MSc and PhD students we have taught or supervised in this area. Their feedback has contributed to making some parts of this book more understandable. Also, we would like to thank the researchers who probably influenced this book since the authors attended their classes or seminars and interacted with them. We will just mention some of them: Professor Jörgen Weibull (Stockholm School of Economics, Sweden), Professor Sylvain Sorin (Université Pierre et Marie Curie – Paris 6), Professor Shmuel Zamir (The Hebrew University of Jerusalem, Israel), Professor Rida Laraki (CNRS – École Polytechnique, Palaiseau, France), and Professor Eitan Altman (INRIA, Sophia Antipolis, France).

Fundamentals of Game Theory

A Very Short Tour of Game Theory

1.1 INTRODUCTION

First of all, what is game theory? It is always difficult to define a theory in a few words without limiting its actual scope and hiding some of the fundamental ideas which originate from it. This is especially true for theories which evolve considerably over time. The essence of game theory is the formal study of *interactions* between several *decision-makers* who can have conflicting or common interests. By interactions, it is meant that what the others do has an impact on each decision-maker (called a player), and what he/she gets from an interactive situation does not only depend on his/her decisions. Although the focus of this book is clearly on wireless games, it will be seen that the meaning of a player can be very broad; it can be a human being, a machine, an automaton, an animal, a plant, etc. Very often, a question arises concerning how game theory differs from optimization. Although some people consider game theory as a part of optimization, it is important to note that games have some features which are not common in classical optimization problems. In contrast to a classical optimization problem where a certain function must be optimized under some constraints, the optimal decision or strategy is generally unclear in interactive situations involving several decision-makers, since no single player completely controls the final outcome. This means that the problem needs to be defined before it can be solved. After reading this book, the reader should be convinced that game theory possesses its own tools and notions, and is enriched with notions coming from different areas, such as economics, biology, and computer science. This leads to concepts we do not encounter in studying convex optimization[1] – for instance cooperative plans, punishments, rationality, risk aversion, and unbeatable strategies.

As said above, the meaning of a player can be very broad, and because of this, games have existed in nature and human societies for a very long time. What is more recent is the appearance of a theory of games. Game theory appeared gradually through the works of Borel (1921), Cournot (1838), Darwin (1871), Edgeworth (1881), Waldegrave (1713), Zermelo (1913) and Ville (1938), but the major works by Nash (1950) are generally considered to be those found which marked the official

---

[1] To elaborate further on this point, note that game theory and distributed optimization obviously have many intersections, but one of the main differences between the two approaches, in their current state, is that the latter assumes that the agents follow some given rules, and are generally coordinated, while in the former the decision is generally left to the player.

birth of the theory of games. Since then, game theory has resounding success in many areas, such as biology, economics, and information sciences. It has been enriched by numerous brilliant game-theorists, such as Allais, Aumann, Debreu, Harsanyi, Maskin, Maynard-Smith, Myerson, Selten, to name just a few. The purpose of the theory of games is to model interactions between players, to define different types of possible outcome for the interactive situation under study (solution concepts), to predict the solution(s) of a game under given information and behavior assumptions, to characterize all possible outcomes of the game, and to design good strategies to reach some of these outcomes.

Before defining the main representations of a game (Sec. 1.3), which corresponds to defining a game mathematically, we now provide four examples, where the ideas of game theory can be used.

1.2 A BETTER UNDERSTANDING OF THE NEED FOR GAME THEORY FROM FOUR SIMPLE EXAMPLES

Before defining the fundamental concepts presented in this chapter, it is useful to consider common situations where game theory is a natural paradigm. We have picked four examples. The first is not related to communication problems, but the other three correspond to models used in the area of wireless networks. All these situations need a tool to predict or understand the outcome of the interactive situation under consideration.

Example 1: A Bankruptcy Problem

The problem is as follows. A man dies, leaving debts totaling more than the value of his estate. Assume that there are three creditors and the debts to them are 100, 200, and 300. How should the estate be divided among the creditors? About two thousand years ago, the Babylonian Talmud gave the answer (reported in the Table 1.1). While the recommendation in the case where the estate is 100 (the estate is divided equally) seems reasonable, those

Table 1.1 A bankruptcy problem and the solution recommended by the Talmud: The estate is divided between three creditors, depending on how much the estate is worth and the claims of the creditors

		Claim		
		100	**200**	**300**
Estate	**100**	$\frac{100}{3}$	$\frac{100}{3}$	$\frac{100}{3}$
	200	50	75	75
	300	50	100	150

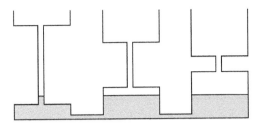

FIGURE 1.1

A physical interpretation (Aumann, 1985) of the game theoretical solution to the bankruptcy problem solved in Aumann and Maschler (1985).

for the cases where it is worth 200 and 300 are very surprising and have baffled many scholars for two thousands years. How can this "mystery" be explained? Many unsatisfying answers were given to this problem, but only in 1985 was a relevant answer derived. It was given by game theorists (namely by Aumann and Maschler (1985)) who proved that this bankruptcy problem can be formulated as a cooperative game for which each solution reported in Table 1.1 corresponds to the nucleolus of the game. We will not go into the details of cooperative games here (see Aumann and Maschler (1985) for more details) but, in order to give some ideas about the solution to the interested reader, we have reported here a physical interpretation of the solution (Aumann, 1985). The estate has to be shared according to Figure 1.1, where the value of the estate is represented by a quantity of a liquid. This is shared between three vessels according to the rule of linked vessels (for each vessel the narrow pipe or small neck has a negligible capacity). This can be thought of as a water-filling solution with channels having a particular (non-linear) structure.

The purpose of reporting the bankruptcy problem here is mainly to show that game theory can be successful where other approaches do not lead to satisfying solutions. However, as the full understanding of this example requires the use of advanced game theoretic notions, simpler examples are necessary to show the usefulness of game theory. We will now consider the examples of the phone conversation, the Braess paradox in routing problems, and power control in interference channels respectively.

Example 2: Phone Conversation

Consider two people A and B who talk to each other by phone. Suddenly, the communication is broken. Each person has two choices: Call or Wait. The question is: What is going to happen? Even though the question is simple, the answer is not. To make the problem simpler, assume that each person can only make one choice (the general case being where a sequence of choices can be made) and the choice of A and B is made simultaneously. Now there are only four outcomes to this situation. Can we predict the effective outcome for certain? To do this it is necessary to make some assumptions. Clearly, the choice made by each person depends on their goal: Does the person want to pursue or finish the conversation? Does he/she want to save money or time? Defining the respective objectives is, however, not sufficient to predict the outcome of the situation. If both players want to pursue the conversation and decide to call, a collision will occur. In fact, it is also necessary to make some behavioral assumptions about the two people and information assumptions about what they know about the situation. For instance, if A does not want to talk to B anymore and B knows this, one could imagine that the outcome is that A waits and B calls. But if A thinks B is going to call, *A* is going to call to cause a collision. Now, if *B* knows that *A* intends to do this trick to reach his objective, *B* can decide to wait; and so on. This example shows how a seemingly trivial communication scenario (which models a collision channel by the way) can lead to complex reasoning. One of the goals of game theory is precisely to model this type of interactive situation, to propose suitable performance criteria for the protagonists, and to formulate information and behavior assumptions.

Example 3: Braess Paradox

The following inequality is well-known in optimization:

$$\max_{x \in \mathcal{A}} f(x) \le \max_{x \in \mathcal{B}} f(x) \qquad (1.1)$$

when $\mathcal{A} \subseteq \mathcal{B}$. Clearly, this holds if the variable is a K-dimensional vector $\underline{x} = (x_1, \dots, x_K)$. It is important to note that it holds because the optimization is performed with respect to whole vector. If the function were optimized with respect to one variable only and the other variables were controlled by an external entity, there is no reason why this equality should still hold. Although this seems trivial and is a well-known fact in game theory, there are real situations where this has not been fully understood. Consider the example given by Braess (1969), which is represented in Figure 1.2. Drivers in *A* want to go from *A* to *B*. On the scenario on the left (scenario 1), they can choose between two paths. Each path comprises two links, each having a certain cost in terms of travel time. For example, if the number of drivers on a given link is represented by x, then the time to travel along this link is $f(x)$ or $g(x)$ depending on the link under consideration. Let us take the original figures used by Braess: $X = 6$ is a number representing the total number of drivers, $f(x) = x + 50$, $g(x) = 10x$, $h(x) = +\infty$. Denote by x_i the quantity representing the number of drivers taking link i. An equilibrium situation is reached when $x_1 = x_2 = 3$. Although the notion of equilibrium would need to be defined rigorously, here we will admit[2] that an equilibrium is reached when the travel time to go from *A* to *B* is identical regardless of the route taken. In scenario 1, the travel time at the (unique) equilibrium point is $D_{1,3} = D_{2,4} = 83$. Now, consider scenario 2 in which an additional link is available. This means that drivers have more options than in scenario 1 (3 possible paths instead of 2). An argument behind adding such a link can be to divert drivers from link 1 if the latter is found to be too crowded. By choosing a cost function $h(x) = x + 10$, the (unique) new equilibrium is reached for $x_1 = 4$, $x_2 = 2$, and $x_5 = 2$. The

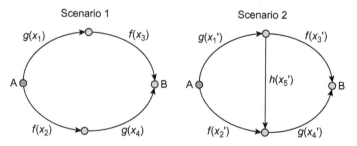

FIGURE 1.2

The routing scenarios considered by Braess.

---

[2]This amounts to admitting an indifference principle.

corresponding travel durations are: $D_{1,3} = D_{2,4} = D_{1,5,4} = 92$. The equilibrium in scenario 2 is worse than the one in scenario 1 even though each driver has more options in the former.[3] This situation is well known in game theory. Real-life examples seem to illustrate this paradox:

- In Stuttgart, Germany, after investments in the road network in 1969, the traffic situation did not improve until a section of the newly-built road was closed for traffic (Knoedel, 1969).
- In 1990, the closing of 42nd street in New York City, USA, reduced the amount of congestion in the area (Kolata, 1990).
- In Seoul, South Korea, there were three tunnels in the city, and one of them needed to be shut down (from 2003) to restore a river and a park. People involved in the project (experts in urban planning in particular) discovered the traffic flow was improved after this change (Vidal, 2006).

At several places in this book the reader will find situations in which enlarging the action space of the players leads to a less efficient outcome. Some examples (e.g., in Chapter 8) will also be provided in which enlarging the observation space of the players leads to a worse outcome.

Example 4: Cognitive Interference Channels

As mentioned in Section 1.1, the importance of multi-user channels for the design of modern communication networks has increased. The interference channel is a simple communication model which is often used where communication systems operating in unlicensed bands or networks with cognitive terminals are assumed. In its original definition (Carleial, 1975, 1978), the interference channel consists of two point-to-point communications operating on the same channel (e.g., on the same transmission frequency band). For the Gaussian interference channel the signal observed by receiver $i \in \{1, 2\}$ is written as $y_i = \sum_{j=1}^{2} h_{ji} x_j + z_i$. For fixed power levels, the different signal-to-interference plus noise ratios (SINR) are given by:

$$\text{SINR}_i = \frac{g_{ii} p_i}{\sigma_i^2 + g_{ji} p_j} \tag{1.2}$$

where $|h_{ji}|^2 \triangleq g_{ji}$, $j = 1$ if $i = 2$, $j = 2$ if $i = 1$, and $p_i \leq P_i^{\max}$ is the transmission power of transmitter i. When the transmitters can sense each other, they can adapt their power level to each other and the SINRs become:

$$\text{SINR}_i = \frac{g_{ii} p_i(p_{-i})}{\sigma_i^2 + g_{ji} p_{-i}(p_i)}. \tag{1.3}$$

Just as in the example of the phone conversation, predicting the pair of SINRs required to make behavior and information assumptions allows one to characterize the reaction of each transmitter $p_i(p_{-i})$. This situation is the one applying to a (power control) game.

These four examples give an outline of the objectives of game theory. As already mentioned, one of its purposes is to analyze interactions between decision-makers. For this purpose, it proposes interaction models, behavior models, and possible information assumptions. Based on this, one of the goals is to predict the outcome of the

---

[3]Here it is useful to mention that these figures result from a purely mathematical analysis based on the model studies by Braess, and do not take into account effects such as induced demand.

interactive situation. Solution concepts are proposed, and natural issues like existence and uniqueness of the solution(s) are considered. For a given wireless scenario,
it may be that no suitable solution concept is available in the current literature of game
theory. This does not necessarily mean that game theory is not the right approach;
it is more likely that the theory needs to be enriched. To this end, the authors would
like to draw to the attention of the reader who is new to game theory, that it is not
only about:

1. rational decision-makers;
2. selfish decision-makers;
3. non-cooperative games;
4. Nash equilibrium as a solution concept;
5. interactions due to the fact that the players' utility depends on the actions of the
   others.

Some elements justifying these assertions are given throughout this chapter, especially in Sec. 1.5.

1.3 REPRESENTATIONS AND CLASSIFICATION OF GAMES

Now we have a rough idea of what game theory is about, the second natural question
is: What is a game exactly? While everyone has an intuitive idea of what a game
is, and knows many examples (e.g., board games, Olympic games, TV games, video
games), game theory provides a mathematical definition of games. The definition
depends on the representation considered. In this section, three[4] existing representations of a game are described, and several classes of games are mentioned. The
representations allow one to see how a game can be generically defined. In order to
go further with the definitions (e.g., by defining the strategies precisely), the class of
games under consideration has to be specified.

1.3.1 Representations of a Game

There are three main types of representation of a game:

- the normal or strategic form;
- the extensive form;
- the coalitional form.

Coalitional games will not be addressed in this book. Indeed, unless explicitly stated
otherwise, **non-cooperative games** will always be considered. This means that it will
always be possible to identify the utility function of a particular player (or at least his
preference order).

---

[4]Here, only the three dominant representations are described. Other representations exist (e.g., the
standard form which is used in the theory of equilibrium selection (Harsanyi and Selten, 2003)) but are
used much more marginally.

1.3.1.1 *The Strategic Form*

> ## Definition 5: Strategic Form Games (1)
>
> A strategic form game is an ordered triplet:
>
> $$\mathcal{G} = \left(\mathcal{K}, \{\mathcal{S}_i\}_{i \in \mathcal{K}}, \succeq_i\right) \qquad (1.4)$$
>
> where $\mathcal{K} = \{1,\dots,K\}$ is the set of players, $\mathcal{S}_i$ is the set of strategies of player i, and $\succeq_i$ is a preference order over the set of action profiles for player i.

The strategic form is the most used form, in both game theory and wireless literature. It is generally more convenient for mathematical analysis and is usable for both discrete and continuous strategy sets. In fact, the most common form relies on the existence of a cost/reward/payoff/utility function for each player; in this book the term "utility" will be used almost everywhere. When it is possible, it is much more convenient to replace the preference order with a real-valued function: the idea is to assign a numerical value to each possible outcome of the game. In wireless communications, the performance metrics (e.g., the quality of service) are generally known, which give the utility functions to be used directly. However, it may be that utility functions are difficult or even impossible to find. More details about utility theory can be found in Fishburn (1970).

First of all, it is interesting to note that, for a given preference order, utility functions do not always exist. The lexicographic order is well-known for ordering words or names but can also be applied to vectors of reals. More specifically, in $\mathbb{R}^2$ we have that $\underline{x} = (x_1, x_2) \succeq \underline{y} = (y_1, y_2)$ if and only if: either $x_1 > y_1$ or $x_1 = y_1$, $x_2 > y_2$. For this order, Debreu has proved the proposition below (Debreu, 1954).

> ## Proposition 6: Debreu (1954)
>
> There exist no utility functions for lexicographic ordering on $\mathbb{R}^2$.

More constructively, there exist theorems based on sufficient conditions under which the existence of a utility function is guaranteed. The two results below can be found in Fishburn (1970) and Debreu (1954) respectively (for a review of used mathematical notions such as connectedness and closedness, the reader is referred to Appendix A.1).

> ## Theorem 7: Existence Theorem of a Utility Function
> ##   (Countable Sets)
>
> There exists a utility function for every transitive and complete ordering on any countable set:
>
> - completeness: $x \succeq y$ or $y \succeq x$ or both;
> - transitivity: "$x \succeq y$ and $y \succeq z$" $\Rightarrow x \succeq z$.

Definition 8: Continuous Ordering

The preference order $\succeq$ is said to be continuous if $\mathcal{B}(x) = \{y \in \mathcal{X} : x \succeq y\}$ and $\mathcal{W}(x) = \{y \in \mathcal{X} : y \succeq x\}$ are closed.

Theorem 9: Existence Theorem of a Utility Function (Continuous Sets)

There exists a utility function for every transitive, complete, and continuous ordering on a continuous set $\mathcal{X} \subset \mathbb{R}^N$, provided $\mathcal{X}$ is non-empty, closed, and connected.

To conclude on the existence of a utility function, we mention a theorem which applies to the preference over lotteries. Indeed, when each strategy or alternative corresponds to a simple probability measure on the consequences in a set $\mathcal{X}$, the expected utility model is considered for computing utilities of the strategies or their associated measures. The following theorem has been developed by von Neumann and Morgenstern (1944).

Theorem 10: Preferences over Lotteries

The complete and transitive preference ordering $\succeq$ over $\Delta(\mathcal{S})$ admits a utility function (expected utility) if and only if $\succeq$ meets the axioms of independence and continuity:

- independence axiom: $x \succ y \Rightarrow (1 - \mu)x + \mu z \succ (1 - \mu)y + \mu z, \ \mu \in]0,1[$;
- continuity axiom: $x \succ y \succ z \Rightarrow \exists \mu \in]0,1[, \ (1 - \mu)x + \mu z \succ y \succ (1 - \mu)z + \mu x$.

From now on, the existence of a utility function will be assumed, which leads us to a slightly modified definition for a strategic form game.

Definition 11: Strategic Form Games (2)

A strategic form game is an ordered triplet:

$$\mathcal{G} = \left(\mathcal{K}, \{\mathcal{S}_i\}_{i \in \mathcal{K}}, \{u_i\}_{i \in \mathcal{K}}\right). \tag{1.5}$$

Let us consider an example of a game in its strategic form. The prisoner's dilemma (PD) is a famous game originally framed by Flood and Dresher working at the RAND corporation (USA) in 1950 and formalized by Tucker in the form presented here. A version of this game can be described as follows (Stanford Encyclopedia, 2007): "Two suspects are arrested by the police. The police have insufficient evidence for a conviction, and, having separated the prisoners, visit each of them to offer the same deal. If one testifies for the prosecution against the other (defects), and the other remains silent (cooperates), the defector goes free and the silent accomplice receives the full 10-year sentence. If both remain silent, both prisoners are sentenced to only six months in jail for a minor charge. If each betrays the other, each receives a 5-year sentence. Each prisoner must choose to betray the other or to remain silent." Assuming this game is played only once, what is the strategic form of this game?

- The players are the two prisoners, who are called 1 and 2: that is, $\mathcal{K}^{PD} = \{1,2\}$.
- The possible choices for the prisoners define the action set for each of them: $\mathcal{A}_i^{PD} = \{\text{Cooperate}, \text{Defect}\}$.
- There is an infinite number of possible utility functions to model this interactive situation (in strategically equivalent ways). For example, let us assume that being free is valued as 4 points for each prisoner, staying in jail for 6 months is valued as 3 points, staying in jail for 5 years is valued as 1 point, and staying in jail for 10 years is valued as 0 points. With this choice, the utility function of player $i \in \{1,2\}$, is given by:

$$u_i^{PD} : \left| \begin{array}{ccc} \mathcal{A}_1^{PD} \times \mathcal{A}_2^{PD} & \to & \{0,1,3,4\} \\ (a_1, a_2) & \mapsto & u_i^{PD}(a_1, a_2) \end{array} \right. . \tag{1.6}$$

The triplet $\left(\mathcal{K}^{PD}, \{\mathcal{S}_i^{PD}\}_{i \in \mathcal{K}}, \{u_i^{PD}\}_{i \in \mathcal{K}}\right)$ is the strategic form of the game. A convenient representation of a strategic form game with finite numbers of actions and players is the matrix form. For the prisoner's dilemma, a matrix form is given in Table 1.2: player 1 chooses the row, player 2 chooses the column and a pair of reals is associated with each pair of choices (namely the values of the utility functions for the two corresponding players).

Often in this book, we will use a wireless version of this game called the forwarder's dilemma (Felegyhazi and Hubaux, 2006). The problem is represented in Figure 1.3. Two source nodes want to send a packet to their respective destination nodes. To reach its destination, the packet has to go through an intermediate node which is actually the other source node. Each source node can decide to forward the packet to the destination of the other source (Forward) or drop it (Drop). The utility of each source is assumed to have the form "utility = benefit − cost" where the benefit is 1 if the packet is forwarded and 0 if it is dropped; the cost of transmitting is $c > 0$. This is shown in tabular form in Table 1.3.

1.3.1.2 The Extensive Form
The key ingredient of the extensive form is a tree, which can be very useful for games played sequentially. The notion of a tree is reviewed here to make this chapter suitably self-contained.

Table 1.2 The Prisoner's Dilemma in a Matrix Form

	Cooperate	Defect
Cooperate	(3, 3)	(0, 4)
Defect	(4, 0)	(1, 1)

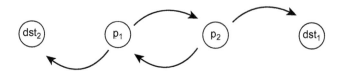

FIGURE 1.3

The communication scenario under study: each source node has to send a packet to a certain destination and has to go through to another source node. For this, the two sources have to decide simultaneously either to forward or to drop the received packet.

Table 1.3 The Forwarder's Dilemma in a Matrix Form		
	Forward	**Drop**
Forward	$(1-c, 1-c)$	$(-c, 1)$
Drop	$(1, -c)$	$(0, 0)$

Definition 12: Tree

A tree is a triplet $(\mathcal{V}, v_{\text{root}}, \pi)$ where:

- $\mathcal{V}$ is set of nodes (or vertices);
- v_{root} is the root of $\mathcal{V}$;
- π is the predecessor function. It verifies: $\forall v \in \mathcal{V}, \exists n \geq 1, \pi^{(n)} = \pi \circ \ldots \circ \pi = v_{\text{root}}$

Definition 13: Standard Extensive Form Games

A standard extensive form game is a 6-uplet:

$$\mathcal{G} = \left(\mathcal{K}, \mathcal{V}, v_{\text{root}}, \pi, \{\mathcal{V}_i\}_{i \in \mathcal{K}}, \underline{u}\right) \tag{1.7}$$

where $\mathcal{K} = \{1, \ldots, K\}$ is the set of players, $(\mathcal{V}, v_{\text{root}}, \pi)$ is a tree, $\{\mathcal{V}_i\}_{i \in \mathcal{K}}$ is a partition of $\mathcal{V}$, and $\underline{u}$ is a result function, namely $\underline{u} = (u_1, \ldots, u_K)$ in the context of non-cooperative games.

This definition shows that a game can be represented by a (game) tree. At each node of the tree, a given player can make a decision, so the next node depends on the decision made. Before justifying the term "standard", consider an example.

Example 14: Extensive Form of a Multiple Access Game

Assume two terminals $\{1, 2\}$ who have two choices: transmit (T) or not transmit ($\overline{\text{T}}$). One of the terminals (say 1) has to make his choice before the other. The terminal playing second knows what has been played. Again, the utility of the terminal is assumed to have the form "utility = benefit − cost" where the benefit is 1 if the terminus uses the medium alone and 0 if both terminals are using it; the cost is $c > 0$ if the terminal transmits and 0 otherwise. This two-stage game is show in extensive form in Figure 1.4. Player 1 makes his decision at

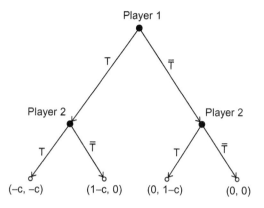

FIGURE 1.4

The multiple access game under extensive form.

the top node (the root) and player 2 makes his decision at one of the nodes which has been selected by player 1. No decision can be made at the terminal nodes, since they correspond to the possible outcomes of the game.

In the example above, it can be noticed that each player knows at which node he is when he makes his decision. This type of game is called a game with *perfect information*. Obviously, this property is not always available. An extensive form of the forwarder's dilemma shows this.

Example 15: Extensive Form of the Forwarder's Dilemma

Compared to the multiple access game, the forwarder's dilemma possesses several special features. As well as different outcomes, the most important feature to be noticed is that the game is played simultaneously. This means that player 2 does not observe what player 1 chooses (and conversely). This is indicated by a dotted line in Figure 1.5. This line is used to link nodes which can be distinguished by player 2 (source 2)[5]. The corresponding set of nodes is called the information set of player 2.

This example shows that the standard extensive form is incomplete, since it does not comprise the description of the information sets of the players. The standard form is therefore the most basic extensive form corresponding to games with perfect information; that is, the information set of every player is reduced to a singleton. To describe simultaneous games, games with limited memory, sequential games where the players do not know the history of the game perfectly, sequential games where randomness intervenes (e.g., where Nature is interpreted as a player choosing certain nodes), etc, a more advanced form is required. This is covered by the next definition.

---

[5] An equivalent extensive form in which the roles of players 1 and 2 were switched could have been used.

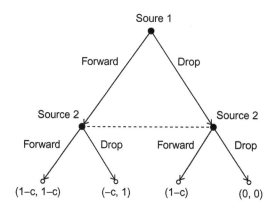

FIGURE 1.5

The forwarder's dilemma under extensive form.

Definition 16: Extensive Form for Games with Imperfect Information

An extensive form game with imperfect information is a 9-uplet:

$$\mathcal{G} = \left(\mathcal{K}, \mathcal{V}, v_{root}, \pi, \mathcal{V}_0, \left\{ q_0^j \right\}_{j \in \mathcal{V}_0}, \{\mathcal{V}_i\}_{i \in \mathcal{K}}, \left\{ \mathcal{W}_i^k \right\}_{k \in \{1, \dots, k_i\}}, \underline{u} \right) \tag{1.8}$$

where $\mathcal{K} = \{1, \dots, K\}$ is the set of players, $(\mathcal{V}, v_{root}, \pi)$ is a tree, $\mathcal{V}_0$ is a set of vertices associated with player 0 (Nature), $\forall j \in \mathcal{V}_0$ q_j^0 represents the transition probability used by player 0 to choose a successor to j, $\{\mathcal{V}_i\}_{i \in \mathcal{K}}$ is a partition of $\mathcal{V}$, $\mathcal{W}_i^k$ corresponds to the partition of $\mathcal{V}_i$ defining the information structure for player i, and $\underline{u}$ is a result function $(\underline{u} = (u_1, \dots, u_K))$.

It can be checked that this form generalizes the standard form by considering Nature to be deterministic and $\mathcal{W}_i^k = \{w_i^k\}$ (singleton) for all $(i, k) \in \{1, \dots, K\} \times \{k_1, \dots, k_K\}$.

1.3.1.3 *The Strategic Form and Extensive Form*

A natural question is how to choose between the strategic and extensive form to model a non-cooperative game. There is no strict rule for this, but some useful observations can be made:

1. The strategic form is generally more convenient than the extensive form for mathematical analysis and continuous action sets.
2. The extensive form is more complete than the strategic form.
3. The extensive form is sometimes more intuitive and allows the game to be better understood.
4. The tree structure of the extensive form can be useful for computer-based analyses.

Based on (1), it can be useful to transform an extensive form game into a strategic form game. In fact, if the transformation used is based on the von Neumann and Morgenstern expected utility (more details can be found, e.g., in Fishburn (1970)), there is a unique strategic form game to represent a given extensive form game. The converse is however not true – there can be several (and even many) extensive forms associated with a given strategic form game (this explains point 2).

1.3.1.4 *The Coalitional Form*

The coalitional form is not described in detail here (for more details see for example Osborne and Rubinstein (1994)). We will merely mention its existence and emphasize the main difference between the two above forms and the coalitional form. For this purpose, only one (common) type of coalitional form is reported here – namely, the characteristic (coalitional) form, introduced by von Neumann and Morgenstern (1944) for coalitional games with transferable utility (TU).

....

Definition 17: Characteristic Form in TU Coalitional Games

A characteristic form game is defined by a pair $(\mathcal{K}, v)$ where $\mathcal{K}$ is the set of players of the game and v is the characteristic function $v : 2^{\mathcal{K}} \to \mathbb{R}$.

The notation $2^{\mathcal{K}}$ is common in game theory. Here, it indicates that the function v assigns a value to a certain subset of players (or coalition). The notation becomes natural when labeling a coalition by a K-dimensional vector with entries equal to 0 or 1. Compared to the strategic and extensive form, which are intimately concerned with the details of the process and rules defining a game, the characteristic form (and more generally any coalitional form) abstracts away from such rules, and instead looks only at a more general description which specifies only what each coalition can get, without saying how.

1.3.2 **Classification of Games**

As for the types of representations, here again the authors do not attempt to be exhaustive. The goal is rather to give to the reader a rough idea of the main features which distinguish games in the different main classes. One of the dangers in classifying games is that some arbitrary frontiers put in place may become meaningless as knowledge of the field progresses. Below is a short list of classes of games:

a. static and dynamic games;
b. non-stochastic and stochastic games;
c. non-cooperative and cooperative games;
d. games with complete/incomplete information;
e. games with perfect/imperfect information;
f. zero-sum games and non zero-sum games.

(a) The distinction between static and dynamic games will become clear in Chapter 3. Although there is not a perfect consensus on this point, static games will be considered in this book as a special case of dynamic games. In fact, dynamic games generally assume that players can extract some information from past moves, observations, and chosen strategies and take this into account to adjust current and future moves. In static games, players have a certain knowledge (information assumptions, behavior assumptions) and this does not change.

(b) One of the key ingredients of stochastic games is the presence of a state in the game (e.g., the channel gain) which evolves (over time most often, and according to a certain stochastic rule). This class of games will also be studied in Chapter 3. In particular, the chapter should allow the reader to distinguish between repeated games (which correspond to a special case of dynamic games where the same static game is repeated) and stochastic games (the latter includes the former as a special case).

(c) Elaborating on repeated games, it is worth mentioning that cooperative plans can be put in place in non-cooperative, repeated games. The point here is that cooperation can exist in non-cooperative games and, more generally, it is not always clear whether players can be said to be selfish or altruistic. Even if the boundary is not always clear, there are still two different approaches or branches in game theory. In non-cooperative games, the individual goals and strategies can be distinguished, whereas in cooperative games this is not always possible. The second branch rather deals with the options available to the group, what coalitions form, how the available global utility is divided (e.g., consider the cost allocation problem (Young, 1994)). A key distinction between the two types of game is to know whether commitments are enforceable or not: in cooperative games commitments are fully binding and enforceable, while they are not in non-cooperative games.

(d) In games with complete information, it is assumed that the data of the game is common knowledge. Considering a strategic form game, this means that the actions available to the players and the utility functions are common knowledge. By that is meant that every player knows the data of the game, every player knows that the other players know the data of the game, every player knows that every player knows the data of the game, and so on, ad infinitum. Games with incomplete information (also known as Bayesian games in game theory (Zamir, 2009)) are models in which the players have only partial information about the game. Although games with incomplete information appear at first glance to be more realistic than games with complete information, both classes of games are important in practice. As can be seen in Chapters 4, 5, and 6, it may happen that solutions predicted in a game with complete information can be observed in a degraded version of the game where only partial information is available. This remark will be developed further in this book; in particular Nash equilibria predicted in a game with complete information can also be observed in a game between learning automata.

(e) The distinction between games with perfect and imperfect information has been already mentioned when defining the extensive form. The key notion used to make this distinction is the history of the game. When all the players know the history of the game perfectly, it is said to be played with perfect information, and imperfect information otherwise.

A few comments are in order. First note that, in its basic definition, the notion of perfect or imperfect information is not necessarily related to the extensive form or to the dynamic aspect of the game. Indeed, one can have repeated games in strategic form with perfect or imperfect information (in Chapter 3 the history of a game is defined mathematically for repeated games). Also, the prisoner's dilemma is a static game with imperfect information. Second, the terminology used in game theory differs from that used in communications. In the latter, "imperfect" usually refers to the situation where a quantity is not known perfectly (e.g., because of intrinsic estimation errors), whereas in game theory the underlying notion is the knowledge of the history of the game. In fact, scenarios where some quantities are estimated correspond to the framework of games with incomplete information.

(f) Zero-sum games are those where the sum of utilities is zero (or a constant). The idea is that if someone wins something, someone else necessarily has to lose. Zero-sum games have been studied extensively in the game theory literature (e.g., see Sorin (2002)). In this book, no specific section is dedicated to this class of games. The main reason for this is that this special structure does not seem to appear very often in wireless games, as indicated by the wireless literature (Special issue, 2008).

1.4 SOME FUNDAMENTAL NOTIONS OF GAME THEORY

In this section, several fundamental game-theory notions are defined. Understanding these concept is necessary for the other chapters of this book, particularly with regard to Nash equilibrium, a solution concept of great importance in game theory. Nash equilibria are defined in this chapter and explained in detail in Chapter 2. From now on, strategic form games will always be considered, unless stated otherwise. The notation $\mathcal{G}$ will be used by default.

1.4.1 Actions, Pure Strategies, Mixed Strategies

To get started, let us define an action profile.

Definition 18: Action Profile

Let $\mathcal{K} = \{1,\ldots,K\}$ be the set of players. Let $\{\mathcal{A}_i\}_{i \in \mathcal{K}}$ be the different sets of actions of these players. The vector of actions $\underline{a} = (a_1,\ldots,a_K) \in \mathcal{A}$, with $\mathcal{A} = \mathcal{A}_1 \times \mathcal{A}_2 \times \ldots \times \mathcal{A}_K$, is called an action profile.

The word "action" has to be distinguished from the word "strategy". The forwarder's dilemma (Felegyhazi and Hubaux, 2006) was presented in such a way that the notion of strategy coincides (possibly through a one-to-one mapping) with that of action, but this is not general at all. In such a (special) case, the strategy used by a player is called a *pure strategy*; that is, the strategy is said to be pure when the player assigns a probability of 1 to one of the options he has. Often, the action is a result of a strategy and can be seen as the translation of a player's decision into the ultimate means to influence the physical world where the games takes place. Roughly speaking, if the player can be modeled by a computer, a strategy can be seen as a program, and the action corresponds to the result of running the computer program (start/stop printing depending on the number of jobs sent to the printer, allocate more or less processing resources to a task, etc.). Mixed strategies generalize the notion of pure strategies.

....

Definition 19: Mixed Strategy

A mixed strategy ρ_i for player i is a probability distribution over his actions (or pure strategies), that is, $\rho_i \in \Delta(\mathcal{A}_i)$ where $\Delta(\mathcal{X})$ is the $(|\mathcal{X}| - 1)$–dimensional unit simplex associated with the set $\mathcal{X} = \{x_1, \ldots, x_{|\mathcal{X}|}\}$: $\Delta(\mathcal{X}) = \left\{ (p_1, \ldots, p_{|\mathcal{X}|}) \in \mathbb{R}_+^{|\mathcal{X}|}, \sum_{\ell=1}^{|\mathcal{X}|} p_\ell = 1 \right\}$. Additionally, by definition, the mixed strategies of the players are assumed to correspond to independent randomizations.

Mixed strategy profiles can be readily defined as follows:

....

Definition 20: Mixed Strategy Profile

Based on definitions 18 and 19, a mixed strategy profile is defined by $\underline{\rho} = (\rho_1, \ldots, \rho_K) \in \Delta(\mathcal{A}_1) \times \Delta(\mathcal{A}_2) \times \cdots \times \Delta(\mathcal{A}_K)$.

The relevance and interpretation of mixed strategies is very well explained in books like Osborne and Rubinstein (1994). This book will only supply a few comments on them, to understand how they are used. First, we can see immediately the difference between a strategy and an action. For example, a mixed strategy might consist in choosing a coin from different coins put on a table, each of them having a given probability to produce head or tail as a result of being flipped. The strategy consists in choosing the coin, but the action is either head or tail. Another familiar game where mixed strategies can be assumed to be used is given by the penalty game (see Figure 1.6). If the goalkeeper could predict the action of the player kicking the ball, he would improve his chances of catching it. However, the action of the shooter randomizes his actions; say Left, Center, and Right to make it simple. Similarly, the goalkeeper has to be unpredictable from the shooter's point of view and uniform randomization appears to be the best distribution to use. As a second remark on mixed strategies, from a mathematical point of view, mixed strategies generalize pure strategies: mixed strategies cover the whole unit simplex, while pure strategies correspond to its vertices. Mixed strategies therefore occupy a set having strong mathematical properties (e.g., it is convex). The corresponding utility functions have, also, nice properties like multi-linearity. All of these properties are very useful for mathematical analysis, for instance to prove the existence of Nash equilibria. In this book, the term "mixed extension" will be used in some places to refer to the corresponding transformed game. In the case of finite action sets, the corresponding definition is as follows:

....

Definition 21: Mixed Extension

Let $\mathcal{G} = (\mathcal{K}, \{\mathcal{A}_i\}_{i \in \mathcal{K}}, \{u_i\}_{i \in \mathcal{K}})$ be a strategic form game, $\mathcal{A}_i$ being the set of actions of player i. The mixed extension of $\mathcal{G}$ is the game $\widetilde{\mathcal{G}} = (\mathcal{K}, \{\Delta(\mathcal{A}_i)\}_{i \in \mathcal{K}}, \{\widetilde{u}_i\}_{i \in \mathcal{K}})$ where:

$$\widetilde{u}_i(\rho_1, \ldots, \rho_K) = \sum_{\underline{a} \in \mathcal{A}} \prod_{j \in \mathcal{K}} \rho_j(a_j) u_i(\underline{a}). \tag{1.9}$$

FIGURE 1.6

The penalty game.

Mixed strategies are not necessarily defined on action sets; they can also be defined on probability spaces. In Chapter 3, general strategies will be defined which correspond to mixed strategies over probability distributions in stochastic games.

1.4.2 Dominant Strategies, Rationality, Pareto-Dominance, Best-Response Correspondence

To explain the notion of dominant strategies, which are related to rationality, we borrow the example of Dixit and Nalebuff (1993).

Example 22: Is Indiana Jones a Rational Doctor?

[Indiana Jones, in the climax of the movie "Indiana Jones and the Last Crusade". Indiana Jones, his father, and the Nazis have all converged at the site of the Holy Grail. The two Joneses refuse to help the Nazis reach the last step, so the Nazis shoot Indiana's dad. Only the healing power of the Holy Grail can save the senior Dr. Jones from his mortal wound. Suitably motivated, Indiana leads the way to the Holy Grail, but there is one final challenge: he must choose between literally scores of chalices, only one of which is the cup of Christ. While the right cup brings eternal life, the wrong choice is fatal. The Nazi leader impatiently chooses a beautiful gold chalice, drinks the holy water, and dies the sudden death that follows the wrong choice. Indiana picks a wooden chalice, the cup of a carpenter. Exclaiming "There's only one way to find out" he dips the chalice into the font and drinks what he hopes is the cup of life. Upon discovering that he has chosen wisely, Indiana brings the cup to his father and the water heals the mortal wound. Indy goofed! Although this scene adds excitement, it is somewhat embarrassing that such a distinguished professor as Dr. Indiana Jones would overlook his dominant strategy. He should have given the water to his father without testing it first. If Indiana has chosen the right cup, his father is still saved. If Indiana has chosen the wrong cup, then his father dies but Indiana is spared. Testing the cup before giving it to his father does not help, since if Indiana has made the wrong choice, there is no second chance – Indiana dies from the water and his father dies from the wound....]

To better understand this, let us model this situation by a game where Indiana plays against Nature. This is what the table below represents. The value associated with the four possible outcomes is a number representing the number of lives saved: Nature is seen as a player having two choices: either give the good chalice to Indiana or the bad one; Indiana has two choices, either take the chosen chalice first himself or give it to his father. From this table it appears that the values of Indiana's utility in the second column are always greater or equal to those in the first column: whatever Nature does, giving the chalice to his father first is a dominant strategy.

		Indiana	
		To him	To his father
Nature	Good	2	2
	Bad	0	1

More formally, a (weakly) dominant strategy is defined as follows:

Definition 23: Dominant Strategy

Let $\mathcal{G} = (\mathcal{K}, \{S_i\}_{i \in \mathcal{K}}, \{u_i\}_{i \in \mathcal{K}})$ be a strategic form game. The strategy $\hat{s}_i$ is a dominant strategy for player i if:

$$\forall \underline{s}_{-i} \in S_{-i}, \forall s_i \in S_i, u_i(\hat{s}_i, \underline{s}_{-i}) \geq u_i(s_i, \underline{s}_{-i}). \tag{1.10}$$

The definition of strictly dominant strategies is obtained by replacing $\geq$ with $>$ in (1.10). Similarly, weakly and strictly dominated strategies can be defined. All of these concepts are useful for predicting the strategies chosen by rational players. A player is said to be *rational* if he maximizes his utility (or expected utility), taking all his knowledge on the interactive situation into account (von Neumann and Morgenstern, 1944; Savage, 1954). This definition is used by the vast majority of authors, and will be sufficient for all the concepts and results in this book. Other definitions exist. For example, in Aumann (1997) rationality is defined mathematically as a limit of an iterative reasoning procedure. As a consequence of the definition chosen in this book, a rational player will not play a dominated strategy, and will always play his dominant strategies if they exist. In some games, identifying dominant and dominated strategies can be sufficient to predict the outcome(s) of the game: the game is said to be *dominance solvable*. For instance, dominance solvability can be used to solve the prisoner's dilemma. To conclude on rationality, it is important to bear in mind that, depending on the game model under consideration, the rationality assumption can be sufficient to study the game (this is typically the case in games between learning automata), whereas rationality must be assumed to be *common knowledge* in others (in conventional repeated games for example). Rationality a common knowledge means that every player knows that the others are rational, and the others know he is rational and so on, ad infinitum[6].

Another important order is Pareto-dominance, which allows two action or strategy profiles to be compared.

....

Definition 24: Weak Pareto-Dominance

The strategy profile $\underline{s}$ Pareto-dominates $\underline{s}'$ if $\forall i \in \mathcal{K}$, $u_i(\underline{s}) \geq u_i(\underline{s}')$.

Most often, strict Pareto-dominance is considered; that is the inequality in the above definition is strict for at least one player.

....

Definition 25: Pareto-Optimum

The strategy profile $\underline{s}$ is a Pareto-optimal profile if it is dominated by no other profile.

Pareto-optimality can be easily illustrated for a two-player game by representing the region of feasible utilities of the game (see Figure 1.7). When utility (resp. cost) functions are considered, a point of the region is Pareto-optimal if there are no points at the North-East (resp. South-West) of the considered point. By deviating from a Pareto-optimal profile, at least one player will lose something. When available, Pareto-optimality appears to a minimum requirement for a point at which

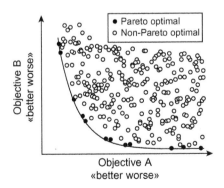

FIGURE 1.7

Illustrating Pareto-optimality in a game with two players minimizing their cost function.

[6]Here again, the notion of common knowledge can be mathematically defined (e.g., see Geanakoplos (1994)) but this definition is not necessary in this book.

a wireless system would operate especially if it involves neither additional signaling cost nor energy cost. Pareto-dominance or Pareto-optimality can sometimes be used to select a given game outcome from several alternatives. Note however, that Pareto-optimality does not necessarily imply having an optimal point in terms of an aggregate or global utility (e.g., the sum-utility).

We will now consider the important notion of best response.

Definition 26: Best Response

The best response (BR) of player i to the (reduced) strategy profile $\underline{s}_{-i}$ is the correspondence given by:

$$\mathrm{BR}_i(\underline{s}_{-i}) = \arg \max_{s_i \in \mathcal{S}_i} u_i(s_i, \underline{s}_{-i}). \tag{1.11}$$

Note that, in general the BR is a correspondence (see Appendix A.1) that is, a set-valued function. If the other players play $\underline{s}_{-i}$, a rational player chooses his action/strategy in the set given by his best response $\mathrm{BR}_i(\underline{s}_{-i})$. The BR allows one to characterize Nash equilibria, which are defined next.

1.4.3 Nash Equilibrium, Dominant Strategy Equilibrium, Correlated Equilibrium, Coarse Correlated Equilibrium

The Nash equilibrium is a concept which was formalized by Nash (1950). To define a Nash point or Nash equilibrium, there is no need to use concepts such as players, rationality, and complete information. To define a Nash equilibrium, only K K-variable functions need to be specified.

Definition 27: Nash Equilibrium

Let $\mathcal{S} = \mathcal{S}_1 \times \mathcal{S}_2 \times \ldots \times \mathcal{S}_K$. Let $u_i : \mathcal{S} \to \mathbb{R}$, $i \in \mathcal{K}$. The vector $\underline{s}^*$ is a Nash equilibrium if:

$$\forall i \in \mathcal{K}, \forall s_i \in \mathcal{S}_i, \ u_i(s_i^*, \underline{s}_{-i}^*) \geq u_i(s_i, \underline{s}_{-i}^*). \tag{1.12}$$

The equilibrium is said to be strict if all the above inequalities are strict. As a strategic game is precisely specified by several multi-variable functions, the Nash equilibrium is a natural concept for strategic form games. In Chapter 2, the importance of this concept will be explained further. A Nash equilibrium can be characterized by the notion of best-response, as follows.

Definition 28: Nash Equilibrium Characterization

Let $\mathcal{G} = (\mathcal{K}, \{\mathcal{S}_i\}_{i \in \mathcal{K}}, \{u_i\}_{i \in \mathcal{K}})$ be a strategic form game. A strategy profile $\underline{s}^*$ is a Nash equilibrium if:

$$\underline{s}^* \in \mathrm{BR}(\underline{s}^*) \tag{1.13}$$

where:

$$\mathrm{BR} : \begin{vmatrix} \mathcal{S} & \to & \mathcal{S} \\ \underline{s} & \mapsto & (\mathrm{BR}_1(\underline{s}_{-1}) \times \mathrm{BR}_2(\underline{s}_{-2}) \times \cdots \times \mathrm{BR}_K(\underline{s}_{-K})) \end{vmatrix} \tag{1.14}$$

This characterization, formalized by Nash (1950), indicates that Nash equilibria are solutions to a fixed point problem. This explains why standard existence theorems are based on topological and geometrical assumptions (see Chapter 2 for more details). In the two definitions above, $\underline{s}$ represents a strategy profile in the broad sense. For instance, it may be a vector of actions, a vector of probability distributions, or a vector of functions. When profiles of pure actions (resp. mixed strategies) are involved, the Nash equilibrium is said to be a *pure (resp. mixed) Nash equilibrium.* A special type of pure/mixed Nash equilibrium is the dominant strategy equilibrium, which is defined as follows:

Definition 29: Dominant Strategy Equilibrium

If each player has a dominant strategy, and each player plays the dominant strategy, then that combination of (dominant) strategies and the corresponding utilities are said to constitute the dominant strategy equilibrium for that game.

As a comment, it can be seen that a mixed Nash equilibrium of a strategic form game corresponds to a pure Nash equilibrium of its mixed extension. By definition (see Def. 19), mixed strategies correspond to independent randomizations; that is the joint probability distribution over the possible action profiles is the product of its marginals. As shown by Aumann (1974), even in the context of non-cooperative games, it is of interest to consider correlated strategies – that is, joint distributions which are not necessarily a product of the marginals. Indeed, even if players are selfish, it may happen that they have access to common information, or that interplayer communication is possible, which is likely to correlate their strategies. It can no longer be assumed that players act independently. This leads us to the notion of correlated equilibria (Aumann, 1974).

Definition 30: Correlated Equilibrium

A correlated equilibrium is a joint probability distribution $q \in \Delta(\mathcal{A})$ which verifies:

$$\forall i \in \mathcal{K}, \forall a_i' \in \mathcal{A}_i, \sum_{\underline{a}\in\mathcal{A}} q(\underline{a})u_i(\underline{a}) \geq \sum_{\underline{a}\in\mathcal{A}} q(\underline{a})u_i(a_i',\underline{a}_{-i}). \tag{1.15}$$

This notion is illustrated in Sec. 1.4.4 and Chapter 3. A simple example of correlated equilibrium is the stop light example. Here two drivers arrive at a cross-roads and can either go or stop depending on the light color (say green or red). By modeling the game in an appropriate manner, the existence of a correlated equilibrium can be demonstrated, i.e., a distribution of recommendations which the drivers have no interest to deviate from in a unilateral manner. An important observation is that for CE, players are assumed to make a decision *after* receiving the recommendation. Another definition of correlated equilibria assumes that players have to decide this before having the recommendation: this is the idea of a *coarse correlated equilibrium* (CCE) (Young, 2004). The definition of a CCE is identical to a CE except that players are assumed to decide whether to follow the recommendation *before* receiving it. In a CCE, all players follow the recommendation because the others also choose to commit. If a single player decides not to commit to follow the recommendations, it experiences a lower (expected) utility. It can be shown (Young, 2004) that the set of CCE includes the set of CE. The notion of CCE will be used in Chapter 6, where a regret matching-based learning scheme is presented.

1.4.4 Illustration of Some Concepts

The purpose of this section is to illustrate some of the notions and concepts presented in the previous sections. In particular, the prisoner's dilemma is considered.

1.4.4.1 *Dominant Strategy in the Prisoner's Dilemma*
First, let us recall the matrix form of the prisoner's dilemma.

	Cooperate	Defect
Cooperate	(3, 3)	(0, 4)
Defect	(4, 0)	(1, 1)

Do the players have a dominant strategy? Indeed, one sees that for Player 1, his second row of utilities $(4, 1)$ is definitely better than the first row $(3, 0)$. This means that, whatever Player 2 does, it is always better to choose Defect, which is therefore a strictly dominant strategy. By symmetry, Player 2 also has such a strategy. The game is therefore dominance solvable: by eliminating the strictly dominated strategies for each of the players, the only remaining profile is (Defect, Defect).

1.4.4.2 *Best Response and Nash Equilibrium in the Prisoner's Dilemma*
If Player i chooses the action C (Cooperate) the best-response of Player $-i$ is to play D. If Player i chooses the action D (Defect) the best-response of Player $-i$ is to play D. From this, the output of the global best-response correspondence BR (see 1.14) equals (D,D) for all the four possible action profiles. The unique fixed-point of this correspondence is clearly (D,D), which is the unique Nash equilibrium of the game. Indeed, it can be checked that the set of correlated equilibria, that of mixed equilibria, and that of pure equilibria are a singleton formed by (D,D).

1.4.4.3 *Pareto-Optimal Points and Braess Paradox in the Prisoner's Dilemma*
Among the four possible action profiles, only the profile (D,D) is not Pareto-optimal. Indeed, for each of the other three profiles, there is no profile such both players can have a better utility.

Although the Braess paradox has been put in evidence for routing problems in Sec. 1.2, it can also be observed in other games. By this, it is meant that enlarging the action set of the players can lead to a worse outcome for all of them. Assume that, in the prisoner's dilemma, the players would be able to play the action C only.

In this scenario, the vector of utilities is $(3,3)$. Now, by allowing the players to choose between C and D (that is to say their set of actions is enlarged), the resulting vector of utilities becomes $(1,1)$ (under appropriate assumptions such as rationality). Therefore, the new outcome is such that every player loses in terms of utility with respect to the first situation where only choice C was available.

1.4.4.4 *Correlated Equilibrium in the Original Example by Aumann*

The prisoner's dilemma is very well-known example, but is a special case in the sense that some notions cannot be emphasized; this is precisely the case of correlated equilibria, as mentioned above. This is the reason why another example is considered, namely the original example by Aumann (1974).

Example 31

Consider the matrix game defined as follows.

	c_1	c_2
r_1	(5, 1)	(0, 0)
r_2	(4, 4)	(1, 5)

This matrix game can easily be shown to have two pure Nash equilibria $((r_1,c_1)$ and $(r_2,c_2))$ and one mixed Nash equilibrium (p_1^*,p_2^*) where p_i^* is a uniform distribution over the two possible actions. The utility vector corresponding to this mixed Nash equilibrium is $(2.5,2.5)$. Now assume a mediator recommends, through a binary public signal (e.g., a coin flip), that the players play either (r_1,c_1) or (r_2,c_2) from a uniform distribution over these two profiles. This distribution can be checked to a correlated equilibrium and leads to a vector of equilibrium utilities $(3,3)$. In fact, by choosing all correlated distributions which can be generated by a public signal, any vector of utilities corresponding to the convex hull of mixed and pure equilibria can be a correlated equilibrium (Haurie and Krawczyk, 2002). Now, if private signals are allowed, it is possible to have equilibrium utilities outside this hull: the idea is that a player is told what to play but is not told what the recommendation to the other player is. Indeed, if a mediator recommends that the players play each of the three profiles (r_1,c_1), (r_2,c_1), (r_1,c_2) with a uniform distribution, the resulting equilibrium utilities are $\left(\frac{10}{3},\frac{10}{3}\right)$.

1.5 MORE ABOUT THE SCOPE OF GAME THEORY

At the end of Sec. 1.2, we mentioned that game theory more than includes the concepts of rational decision-makers, selfish decision-makers, non-cooperative games, Nash equilibrium as a solution concept, and interactions due to the fact that the players' utility depends on the actions of the others. Now that these fundamental ideas have been introduced, it is possible to elaborate a little further on this comment and show why this would be a narrow-minded view of what game theory is about. Bear in mind that even if the theory is inadequate, this does not necessarily mean that game theory is not suitable. It might mean that new concepts need to be invented and the theory enriched with new interaction models. This scenario seems likely to happen in the wireless community if game theory remains as a major paradigm for building modern communications systems.

1.5.1 About Rationality

As far as this book is concerned, two questions are worth asking here. Are players rational in wireless communications? Is rationality always required in game theory?

Often in this book the terminals (e.g., the transmitters) will be considered as the players. From this standpoint, the player is a machine. But the identification of the players is a little more subtle than this. In fact, machines are designed by engineers and the latter follow some rules or recommendations e.g., following a certain policy of an operator, which may in itself be subject to that of the regulator. Current cellular phones can be considered as a special type of player, but this does not mean that they implement the best strategies in terms of spectrum usage, power control, etc. The goal here is not to drown the reader in a philosophical debate but rather to show that the definition of the players is not always obvious and rationality is not automatically present. For implementing advanced game theoretic algorithms in real systems, identifying the players plays an important role in modeling the strategic structure of the environment of a terminal. Additionally, wireless networks also have some specific features, which makes the problem of rationality even more complicated. In heterogeneous networks where players optimize different performance criteria, some players can be perceived as non-rational from a given other player's point of view. Of course, this depends on the knowledge of the considered player. But if two transmitters interact with each other by observing their respective SINR only, the one maximizing his energy-efficiency will perhaps be perceived as non-rational by the one maximizing his transmission rate and vice versa. Finally, even if the rationality assumption is valid, assuming it is common knowledge seems to be more problematic, especially if the network is not designed by the same entity. Now, let us focus on the second question raised above. The answer is in this case less subjective and context-dependent.

Clearly, as in decision theory, the analysis in game theory is from a rational rather than a psychological or sociological point of view. However, this does not necessarily mean that players always must be strictly rational. For instance, risk aversion is well known in game theory (Driesen, 2010); risk averse players do not necessarily maximize their utility in comparison with risk-neutral or risk-seeking players. As a second example, in evolutionary games, full rationality is not required (see Chapter 3 for more details). Evolutionary games, and game theory in general can be used to understand interactions between entities which might be considered as "non-intelligent", such as certain animals, trees, or plants (e.g., see Schuster et al. (2008) where game theory is applied to important problems in biophysics and biochemistry). A third example is automata games. In games between learning automata (Chapters 5 and 6), the assumption that "rationality is common knowledge" is not needed, and automata, although well designed, are not necessarily rational in the sense of von Neumann and Morgenstern (1944) or Savage (1954).

Information in a game is said to be *common knowledge* if all of the players know it, and all of the players know that all other players know it, and all other players know that all other players know that all other players know it, and so on ad infinitum. This is much more than simply saying that something is known by all, but also implies that the fact that it is known is also known by all, etc. Aumann and Shapley (1976) produced one of the first purely formal accounts of the concept of common knowledge in game theory. The common knowledge of rationality implies that every player is motivated by maximizing his own utility. If every player always maximizes his utility, and we are thus able to perfectly calculate the possible probabilistic result of every action, we have *hyper-rationality*.

Players can be said to be with limited or bounded rationality because the information they have may be limited, because of the presence of noise, because they may have a finite amount of time to make decisions, or because of intrinsic limitations in terms of computational capacity (Simon, 1972). This contrasts with the concept of rationality as optimization. The players can sometimes only obtain a satisfactory solution rather than the optimal one, or can make a decision with mistakes and noises, etc. Thus, they do not necessarily optimize their own performance, and the common knowledge assumption need not be satisfied. As a result, a player cannot anticipate the game outcome correctly, and an exact reaction to the others players is difficult.

In his paper entitled *Rationality and bounded rationality*, Aumann (1997) pointed out the following:

> [There are several objections to such models. First, casual empiricism or even just simple introspection leads to the conclusion that even in quite simple decision problems, most economic agents are not in fact maximizers, in the sense that they do not scan the choice set and consciously pick a maximal element from it. Second, such maximizations are often quite difficult, and even if they wanted to, most people including economists and even computer scientists would be unable to carry them out in practice. Third, polls and laboratory experiments indicate that people often fail to conform to some of the basic assumptions of rational decision theory [...]].

The concept of limited rationality or bounded rationality is still a challenging conceptual problem in the area of game theory. Game theorists are currently developing a meaningful formal definition of rationality in situations in which calculation and analysis themselves are costly and/or limited, and not necessarily well-defined in contrast to the standard in which an absolute maximum is chosen from among the set of feasible alternatives, no matter how complex a process that maximization may be. The alternatives themselves involve bounded rationality, but the process of choosing them does not. In the context of bounded rationality, the evolutionary approach has been proposed as a potential key to general solution.

Another way to model bounded rationality within the context of stochastic games is to suppose that players are restricted to using strategies which are implementable by a machine (a finite automaton, a Turing machine, etc.). To give a more precise idea of what bounded rationality can be, we present a model of bounded rationality introduced by Nagel (1995) called level-k rationality. It is recursively defined as follows:

....

Definition 32

Player i is level-k boundedly rational if:

- player i chooses an i.i.d. random action a_0, whenever $k = 0$;
- player i chooses a strategy that is optimal under the belief that all other players are level-$(k-1)$ boundedly rational.

One interpretation of this concept is that level-k boundedly rational players have biased beliefs about the rationality of other players; namely that the other players are level-$(k-1)$ boundedly rational, and that they are playing optimally given those beliefs. A different interpretation is that a level-k boundedly rational player has a limited processing capacity to compute iterated best-response steps. A level-1 boundedly rational player calculates the best response on some arbitrary action of the other players. A level-2 boundedly rational player calculates the best response given that the other players play a best response on some arbitrary action; a level-3 boundedly rational player calculates the best response given that the other players play a best response to players best responding to some arbitrary action, etc. Hence, a level-k boundedly rational player makes k iterative steps of best response computing, and then stops computing. This interpretation is called *bounded processing capacity*. What is the behavior of a level-k boundedly rational player in a two-player game? The process can be described as follows:

- Start with an action profile $\{p_j(0)\}_{j \in \{1,2\}}$ where each $p_j(0)$ represents the action of player j.
- For $k = 1$,

$$\begin{cases} p_1(t+1) \in \mathrm{BR}_1(p_2(t)), \\ p_2(t+1) \in \mathrm{BR}_2(p_1(t)), \end{cases}$$

- $k \geq 2$

$$\begin{cases} p_1(t+1) \in \mathrm{BR}_1 \circ \mathrm{BR}_2(p_1(t-1)) = (\mathrm{BR}_1 \circ \mathrm{BR}_2) \circ (\mathrm{BR}_1 \circ \mathrm{BR}_2)(p_1(t-3)), \\ p_2(t+1) \in \mathrm{BR}_2 \circ \mathrm{BR}_1(p_2(t-1)) = (\mathrm{BR}_2 \circ \mathrm{BR}_1) \circ (\mathrm{BR}_2 \circ \mathrm{BR}_1)(p_2(t-3)) \end{cases}$$

1.5.2 About Selfishness

We already know that game theory both considers non-cooperative and cooperative games. But do the players have to be assumed selfish? The answer is no. In fact, there are situations where modeling players as selfish decision-makers leads to incorrect predictions. As mentioned by Levine (1998), some experiments cannot be explained by using this assumption. For instance, in public goods contribution games, players may make a costly donation to a common pool that provides a social benefit which is greater than the contribution. Because of the free rider problem, it is typically a dominant strategy not to contribute anything. Nevertheless, with as many as ten or more players, some players do contribute to the common pool. In this book, only games with selfish players are studied because this seems to be, in general, the most relevant assumption in wireless games. However, it is useful to bear in mind that this assumption is not necessary and there exists a literature dedicated to the problem of altruism in games (e.g., see Geanakoplos et al. (1989) and Rabin (1993)).

1.5.3 About Interaction

A game corresponds to a situation where (autonomous) decision-makers interact. Is there interaction in wireless networks? The answer depends on the scenario. Often interaction does exist because terminals share common resources. For example, when users exploit the same band at the same time, in the same geographical area, some multiuser interference occurs. Decisions made by the players are inter-dependent, in the sense that their performance metric depends on what the other players do: this is the interaction through actions. It is useful to know that there can be inter-action between players even if the player's utility does not directly depend on the actions of the others. Typically, there can be interaction through observations (e.g., see Rosenberg et al. (2007)). In everyday life, one can picture two mayors who have to choose a policy for the city they are respectively in charge of. Even if the chosen poli-cies have independent consequences, the fact that the two mayors observe each other can induce a correlation between them and lead to new policies, which are possibly more efficient (in terms of utility) than those obtained when no observation is possible.

1.5.4 About the Solution Concepts

While the Nash equilibrium is a key concept in game theory (more details are pro-vided in Chapter 2), it is not the only existing solution concept for a game. Indeed, in addition to providing interaction and behavioral models, game theory tries to provide solution concepts and answer to the question whether these solutions can be effec-tively observed in a given scenario. The importance of solution concepts is that in the interactive multi-person context, the very meaning of "optimal decision" is unclear, since in general, no one player completely controls the final outcome. Below, a short list of solution concepts is provided; most of them are used in this book (those which are explained briefly or not at all are marked by an asterisk ★).

1. Pure Nash equilibrium.
2. Mixed Nash equilibrium.
3. Dominant strategy Nash equilibrium (★).
4. Correlated equilibrium.
5. Coarse correlated equilibrium.
6. Strong equilibrium.
7. Robust equilibrium (★).

8. Logit equilibrium.
9. ϵ-Nash equilibrium.
10. Nash equilibrium refinements: trembling hand perfect equilibrium ($\star$), proper equilibrium ($\star$).
11. Bayesian equilibrium.
12. Wardrop equilibrium ($\star$).
13. Stackelberg equilibrium.
14. Evolutionary stable solution.
15. Satisfaction equilibrium.
16. Maxmin strategy profiles ($\star$).
17. Bargaining solutions: Nash solution, egalitarian solution ($\star$), Kalai-Smorodinsky solution ($\star$).
18. Cooperative games values: Shapley value, Harsanyi value ($\star$).
19. Pareto optimum.
20. Social optimum.

1.5.4.1 Comments on the Concepts of Pure, Mixed, and Correlated Equilibria

The notions of pure, mixed, and correlated Nash equilibria have been presented in Sec. 1.4. Depending on the game, one type of Nash equilibrium can be more appropriate than others. For instance, the following considerations can be made: a pure Nash equilibrium is a special case of a mixed Nash equilibrium, which is itself a special case of a correlated equilibrium. Correlated equilibria (resp. mixed Nash equilibria) are therefore more likely to occur than mixed Nash equilibria (resp. pure Nash equilibria). In finite games, it may happen that there is no pure Nash equilibrium and therefore only mixed or correlated equilibria are relevant. Mixed strategies have many useful interpretations (Osborne and Rubinstein, 1994), but they may not be implementable in some scenarios; correlated equilibria can be seen as an interesting concept in wireless games where some additional signaling is available to the terminals. Correlated equilibria can be more applicable in terms of a modeling situation where some terminals access signals which may correlate their decisions. Alternatively, additional signaling can be generated deliberately (e.g., by a base station) to obtain good trade-offs between global efficiency and signaling cost (for application examples, see Bonneau et al. (2006) and Chapter 3).

1.5.4.2 Comments on the Concept of Strong Equilibrium

Pure, mixed, and correlated equilibria are resilient to single deviations. If two (or more) players deviate in a coordinated manner, they can possibly obtain a higher utility than the one at the equilibrium. If Nash or correlated equilibria are found to be not robust enough in a given game, stronger requirements in terms of robustness may be sought. This is precisely what *strong equilibrium* is about. A strong equilibrium (Aumann, 1959) is a strategy profile $\underline{s}^*$ from which no coalition (of any size) can deviate and improve the utility of every member of the coalition (denoted by $\mathcal{C}$), while possibly lowering the utility of players outside the coalition (the corresponding set is denoted by $-\mathcal{C}$):

$$\forall i \in \mathcal{K}, \forall \mathcal{C} \subset \mathcal{K}, \forall \underline{s}_{\mathcal{C}}, \ u_i(\underline{s}^*) \geq u_i(\underline{s}_{\mathcal{C}}, \underline{s}^*_{-\mathcal{C}}). \tag{1.16}$$

A less strong requirement is to consider *k-strong equilibria*. In this case, the solution is stable to $k \geq 1$ deviations. Obviously, strong and k-strong equilibria are very desirable solutions. The question is whether such solutions exist. Here is one of the rare existence theorems for such a solution concept.

....

Theorem 33: Characterization of Strong Equilibria

Any strong equilibrium is a Pareto-optimal Nash equilibrium point and vice versa.

This means that if a Nash equilibrium can be found, and confirmed to be socially optimal, it is also a strong equilibrium. Below we provide an example of a non-cooperative game that has strong equilibria.

Example 34: Collision Channel Game

Consider the following collision game with two users. If the two users decide to transmit simultaneously, there is a collision. If one of the users transmits (T) and the other stays quiet ($\overline{T}$) then there is successful reception for the transmitter. If both stay quiet, their utility is zero. Table 1.4 provides a matrix form of this game. User 1 chooses a row and user 2 chooses a column.

Table 1.4 The Collision Channel Game in a Matrix Form

	T	$\overline{T}$
T	(0, 0)	(1, 0)
$\overline{T}$	(0, 1)	(0, 0)

The action profiles $(T, \overline{T})$ and $(\overline{T}, T)$ are both Nash equilibria and Pareto optimal points. These equilibria are also strong equilibria (but are not strictly strong equilibria).

1.5.4.3 *Comments on the Concept of Robust Equilibrium*

Robustness against deviators (called resilience) is an attractive feature present in strong equilibria, but it does not give any incentive to "loyal" players who stick to the equilibrium strategies. In order to protect loyal players, the notion of immunity can be added (Halpern, 2008). An equilibrium is said to be (d, ℓ)-robust if it is resilient to d deviators and immune for ℓ loyal players. This generalizes the notion of NE and strong equilibrium. It can be shown that the existence of a mediator, that is, a terminal which sends suitable recommendations, is sufficient to induce RE in the mediator-assisted game; under some conditions, mediation can be induced by direct communications between the players. This concept will not be studied in this book.

1.5.4.4 *Comments on the Concept of Logit Equilibrium*

The concept of a logit equilibrium will be used in Chapters 5 and 6, since it corresponds to the limiting behavior of the logit or Boltzmann-Gibbs learning (BGL) when the latter converges. The BG probability distribution associated with a vector $\underline{v} = (v_1,\ldots,v_M)$ with positive entries $v_m \geq 0$ is given by:

$$\forall m \in \{1,\ldots,M\}, \quad \beta_\alpha(x_m) = \frac{e^{\alpha x_m}}{\sum_{n\in\{1,\ldots,M\}} e^{\alpha x_n}} \tag{1.17}$$

where $\alpha \geq 0$ is a parameter (α represents temperature in physics). The idea of a logit equilibrium is that every player obtains the probability to assign to a given action by considering the mixed strategies played by the others ($\underline{x}_{-i}$) and taking the vector of utilities associated with each of his actions as the argument of $\beta(.)$. For player $i \in \{1,2,\ldots,K\}$:

$$\beta_{i,\alpha_i}(a_m) = \frac{e^{\alpha_i u_i(e_{a_m},\underline{x}_{-i})}}{\sum_{n\in\{1,\ldots,|A_i|\}} e^{\alpha_i u_i(e_{a_n},\underline{x}_{-i})}} \tag{1.18}$$

where a_m is the mth possible action of player i, α_i a given parameter (which will be interpreted as a rationality level in Chapters 5 and 6), and e_{a_m} the probability distribution for choosing the action a_m deterministically. The BG distribution therefore corresponds to a certain type of response for a player, which is called the smooth best response. A logit equilibrium (LE) is merely an intersection point of all these smooth best responses.

....

Definition 35: Logit Equilibrium

Consider the game $\mathcal{G} = \left(\mathcal{K}, \{u_j\}_{j\in\mathcal{K}}, \{A_j\}_{j\in\mathcal{K}}\right)$ and let the vector $\underline{x}^* = (x_1^*,\ldots,x_K^*) \in \Delta(A_1) \times \cdots \times \Delta(A_K)$. Then, $\underline{x}^*$ is a logit equilibrium with parameter $\underline{\alpha} = (\alpha_1,\ldots,\alpha_K)$ if for all $i \in \mathcal{K}$, it holds that:

$$x_i^* = \beta_{i,\alpha_i}\left(u_i\left(e_1,\underline{x}_{-i}^*\right),\ldots,u_i\left(e_{|A_i|},\underline{x}_{-j}^*\right)\right). \tag{1.19}$$

The logit equilibrium is a special type of ϵ-Nash equilibrium, or a Nash equilibrium in a perturbed game (this is explained in Chapter 5).

1.5.4.5 *Comments on the Concept of ϵ-Nash Equilibrium*

An action profile is an ϵ-Nash equilibrium if no player can gain more than ϵ by deviating unilaterally.

....

Definition 36: ϵ-Nash Equilibrium

Let $\epsilon \geq 0$. The profile $\underline{a}^*$ is an ϵ-Nash equilibrium of $\mathcal{G} = \left(\mathcal{K}, \{u_j\}_{j\in\mathcal{K}}, \{A_j\}_{j\in\mathcal{K}}\right)$ if:

$$\forall i \in \mathcal{K}, \forall a_i' \in A_i, \quad u_i(\underline{a}^*) \geq u_i(a_i',\underline{a}_{-i}^*) - \epsilon. \tag{1.20}$$

1.5.4.6 *Comments on the Concept of Refined Nash Equilibrium*

In its original definition, the Nash equilibrium does not take into account the fact that some players may choose unintended strategies, albeit with negligible probability. Nash equilibria are not necessarily robust against small "errors" made by some players. This lack of robustness to such effects can explain why theoretically predicted Nash equilibria may never be observed in some scenarios. To tackle this problem, refinements of Nash equilibria have been proposed (such as the concepts of perfect equilibrium (Selten, 1975), proper equilibrium (Myerson, 1978), essential Nash equilibrium components (Jiang, 1963), strategically stable sets (Kohlberg and Mertens, 1986), sets closed under rational behavior (Basu and Weibull, 1991)). In this short section, only the perfect equilibrium is presented.

The (trembling hand) perfect equilibrium is a refinement of the Nash Equilibrium developed by Selten (1975). A trembling hand perfect equilibrium is one that takes the possibility of off-the-equilibrium play into account. To define it, let us consider the mixed extension of a finite strategic game $\mathcal{G} = \left(\mathcal{K}, \{\mathcal{A}_i\}_{i \in \mathcal{K}}, \{u_i\}_{i \in \mathcal{K}}\right)$. In the extended game $\widetilde{\mathcal{G}}$, only strictly mixed strategies are allowed.

Definition 37: ϵ-Perfect Equilibrium

An ϵ-perfect equilibrium $\mathcal{G}$ is a strictly mixed strategy $\underline{x}^\epsilon$ such that, for each player i, $x_i^\epsilon \in \arg\max_{p_i} \widetilde{u}_i(p_i, \underline{x}^\epsilon_{-i})$ subject to $x_i^\epsilon(a_i) \geq \epsilon(a_i)$ for some $\{\epsilon(a_i)\}_{i \in \mathcal{K}, a_i \in \mathcal{A}_i}$ where $0 < \epsilon(a_i) < \epsilon$.

Definition 38: Perfect Equilibrium

A (trembling-hand) perfect equilibrium is any limit of ϵ-constrained equilibria as ϵ goes to zero.

From Selten (1975), it is known that at least one perfect equilibrium exists in any finite game. The example below illustrates this.

Example 39

Consider the matrix form game with 2 players, represented by Table 1.5.

Table 1.5 A Game Illustrating the (Trembling Hand) Perfect Equilibrium Notion.

	c_1	c_2
r_1	(1, 1)	(2, 0)
r_2	(0, 2)	(2, 2)

This game has two pure Nash equilibria (r_1, c_1) and (r_2, c_2). Only the second one is perfect. Indeed, it can be confirmed that for the first equilibrium, if Player 1 (resp. 2) plays r_1 (resp. c_1) with probability $1 - \epsilon$ and r_2 (resp. c_2) with probability ϵ, Player 2 has no interest in deviating from his equilibrium action c_1 (resp. r_1). However, for the second equilibrium, if Player 1 (resp. 2) plays r_2 (resp. c_2) with probability $1 - \epsilon$ and r_1 (resp. c_1) with probability ϵ, Player 2 (resp. 1) gets a better expected utility by deviating from his equilibrium action c_2 (resp. r_2).

1.5.4.7 *Comments on the Concept of Bayesian Equilibrium*

This concept is explained in Chapter 4 and illustrated in Chapter 8. At this point, it is worth mentioning that the Bayesian equilibrium is a Nash equilibrium for expected utilities. The way the expectation is calculated depends on the knowledge of the players (through conditional probabilities, beliefs, etc.). The Bayesian equilibrium is used in games with incomplete information, where players do not have a complete knowledge of the game parameters (the channel state typically).

1.5.4.8 *Comments on the Concept of Wardrop Equilibrium*

The Wardrop equilibrium was introduced by Wardrop (1952) in the framework of transport problems in which the number of users is generally very large. Its definition is based on two principles, called Wardrop's principles.

Wardrop's first principle. The journey times in all routes actually used are equal and less than those which would be experienced by a single vehicle on any unused route.

Wardrop's second principle. At equilibrium the average journey time is minimum.

This first principle is usually referred to as the user equilibrium. Each user non-cooperatively seeks to minimize his cost of transportation, and a user-optimized equilibrium is reached when no user can reduce his transportation cost through unilateral deviation. The second principle implies that each user behaves cooperatively in choosing his own route, so as to ensure the most efficient use of the whole system. Traffic flows satisfying the second principle are generally deemed "system optimal".

What is the connection between Wardrop and Nash equilibria? The Wardrop equilibrium can be seen as a special case of a Nash equilibrium, in which the number of players is infinite (continuum of players) and all the players have identical action sets and cost functions. Nash himself established this in 1951 for the case of a multi-class of players. The Nash equilibrium condition is equivalent to the following: for any player i, any action $a_i \in \mathcal{A}_i$,

$$x_i^*(a_i) > 0 \Longrightarrow \tilde{u}_i(\mathbf{e}_{a_i}, \mathbf{x}_{-i}^*) = \max_{a_i' \in \mathcal{A}_i} \tilde{u}_i(\mathbf{e}_{a_i'}, \mathbf{x}_{-i}^*) = \max_{\mathbf{x}_i' \in \mathcal{X}_i} \tilde{u}_k(\mathbf{x}_k', \mathbf{x}_{-k}^*) \tag{1.21}$$

which says exactly that:

$$\forall\, i \in \mathcal{K},\ \mathrm{supp}(\mathbf{x}_i^*) := \{a_i \mid x_i^*(a_i) > 0\} \subseteq \arg\max_{\mathcal{X}_i} \tilde{u}_i(., \mathbf{x}_{-i}^*). \tag{1.22}$$

This clearly shows that equilibria defined by the two Wardrop principles are Nash equilibria. Now, in the case of a single population of players, equilibria satisfying the first principle are Nash equilibria. Since routing is a fundamental part of communication networks, it was natural to expect the concepts of WE and of NE to appear in routing games. The natural analogy between packets in a network and cars on the road was not satisfactory for applying the Wardrop concept to networking, since packets, unlike cars, do not choose their routes. These are determined by the network routing protocols. Moreover, for a long time routing issues did not occur in wireless networking, as it was restricted to the access part of a network where each terminal is associated with a given BS. Routing has become relevant to wireless networking since ad-hoc networks were introduced; e.g., see the pioneering work of Gupta and Kumar (1997). However, game theory has penetrated into ad-hoc networks in a completely unexpected way. Indeed, to our knowledge, it was in this context that non-cooperative game theory appeared for the first time as a basic powerful tool for solving problems that do not involve competition. Gupta and Kumar (1997) proposed a routing algorithm for packets that had to satisfy some properties: packets should be routed by the network so as to follow the shortest path. It is thus not the average delay in the network that is minimized; the design objectives are to find routing strategies that (i) equalize the delays of packets that have the same source and destination, so as to avoid the re-sequencing delays that are quite harmful for real-time traffic as well as for the TCP protocol for data transfers, and (ii) make these delays minimal, i.e., do not use routes that have delays larger than the minimal ones. The authors designed a network routing protocol that achieved these two objectives.

However, these objectives are in fact the WE definition, arising in a completely non-competitive context. It was only ten years later that the WE penetrated into ad-hoc networks in a competitive game context. The problem of competitive routing in massively dense ad-hoc networks was introduced in Jacquet (2004). The author observed that as the density of mobiles grows, the shortest paths tends to converge on to some curves that can be described using tools from geometrical optics in non-homogeneous media. Later, the electromagnetic field was used to approximate the routes of packets in massively dense ad-hoc networks (see Toumpis (2006) and references therein). Only later, in Altman et al. (2008) and Silva et al. (2008) the natural concept of WE was introduced in massively dense ad-hoc networks as a natural tool that deals with the competitive interactions between mobiles. This is a special version of WE defined on the continuum limit of a network, whose links and nodes are so dense that they are replaced by a plain.

1.5.4.9 *Comments on the Concept of the Stackelberg Equilibrium*

Stackelberg (1934) introduced a model of competition between two economic actors, where one actor moves first and the second one reacts. A game where one player (called the leader) has to move first and knows that his action is observed by a rational player, and then the other player (called the follower) observes the played action and reacts to it, is called a two-player Stackelberg game. A Nash equilibrium in a game of this structure is called a Stackelberg equilibrium. This notion can be generalized to K-player games with a given hierarchy structure in terms of observation, as illustrated in Chapter 2 in the context of energy-efficient power control games.

1.5.4.10 *Comments on the Concept of an Evolutionary Stable Solution*

Instead of Nash equilibria in classical game theory, evolutionary game theory uses the concepts of evolutionary stability: the unbeatable state concept (Hamilton, 1967), or the evolutionarily stable state (ESS) concept (Smith and Price, 1973). A strategy is *unbeatable* if it is resilient to any fraction of deviants of any size. A strategy is *evolutionarily stable* if a whole population using this strategy cannot be invaded by a small fraction of deviants (mutants). Interestingly, rationality is not required to observe such a solution, and it is especially relevant for large networks where robustness to single deviations can be insufficient. All these notions will be defined formally in Chapter 3 and illustrated in Chapter 4.

1.5.4.11 *Comments on the Concept of Satisfaction Equilibrium*

Although the NE is an attractive solution concept, from a practical point of view, a wireless operator, a service provider, or even an end-user, is often more interested in guaranteeing a certain minimum quality of service level (e.g., for voice or multimedia services) rather than in reaching the highest achievable performance. In such scenarios, using the NE as a solution concept might fail to model the real behavior of wireless communication networks. As a result, in the presence of minimum QoS requirements, a more suitable solution concept for non-cooperative games is the generalized NE (GNE) that was proposed by Debreu (1952) and later by Rosen (1965). Within a wireless network, a GNE is a game outcome in which all transmitters select their action in such a way that their performance cannot be improved by unilateral deviations and, at the same time, certain QoS levels can be guaranteed. This concept is of particular interest for scenarios in which one needs not only a stable state of the game, but also certain guaranteed QoS levels. However, depending on the QoS metrics and the network topology, the GNE might not exist (Lasaulce et al., 2009a). Even when it exists, a transmitter might still end up attaining the highest achievable performance, which, in several scenarios, turns out to be costly in terms of energy consumption, spectral efficiency, etc. This leads us to the idea of satisfaction equilibrium: a player is said to be satisfied if he plays a strategy which satisfies his constraints. Once a player satisfies his own constraints he has no interest in changing his strategy, and, thus, an equilibrium is observed if all players are simultaneously satisfied. We refer to this solution as satisfaction equilibrium and we define it as follows:

Definition 40: Satisfaction Equilibrium

Let $\mathcal{G}$ be a strategic form game and $f_1, f_2, \ldots, f_K$ be K set-valued satisfaction functions. The strategy profile $\underline{s}^* = (s_1^*, s_2^*, \ldots, s_K^*)$ is an SE if:

$$\forall i \in \mathcal{K}, \quad s_i^* \in f_i(\underline{s}_{-i}^*). \tag{1.23}$$

Note that if one defines the correspondence f_i, for all $i \in \mathcal{K}$, as $f_i(\underline{s}_{-i}) = \{s_i \in \mathcal{S}_i : u_i(s_i, \underline{s}_{-i}) \geqslant \gamma_i\}$, where γ_i is the minimum utility level required by player i, then Def. 40 coincides with the definition of SE provided in Ross and Chaib-draa (2006). Finally, note that an SE can be re-interpreted as an NE in an auxiliary game where the new player's utility (say v_i) equals 1 if he is satisfied and 0 otherwise, justifying why the term "equilibrium" is used. However, there is no equivalence in terms of solutions between the original game and this auxiliary game.

1.5.4.12 *Comments on the Concept of Maxmin Strategy Profiles*

By definition, a Nash equilibrium is a profile of strategies which is stable to single deviations. In this sense, the concept of Nash equilibria possesses some stability properties. As mentioned in Sec. 1.5.1, players do not always (or may not necessarily want to) maximize their utility. Many people who arrive much earlier than their train or bus leaves could maximize their expected utility by decreasing the margin they take to catch their train or bus with high probability. Nash equilibria do not capture the notion of risk. By contrast, the maxmin solution is a very simple concept, which employs the latter notion (more advanced concepts and different approaches of risk aversion exist, e.g., see Driesen (2010)). Let us illustrate this concept with a simple example (Laraki and Zamir, 2003).

Example 41

Consider the game described by Table 1.6: player 1 chooses the row and player 2 chooses the column. The unique pure Nash equilibrium of this matrix game is the action profile (r_3, c_2). However, if for some reason, player 2 chooses column c_1, player 1 gets a very bad utility (namely -100). As a consequence, playing at the Nash equilibrium is very dangerous from the standpoint of player 1. Such a player may want to play the action r_1 which provides him with a utility equal to 2 whatever the other player does.

Table 1.6 A Game Illustrating the Notion of Maxmin Solution (Laraki and Zamir, 2003)		
	c_1	c_2
r_1	(3, 1)	(2, 2)
r_2	(0, 8)	(0, −1)
r_3	(−100, 2)	(3, 3)

This utility level is precisely the maxmin level of player 1. It is obtained as follows: for each action of player 1, player 1 assumes that the other player chooses the worst action in terms of player 1's utility:

- if player 1 chooses r_1, player 2 chooses c_2 ($u_1(r_1, c_2) = 2$);
- if player 1 chooses r_2, player 2 chooses c_1 or c_2 ($u_1(r_2, c_2) = u_1(r_2, c_2) = 0$);
- if player 1 chooses r_3, player 2 chooses c_1 ($u_1(r_2, c_1) = -100$).

Clearly $\max\{2, 0, -100\} = 2$. Clearly the maxmin level of player 2 can be found, which is $\max\{1, -1\} = 1$. The latter utility is guaranteed by playing c_1. Therefore if the maxmin profile (r_1, c_1) is played, player 1 is certain to obtain at least 2 and player 2 is certain to obtain at least 1.

1.5.4.13 *Comments on the Concept of Nash Bargaining Solution*

We have already mentioned the importance of the Nash equilibrium, which is one building block of game theory. As will be discussed in Chapter 2, Nash equilibria can not only be interpreted as possible outcomes of interactive situations involving rational players but also as a result of some evolutionary process. A second ground-breaking contribution by Nash is the Nash bargaining solution. Different bargaining solutions exist, but only the one by Nash is discussed here. One important idea of bargaining theory, initiated by the Nash bargaining solution for two-player games, is that solution concepts of cooperative games (which are generally based on an abstract approach of the problem) can be re-interpreted as non-cooperative games (involving much more detail and based on the idea of individual utility maximization from the knowledge/belief of the players). The concern of the corresponding literature is to construct non-cooperative bargaining games that sustain various cooperative solution concepts as their equilibrium outcomes. As far as wireless networks are concerned, the Nash bargaining solution can be seen as a way of inducing a cooperative/efficient/fair solution. For example, it was used about 20 years ago to obtain fair solutions to flow control problems in communication networks (Mazumdar et al., 1991). More recently, it has been exploited; e.g., in Touati et al. (2006) for solving bandwidth allocation problems, in Boche et al. (2007), for having weighted proportional fairness in resources allocation in wireless networks, and in Larsson and Jorswieck (2008) to obtain a cooperative solution in multiple input, single output, interference channels. Mainly based on the idea of having a solution which is fair and more efficient then the one obtained without bargaining or agreement, the Nash bargaining solution is characterized by a set of axioms regarding the desired outcome:

1. Pareto optimality (PO);
2. individual rationality (IR);
3. invariance to positive affine transformations (IPAT);
4. independence of irrelevant alternatives (IIA);
5. symmetry (S).

To define these axioms, some definitions are in order.

Definition 42: Bargaining Problem

A bargaining problem is a pair $(\mathcal{F}, \underline{v})$ where $\mathcal{F} \in \mathbb{R}^2$ is a closed and convex set, $\underline{v} = (v_1, v_2) \in \mathbb{R}^2$, and such that the set $\mathcal{F} \cap \{(u_1, u_2) \in \mathbb{R}^2 : u_1 \geq v_1, u_2 \geq v_2\}$ is bounded and non-empty.

Following the terminology used by Nash, $\mathcal{F}$ represents the set of deals available to the bargainers, while $\underline{v}$ is a point representing the status quo.

Definition 43: Bargaining Solution

Let $\mathcal{B}$ be the set of all bargaining problems. A bargaining solution is an application which assigns to each bargaining problem a unique solution that is

$$\underline{\mu}: \begin{array}{ccc} \mathcal{B} & \to & \mathbb{R}^2 \\ (\mathcal{F}, \underline{v}) & \mapsto & \underline{\mu}(\mathcal{F}, \underline{v}) = (\mu_1, \mu_2) \end{array}. \tag{1.24}$$

Definition 44: Symmetric Bargaining Problem

A bargaining problem $(\mathcal{F}, \underline{v})$ is symmetric if:

- $v_1 = v_2$;
- $(u_1, u_2) \in \mathcal{F} \Leftrightarrow (u_2, u_1) \in \mathcal{F}$.

From these definitions, the above axioms can be expressed mathematically as follows:

1. PO: $\forall \underline{u} \in \mathcal{F}, \underline{u} \geq \underline{\mu}(\mathcal{F}, \underline{v}) \Rightarrow \underline{u} = \underline{\mu}(\mathcal{F}, \underline{v})$;
2. IR: $\underline{\mu}(\mathcal{F}, \underline{v}) \geq \underline{v}$;
3. IPAT: let the pair $(\mathcal{F}', \underline{v}')$ be defined as the result of an affine transformation applied to $(\mathcal{F}, \underline{v})$: $\mathcal{F}' = \{(u'_1, u'_2)\mathbb{R}^2 : u'_i = \alpha'_i u_i + \beta_i, (u_1, u_2) \in \mathcal{F}, \alpha_i > 0, \beta_i \in \mathcal{R}\}$. Then $\underline{\mu}(\mathcal{F}', \underline{v}' = (\alpha_1 \mu_1 + \beta_1, \alpha_2 \mu_2 + \beta_2)$;
4. IIA: let $\mathcal{F}' \subseteq \mathcal{F}$. Then, $\underline{\mu}(\mathcal{F}, \underline{v}) \in \mathcal{F}' \Rightarrow \underline{\mu}(\mathcal{F}', \underline{v}) = \underline{\mu}(\mathcal{F}, \underline{v})$;
5. S: if $(\mathcal{F}, \underline{v})$ is symmetric then $\mu_1 = \mu_2$.

The following theorem exploits these axioms.

Theorem 45: Nash Bargaining Solution

There exists a unique bargaining solution verifying the axioms 1–5. This solution is the unique Pareto-optimal utility profile, satisfying:

$$\underline{\mu}(\mathcal{F}, \underline{v}) \in \arg \max_{\underline{u} \in \mathcal{F}, \underline{u} \geq \underline{v}} (u_1 - v_1)(u_2 - v_2). \qquad (1.25)$$

This solution concept can be extended to K-players for $K \geq 2$ but it does not take into account the fact that some coalitions of players can form, and deviate from the solution.

1.5.4.14 *Comments on the Concept of Value*

The most famous type of value is the one invented by Shapley (1964). Like the Nash bargaining solution, the Shapley value is based on several axioms, and can be sustained via non-cooperative games. Indeed, it can be interpreted as a measure for evaluating a player's influence on a cooperative game, and also as a rule for allocating collective benefits or costs.

1.5.4.15 *Comments on the Concept of Pareto-Optimality*

Pareto-optimality was defined in Def. 25 and does not call for particular comments at this point. This concept is more a property which is desirable for other solution concepts, such as the Nash equilibrium. For instance, Pareto-optimality can serve as a mean of selecting Nash equilibria. When utility regions are considered, the set of all Pareto-optimal points is called the *Pareto frontier*.

1.5.4.16 *Comments on the Concept of Social Optimum*

A common way of measuring the performance of a system composed of several entities is to consider the sum of individual performance metrics. This is especially important in decentralized systems, where these entities make their decisions autonomously. In the literature of non-cooperative games (whose terminology is influenced by economics), the corresponding quantity is called social welfare (Arrow, 1963):

Definition 46: Social Welfare

The social welfare of a game is defined as the sum of the utilities of all players:

$$w = \sum_{i=1}^{K} u_i. \tag{1.26}$$

A *social optimum* is therefore a maximum of the social welfare. Any social optimum is Pareto optimal (Arora, 2004). However, the converse is not always true and only holds under some sufficient conditions. In particular, if the region of feasible utilities is convex, the converse is true (Athan and Papalambros, 1996). The notion of social welfare is used throughout this book, in particular to define the price of anarchy (Chapter 2).

Playing with Equilibria in Wireless Non-Cooperative Games

2.1 INTRODUCTION

As seen in Chapter 1, equilibria, particularly Nash equilibria, are of paramount importance in non-cooperative games. Intrinsically, Nash points are just vectors which satisfy a given property which involves $K \in \mathbb{N}^*$ functions (or mappings), and are, a priori, defined independently of the concept of a game. Since their original definition by Nash (1950) several fundamental interpretations of these points have been made, the most famous of these (made by Nash himself) is that they can be the solutions of non-cooperative games involving rational players. In these games, which can be static or dynamic, with complete or incomplete information, with perfect or imperfect observation, players are generally assumed to have a good knowledge of the structure of the game (e.g., they know the number of players).

Remarkably, there are other interpretations of Nash equilibria which correspond to scenarios that have less stringent information and behavior assumptions for the players. One of the most important of these scenarios is that NE can be the result of an evolution or learning process. Roughly speaking, the main idea behind this is that a NE predicted in a static game with full information can also be observed in a dynamic game with partial information. The best response dynamics and learning reinforcement algorithms are simple examples of dynamics which can lead to Nash equilibria; they will be presented in detail in Chapters 5 and 6. The iterative water-filling algorithm introduced in Yu et al. (2004) in the context of distributed power allocation in multiple access channels corresponds exactly to the best-response dynamics based on the actions (namely the power allocation policies) carried out by the other transmitters; say $\underline{a}_{-i}$, then transmitter i plays his best response, player $j \neq i$ then updates his action, and the others do so in a sequential manner. This procedure can be proved to converge on the pure Nash equilibrium that would be obtained in a one-shot game played by players having complete information (Bertsekas, 1995).

In the context of the best-response dynamics, players need to observe the actions played by the others; however, reinforcement algorithms (e.g., Mosteller and Bush (1955) and Sastry et al. (1994b)) only assume the knowledge of the instantaneous value of the utility functions at each stage of the game. More precisely, there is no need to assume that rationality is common knowledge and each player knows the structure of the game. This approach is close in spirit to the mass-action interpretation made by Nash in his thesis where he mentioned:

> *"It is unnecessary to assume that the participants have full knowledge of the total structure of the game, or the ability and inclination to go through any complex reasoning processes."*

But the participants are supposed to accumulate empirical information on the relative advantages of the various pure strategies at their disposal. To be more precise, we assume a population (in the sense of statistics) of participants for each position of the game. Let us also assume that an "average play" of the game involves n participants selected at random from the population, and that there is a stable average frequency with which each pure strategy is employed by the average member of the population. Since there is no collaboration between individuals playing in different positions of the game, the probability that a particular n-tuple of pure strategies will be used in a play should be the product of the probabilities of each of the players. [...] Thus the assumption we made in this mass-action interpretation leads to the conclusion that the mixed strategies representing the average behavior in each of the populations form an equilibrium point. The considerable development of learning theory in games, evolutionary games, etc., has shown how relevant this interpretation is.

These considerations show how Nash equilibria appear to be a good solution concept in various interactive situations. This justifies, in part, why so many papers in the wireless literature are dedicated to Nash equilibrium analysis. This chapter provides some elements to help the reader to conduct her/his own analysis of Nash equilibria in wireless games. Typical issues are the problems of existence, uniqueness, selection, efficiency, and also the convergence properties of the game under study, and important questions to be answered in a conflicting or cooperative situations are: Is there an outcome to this situation? If yes, is it unique? If it does not exist, what can be done? On the other hand, if there are several possible outcomes, how to select one of them? If the selected or effective outcome is found to be inefficient in a given sense of efficiency measure, how can the game be modified to improve its efficiency? This chapter will be structured according to this list of questions, looking at situations where the outcomes of interest are Nash equilibria. As an important comment, note that, unless stated otherwise, all the results will hold for any strategic form game. This means that the results can be directly applied to one-shot games in their strategic form, but also to dynamic games if the latter can be written under this form. However, for the sake of simplicity, the wireless examples used to illustrate the theorems will concern one-shot games exclusively.

2.2 EQUILIBRIUM EXISTENCE

Mathematically speaking, proving the existence of a Nash equilibrium amounts to proving the existence of a solution to a fixed-point problem (Nash, 1950); this is why so much effort is made to derive fixed-point theorems (e.g., see Border (1985) and Smart (1974)) by game theorists. Therefore, it will not be surprising to see that equilibrium existence theorems are based on the topological properties of the strategy sets of the players, and the topological and geometrical properties of their utility functions. This explains why the respective works by Lefschetz (1929), Hopf (1929), Brouwer (1912) and Kakutani (1941) on fixed-point theorems (FPT) have had an important impact in game theory.

To prove the existence of an equilibrium for a problem at hand, one may always attempt to derive an FPT, which is the most general method. However, there are many scenarios in usual channel models and performance metrics for which existing theorems are sufficient. For example, Shannon transmission rates or rate regions quite often have desirable convexity properties that are needed in standard known theorems for equilibrium existence. The purpose of this section is to give details of certain of these useful existence theorems,[1] and mention some examples where they have been applied in the literature of wireless communications. The structure of this section follows the chart in Figure 2.1.

2.2.1 Better-Reply Secure and Quasi-Concave Games

Before stating the corresponding theorem (developed in Reny (1999)), we first define the concepts of better-reply security and quasi-concavity (other mathematical notions, such as concavity and compactness are reviewed in the Appendices). As in Chapter 1, we will use the notation $\underline{s} = (s_1, \ldots, s_K)$ and $\underline{u} = (u_1, \ldots, u_K)$ to refer to strategy and utility profiles respectively.

....

Definition 47: Better-Reply Secure Games

Let $\mathcal{G} = (\mathcal{K}, \{S_i\}_{i \in \mathcal{K}}, \{u_i\}_{i \in \mathcal{K}})$ be a strategic game. Let $\underline{s}'$ be a strategy profile which is not a Nash equilibrium of $\mathcal{G}$, and $\underline{u}'$ the corresponding utility profile. The game $\mathcal{G}$ is better-reply secure (BRS) if, whenever the pair $(\underline{s}', \underline{u}')$ is in the closure of the graph of its utility profile function, then for some player i, there exists $\bar{s}_i$ such that for all $\underline{s}_{-i}$ close to $\underline{s}'_{-i}$, $u_i(\bar{s}_i, \underline{s}_{-i}) > u'_i$.

Better-reply secure games can be seen as a generalized version of continuous games in which utility functions are continuous with respect to the strategy profile. An historical example of better-reply secure games is the Bertrand competition (Bertrand, 1883). Considering games which are not continuous in wireless communications might be surprising, but in fact this scenario can happen (see for example, the base station location game treated in Mériaux et al. (2011a)) and will probably be discussed more and more often in the wireless literature.

Now let us define quasi-concavity.

---

[1] In the whole chapter the strategy sets are always assumed to be non-empty. For refinements concerning this assumption, the reader is invited to consult the references associated with the relevant theorems.

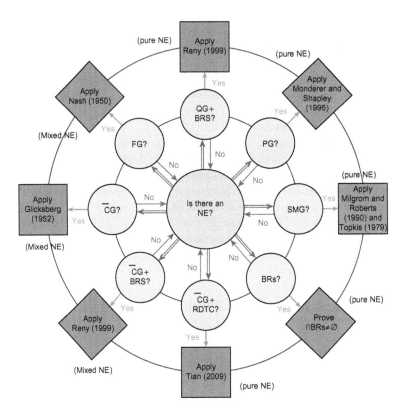

FIGURE 2.1

A (non-exhaustive) methodology for proving the existence of a Nash equilibrium in strategic form games. The meaning of the used acronyms is as follows: QG (quasi-concave games), BRS (better-reply secure; Reny, 1999), PG (potential games; Monderer and Shapley, 1996), SMG (S-modular games; Milgrom and Roberts, 1990; Topkis, 1979), BRs? (Are the best-responses known?), $\overline{\text{C}}$G (compact games; Glicksberg, 1952), RDTC (recursive diagonally transferable continuity; Tian, 2009), FG (finite games; Nash, 1950).

Definition 48: Quasi-Concave Functions

A function ψ is quasi-concave on a convex set $\mathcal{S}$ if, for all $\alpha \in \mathbb{R}$, the upper contour set $\mathcal{U}_\alpha = \{x \in \mathcal{S}, f(x) \geq \alpha\}$ is convex. We will use the acronym QG to refer to games in which utilities are quasi-concave.

At this point, the Reny existence theorem for pure NE (Reny, 1999) can be stated.

Theorem 49: Reny (1999)

Let $\mathcal{G}$ be a strategic form game. If: $\forall i \in \mathcal{K}$: $\mathcal{S}_i$ is a compact and convex set; $u_i(\underline{s})$ is quasi-concave in s_i; and $\mathcal{G}$ is better-reply secure, then the game $\mathcal{G}$ has at least one pure NE.

Whereas this theorem has been applied in the game theory literature (e.g., for auctions), it seems that it has not yet been applied in the wireless literature. A less general version, obtained by assuming continuous utility functions, has been studied. This is the purpose of the next section.

2.2.2 Continuous Quasi-Concave Games

A very useful existence theorem is stated in Fudenberg and Tirole (1991), resulting from the contributions of Debreu (1952), Fan (1952) and Glicksberg (1952). The Debreu-Fan-Glicksberg theorem, as in Fudenberg and Tirole (1991), can be applied if the users' utilities are both continuous and quasi-concave, and some other conditions are satisfied.

....

Theorem 50: Debreu (1952)

Let $\mathcal{G}$ be a strategic form game. If: $\forall i \in \mathcal{K}$: $\mathcal{S}_i$ is a compact and convex set and $u_i(\underline{s})$ is a continuous function in the profile of strategies $\underline{s}$ and quasi-concave in s_i, then the game $\mathcal{G}$ has at least one pure NE.

A special case of this theorem is obtained by assuming the utility functions to be concave[2]. In this respect, Theorem 1 by Rosen (1965) for concave K-person games can be seen as a corollary of Theorem 50. As a comment, note that we will use the term "concave games" but not the term "convex games", which can be confused with games where strategy sets are convex.

Example 51

For the distributed energy-efficient power control problem introduced in Goodman and Mandayam (2000), $\mathcal{G}$ is defined as follows. The strategy (which is an action in the setup under consideration) of user $i \in \mathcal{K}$ is his instantaneous transmit power $p_i \in [0, P_i^{\max}]$ and his utility is $u_i(\underline{p}) = \frac{f(\mathrm{SINR}_i)}{p_i}$ where $f : \mathbb{R}^+ \to [0,1]$ is a sigmoidal[3] efficiency function (e.g., the packet success rate) and SINR_i is the signal-to-interference plus noise ratio for user i. This game can be shown to be quasi-concave and thus has at least one pure NE.

Of course, there is no reason why utilities used in wireless games should always be quasi-concave. For instance, in the case of the energy-efficient power control game we have mentioned, the quasi-concavity property is lost when the utility is modified into $\tilde{u}_i(\underline{p}) = u_i(\underline{p}) + \alpha p_i$, which corresponds to implementing a linear pricing technique (Saraydar et al., 2002). If a game can be shown to be non-quasi-concave, it still can have some properties that ensure the existence of a pure NE. This is exactly the case if one has to deal with a potential game (PG; Monderer, 1996) or an S-modular game (SMG; Milgrom and Roberts, 1990; Topkis, 1998). S-modular games include

---

[2]Concavity implies quasi-concavity. However, note that the sum of quasi-concave functions is not necessarily quasi-concave and quasi-concavity is generally not preserved by adding an affine function.
[3]A sigmoidal or S-shaped function is a function which is convex up to a point and then becomes concave from this point on.

sub-modular games (Topkis, 1979) and super-modular games (Milgrom and Roberts, 1990; Topkis, 1998).

The following discussion will give the definition, an existence theorem, and an example of a game for these types of games. The notation $\underline{a} \geq \underline{b}$ will mean that each entry of the vector $\underline{a}$ is greater than or equal to the entry having the same index in the vector $\underline{b}$.

2.2.3 Potential Games

Potential games have been introduced by Monderer and Shapley in Monderer (1996). They proposed four types: weighted potential games, exact potential games, ordinal potential games, and generalized potential games. Since then, other types of potential games have been introduced. For instance, Voorneveld introduced best-response potential games in Voorneveld (2000). As far as the existence of a Nash equilibrium is concerned, distinguishing between these types of potential games is not relevant. The reasons for their inclusion are twofold: depending on the game under consideration, proving that the game is potential can be easier for a given type; and, although existence is guaranteed for all the mentioned types, they do not have the same properties in general (especially in terms of convergence).

··

Definition 52: Exact Potential Games (EPG)

The game $\mathcal{G}$ is an exact potential game if there exists a function ϕ such that:

$$\forall i \in \mathcal{K}, \forall \underline{s} = (s_i, \underline{s}_{-i}) \in \mathcal{S}, \forall s_i' \in \mathcal{S}_i, u_i(\underline{s}) - u_i(s_i', \underline{s}_{-i}) = \phi(\underline{s}) - \phi(s_i', \underline{s}_{-i}). \qquad (2.1)$$

As in physics, there can be an infinite number of potential functions. What generally matters is whether one of them can be found. Alternatively, one of the following conditions can be tested:

- the weighted potential game (WPG) condition: there exists a function ϕ such that:

$$\forall i \in \mathcal{K}, \forall \underline{s} = (s_i, \underline{s}_{-i}) \in \mathcal{S}, \forall s_i' \in \mathcal{S}_i, \; u_i(\underline{s}) - u_i(s_i', \underline{s}_{-i}) = w_i \left[\phi(\underline{s}) - \phi(s_i', \underline{s}_{-i}) \right]. \qquad (2.2)$$

  for some vector of positive numbers $\underline{w} = (w_1, w_2, \dots, w_K)$;
- ordinal potential game (OPG) condition: there exists a function ϕ such that:

$$\forall i \in \mathcal{K}, \forall \underline{s} = (s_i, \underline{s}_{-i}) \in \mathcal{S}, \forall s_i' \in \mathcal{S}_i, \; u_i(\underline{s}) - u_i(s_i', \underline{s}_{-i}) > 0 \Leftrightarrow \phi(\underline{s}) - \phi(s_i', \underline{s}_{-i}) > 0; \qquad (2.3)$$

- generalized potential game (GPG) condition: in the ordinal potential game condition, replace the equivalence $\Leftrightarrow$ by an implication $\Rightarrow$;
- best-response potential game (BRPG) condition: there exists a function ϕ such that:

$$\arg\max_{s_i \in \mathcal{S}_i} u_i \left(s_i, \underline{s}_{-i} \right) = \arg\max_{s_i \in \mathcal{S}_i} \phi \left(s_i, \underline{s}_{-i} \right). \qquad (2.4)$$

By denoting Γ^{XPG} as the set of strategic potential games with $X \in \{E, W, O, G, BR\}$, we have that $\Gamma^{EPG} \subseteq \Gamma^{OPG} \subseteq \Gamma^{GPG}$ and $\Gamma^{EPG} \subseteq \Gamma^{OPG} \subseteq \Gamma^{BRPG}$ (see Monderer (1996) and Voorneveld (2000)). In any case, it is important to note that the potential function is independent of the user index: for every player, ϕ allows one to quantify the impact of a unilateral deviation on all the users' utilities in exact potential games while it gives the sign of the difference of utilities in ordinal potential games. This is reminiscent to the definition of a potential in physics. For example, when there exists a potential function ϕ for the field $\underline{E}$ we have that $\underline{E} = -\text{grad}(\phi)$. This can mean that the N-dimensional problem (with $N = \dim \underline{E}$) can be studied through a one-dimensional problem where ϕ is exploited (e.g., think about the calculation of the work of a force when the force is gravitational). In this respect, in certain games with a continuum of players (e.g., see Sandholm (2011)), the exact potential game has precisely the form $\underline{u} = -\text{grad}(\phi)$. Finally, note that when the type of potential game is not specified, exact potential games are generally considered; this is the convention we adopt in this book.

For EPG/WPG/OPG/GPP/BRPG, we have the following existence theorem (Monderer, 1996; Voorneveld, 2000).

....

Theorem 53: Monderer–Shapley–Voorneveld

If $\mathcal{G}$ is a potential game with a finite number of players and:

- either non-empty compact strategy sets and continuous utilities
- or finite non-empty strategy sets,

then it has at least one pure NE.

We see that Theorem 53 is very useful if a potential function can be found. If such a function cannot be easily found, other results can be used to verify that the game is potential. As mentioned in Monderer (1996), a very simple case where it is too easy to verify if the game is potential is one where the strategy sets are intervals of $\mathbb{R}$. In this case we have the following theorem:

....

Theorem 54: Monderer (1996)

Let $\mathcal{G}$ be a game in which the strategy sets are intervals of real numbers. Assume the utilities are continuously differentiable twice. Then $\mathcal{G}$ is an exact potential game if and only if:

$$\forall (i,j) \in \mathcal{K}^2, \frac{\partial^2 (u_i - u_j)}{\partial s_i \partial s_j} = 0. \tag{2.5}$$

While potential games were fully formalized in 1996 by Monderer and Shapley (Monderer, 1996), they seem to have been applied in the wireless literature for the first time in 2002. In their study, Neel et al. (2002) show that potential games can be applied to cognitive radio for problems like the distributed power control problem. Here, we mention two examples; the one of Scutari et al. (2006) and the one treated in Perlaza et al. (2009b).

Example 55

In Scutari et al. (2006) the authors formulate a constrained power control problem as a game where each user wants to minimize his transmit power p_i, subject to the constraints $f_i(\text{SINR}_i) \geq \gamma_i$ and $p_i \in [0, P_i^{\max}]$. Considering without loss of generality $u_i(\underline{p}) = \log(p_i)$ for the users's cost functions, it can be verified that $\phi(\underline{p}) = \sum_{i=1}^{K} \log(p_i)$ is a potential function for this game.

Example 56

In Perlaza et al. (2009b), the authors study distributed parallel multiple access channels when Shannon rates are considered for the transmitters' utilities. Each transmitter $i \in \{1,...,K\}$ (e.g., a mobile station) can connect to several receivers $s \in \{1,...,S\}$ (e.g., base stations), each of them using a given band B_s. All of the receivers are assumed to be connected and using a total band which equals to $B = B_1 + \cdots + B_S$. If the transmitters' utility is chosen to be the achievable Shannon rate, the game can be shown to be an exact potential game. Indeed, if the utility function is:

$$u_i(\underline{p}_i, \underline{p}_{-i}) = \sum_{s=1}^{S} \frac{B_s}{B} \log_2\left(1 + \text{SINR}_{i,s}\right) \text{ [bps/Hz]}, \tag{2.6}$$

where $\underline{p}_i$ is the power vector associated with the power allocation policy of transmitter i, $\text{SINR}_{i,s}$ is the signal to interference plus noise ratio (SINR) seen by player i on his channel s, i.e.

$$\text{SINR}_{i,s} = \frac{p_{i,s}|h_{i,s}|^2}{\sigma_s^2 + \sum_{j \neq i} p_{j,s}|h_{j,s}|^2} \tag{2.7}$$

then the function:

$$\phi(\underline{p}_1,...,\underline{p}_K) = \sum_{s=1}^{S} \frac{B_s}{B} \log_2\left(\sigma_s^2 + \sum_{i=1}^{K} p_{i,s}|h_{i,s}|^2\right). \tag{2.8}$$

is a potential function (σ_s^2 and $h_{i,s}$ respectively correspond to standard notations for the noise level at receiver s and the channel gain for the link between i and s). Note that the game is potential in the cases where the action space is compact (receiver sharing games) and in which it is discrete (receiver selection games).

2.2.4 S-Modular Games

Now we turn our attention to another important type of game: the S-modular game. Interestingly, very simple results exist for this type of game. They are defined as follows:

Definition 57: S-Modular Games

The strategic game $\mathcal{G}$ is said to be super-modular (resp. sub-modular) if: $\forall i \in \mathcal{K}$, $\mathcal{S}_i$ is a compact subset of $\mathbb{R}$; u_i is upper semi-continuous in $\underline{s}$; $\forall i \in \mathcal{K}, \forall \underline{s}_{-i} \geq \underline{s}'_{-i}$ the quantity $u_i(\underline{s}) - u_i(s_i, \underline{s}'_{-i})$ is non-decreasing (resp. non-increasing) in s_i.

The definition is easy to understand in specific contexts like the one of distributed power control. In such a context, super-modularity means that if all the other transmitters $(-i)$ increase their power, transmitter i also has an interest in increasing his.

Furthermore if all utility functions of the game are differentiable twice, there is a simple characterization of S-modular games.

Definition 58: Characterization of S-Modular Games

If $\forall i \in \mathcal{K}$, u_i is twice differentiable, then the game $\mathcal{G}$ is super-modular (resp. sub-modular), if and only if:

$$\forall (i,j) \in \mathcal{K}^2, i \neq j, \frac{\partial^2 u_i}{\partial s_i \partial s_j} \geq 0 \text{ (resp. } \leq 0). \tag{2.9}$$

One of the nice properties of S-modular games is that they do not require convexity assumptions on the utilities to ensure the existence of an NE. The following theorem for S-modular games can be found in Topkis (1979) and follows from the Tarski fixed-point theorem (Tarski, 1955).

Theorem 59: Topkis (1979)

If $\mathcal{G}$ is an S-modular game, then it has at least one pure NE.

Example 60

A first example of super-modular games can be found in Saraydar et al. (2002), where the authors use a linear pricing technique to improve the energy-efficiency of the NE of the distributed power control game. The corresponding utilities are $\forall i \in \mathcal{K}, \bar{u}_i(p) = u_i(p) + \alpha p_i$ which are not quasi-concave, as already mentioned. Super-modularity can be confirmed over a certain interval for the transmit power which is defined in Saraydar et al. (2002).

Example 61

A second simple example of a super-modular game is the special case of power allocation games, addressed in Mochaourab and Jorswieck (2009). The authors studied a 2-band 2-transmitter interference channel (Figure 2.2). On the first band (called the protected band) there is no interference, while on the second band (called the shared band), there

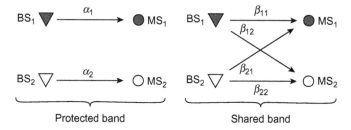

FIGURE 2.2

A special case of multi-band interference channels for which Shannon rate-efficient power allocation games are super-modular.

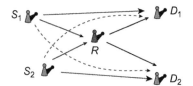

FIGURE 2.3

The 2-transmitter and 2-receiver interference relay channel.

can be some interference at the receivers. This scenario is represented in Figure 2.3. The quantities α_i, β_{ij} represent the channel gains associated with the different links. By denoting π_i (resp. $\overline{\pi} = 1 - \pi_i$) the fraction of power base station i allocated to the protected band (resp. shared band) and ρ the signal-to-noise ratio, the Shannon transmission rates of the two transmitters are given by:

$$u_1(\pi_1, \pi_2) = \log_2 (1 + \rho\alpha_1\pi_1) + \log_2 \left(1 + \frac{\rho\beta_{11}\overline{\pi}_1}{1 + \rho\beta_{21}\pi_2}\right)$$

$$u_2(\pi_1, \pi_2) = \log_2 (1 + \rho\alpha_2\overline{\pi}_2) + \log_2 \left(1 + \frac{\rho\beta_{22}\pi_2}{1 + \rho\beta_{12}\overline{\pi}_1}\right).$$

(2.10)

The characterization provided in Definition 58 directly applies here.

Other examples of sub-modular games can be found in Yao (1995) and Altman and Altman (2003). To conclude, we will mention that, even though super-modular and sub-modular games are defined in a similar way (leading to a common existence result), they can have markedly different properties. As a specific example, it can be mentioned that the convergence properties of Nash equilibria in super-modular games are generally stronger than those in sub-modular games.

It is interesting to notice that in the (non-exhaustive) list of theorems stated so far, none requires the best responses of the players to be explicate (after Chapter 1, the best response –BR– of player $i \in \mathcal{K}$ corresponds, by definition, to the set of strategies $BR_i(\underline{s}_{-i})$ maximizing the utility of user i when the rest of the world plays $\underline{s}_{-i}$:

$\mathrm{BR}_i(\underline{s}_{-i}) = \arg\max_{s_i} u_i(s_i, \underline{s}_{-i}))$. In general, BR_i can be a correspondence but it is a function in many wireless games already addressed in the current literature. If the BRs can be explicated, the existence proof boils down to proving that the BRs have a non-empty intersection, which can be very simple in some scenarios. For instance, in the power allocation (PA) game of Belmega et al. (2009a), where the authors study interference relay channels, the best responses of the users are piecewise affine functions in the case of their amplify-and-forward protocol. The existence and uniqueness issues are quite simple to analyze in such a case.

2.2.5 Recursive Diagonally Transferable Continuous Games

All the existence theorems and results given so far provide sufficient conditions under which the existence of at least one pure Nash equilibrium is guaranteed. In fact, there exists a recent result by Tian (2009) which also provides a necessary and sufficient condition for a game to have a pure Nash equilibrium, which is relatively rare in the literature of game theory. To state the corresponding theorem, a few definitions are in order.

Definition 62: Aggregator Function

Let $\mathcal{G} = (\mathcal{K}, \{\mathcal{S}_i\}_{i \in \mathcal{K}}, \{u_i\}_{i \in \mathcal{K}})$, a strategic form game. The aggregator function is defined by:

$$U : \begin{vmatrix} \mathcal{S}^2 & \to & \mathbb{R} \\ (\underline{y}, \underline{x}) & \mapsto & \sum_{i=1}^{K} u_i(y_i, \underline{x}_{-i}). \end{vmatrix} \qquad (2.11)$$

Definition 63: Recursive Upsetting

A strategy profile $\underline{y}^0$ is said to be recursively upset by $\underline{z}$ if there exists a finite set of profiles $\{\underline{y}^1, \underline{y}^2, \dots, \underline{y}^{m-1}, \underline{y}^m\}$ such that $\forall j \in \{1, 2, \dots, m\}$, $U(\underline{y}^j, \underline{y}^{j-1}) > U(\underline{y}^{j-1}, \underline{y}^{j-1})$ where $\underline{y}^m = \underline{z}$ by convention.

Definition 64: Recursive Diagonally Transferable Continuity (RDTC)

A game $\mathcal{G} = (\mathcal{K}, \{\mathcal{S}_i\}_{i \in \mathcal{K}}, \{u_i\}_{i \in \mathcal{K}})$ is RDTC if $U(\underline{y}, \underline{x}) > U(\underline{x}, \underline{x})$ implies that $\exists \underline{y}^0 \in \mathcal{S}$ and a neighborhood $\mathcal{V}_{\underline{x}}$ such that:

$$\text{"}\underline{z} \in \mathcal{A} \; upsets \; \underline{y}^0 \text{"} \Rightarrow \text{"}U(\underline{z}, \mathcal{V}_{\underline{x}}) > U(\mathcal{V}_{\underline{x}}, \mathcal{V}_{\underline{x}})\text{"}.$$

Based on these definitions, the following theorem can be stated (Tian, 2009).

Theorem 65: Tian (2009)

Assume that the strategy sets of $\mathcal{G}$ are compact. Then $\mathcal{G}$ possesses a pure Nash equilibrium if and only if it verifies the RDTC condition.

At the time of writing, this theorem has not yet been applied in wireless games. It should, however, be useful for proving a wireless game has no Nash equilibrium, by exploiting the fact that the RDTC condition is necessary. For the sufficiency part, the theorem seems to be more difficult to exploit. Compared to conditions such as quasi-concavity of the utility functions, we see that the price to be paid to have a general existence condition is the loss in terms of applicability of the result.

2.2.6 Existence of Mixed Nash Equilibria

All the results stated so far concern the existence of pure Nash equilibria. To conclude this section, we mention three simple cases where the existence of at least one mixed Nash equilibrium is guaranteed. The first theorem was introduced by Reny (1999).

> ### Theorem 66: Compact and Better-Reply Secure Games (Reny, 1999)
> Assume a strategic form game $\mathcal{G} = (\mathcal{K}, \{\mathcal{S}_i\}_{i\in\mathcal{K}}, \{u_i\}_{i\in\mathcal{K}})$. If $\mathcal{S}_i$ is compact and convex, and $\mathcal{G}$ is better-reply secure, then there is at least one mixed Nash equilibrium.

The second theorem provided here is due to Glicksberg (1952) and also concerns compact games.

> ### Theorem 67: Compact Games (Glicksberg, 1952)
> Assume a strategic form game $\mathcal{G} = (\mathcal{K}, \{\mathcal{S}_i\}_{i\in\mathcal{K}}, \{u_i\}_{i\in\mathcal{K}})$. If $\mathcal{S}_i$ is compact and u_i is continuous in $\underline{s} \in \mathcal{S}$, then there is at least one mixed Nash equilibrium.

Interestingly, the strategy sets or spaces do not need to be convex, which means that this theorem is not a corollary of the existence theorem by Reny for compact and better-reply secure games (the latter is only more general in terms of utility continuity assumptions). It can be confirmed that the assumptions of the Glicksberg existence theorem hold for all the examples of wireless games given in this section.

> ### Theorem 68: Finite Games (FG) (Nash, 1950)
> Assume a strategic form game $\mathcal{G} = (\mathcal{K}, \{\mathcal{S}_i\}_{i\in\mathcal{K}}, \{u_i\}_{i\in\mathcal{K}})$. If the set of strategy profiles $\mathcal{S} = \mathcal{S}_1 \times \mathcal{S}_2 \times \cdots \times \mathcal{S}_K$ is finite, then there is at least one mixed Nash equilibrium.

Although this theorem was derived by Nash before the Debreu-Fan-Glicksberg existence theorem, mathematically the former is a special case of the latter. This is due to the fact that the mixed strategy of player $i \in \mathcal{K}$, i.e., the probability distribution q_i used by player i, belongs to a compact and convex set $\Delta(\mathcal{S}_i)$ and his (averaged) utility (i.e., the utility of the mixed extension game) is continuous with respect to the profile of strategies. The Nash existence theorem can be applied not only to any static wireless game which is finite (e.g., a game with a finite number of transmitters who have to select a modulation order) but also to dynamic wireless games. For example, a repeated power control game with a finite number of transmitters, a finite number of power levels, and a finite number of stages (e.g., time slots) has necessarily at least one mixed Nash equilibrium.

Conclusion. In this section we have seen some ways of proving the existence of pure Nash equilibria. Essentially the methodology is as follows:

1. Analyze the properties of the game and apply and exploit an existing theorem.
2. Otherwise, come back to the definition of an NE by expressing the best responses and studying their intersection points.
3. Finally, derive a fixed-point theorem (or prove a variational inequality) specific to the game under investigation.

Of course, it may be that there is no pure Nash equilibrium. In this case, a possible solution is to consider other types of equilibria, for example mixed equilibria, correlated equilibria, or evolutionary equilibria. Another solution could be to modify the game to give it a Nash equilibrium, which is in spirit closer to the mechanism design approach (e.g., see Myerson (2008)).

2.2.7 A Note on the Existence of Nash Equilibria in Extensive-Form Games

As mentioned in Chapter 1, only the strategic form is discussed in this book. Therefore, for specific existence results related to the extensive form, the reader is invited to consult references such as Hart (1992). Here, we will just mention one well-known result due to Kuhn (1953).

Theorem 69: Kuhn (1953)

Every finite game of perfect information has at least one pure Nash equilibrium.

We already know that every finite game has at least one mixed Nash equilibrium. Here, we add that the assumption of perfect information guarantees the existence of a pure Nash equilibrium. Note that, when the equivalence exists, the notion of perfect information also applies to strategic form games, but it can be more easily verified for extensive-form games (where information sets are reduced to singletons). For example, when expressing the forwarder's dilemma under the extensive form, it is immediately seen that the information set of one the players has two elements and therefore the condition of perfect information is not verified.

2.2.8 Other Classes of Games Possessing at Least one Pure Nash Equilibrium

To make this section suitably self-contained, we conclude it by providing a list of special cases where the existence of a pure NE is guaranteed.

- Dominance solvable games (see Chapter 1 for a definition).
- Common interest or team games. They are also called identical interest games. In such a game, the players' utility functions $\{u_i\}$ are chosen to be the same: $u_i(\underline{a}) = \gamma(\underline{a})$ for $\gamma: \prod_i \mathcal{A}_i \rightarrow \mathbb{R}$. This is a particular case of exact potential games.
- Dummy games. A player i is a dummy player if he has no effect on the outcome of the game. The utility function is insensitive to player i's action: $u_i(\underline{a}) = g_i(\underline{a}_{-i})$ for a certain function g_i defined over $\prod_{i' \neq i} \mathcal{A}_{i'}$.
- Coordination games (Cooper, 1998). They capture the notion of positive externalities; i.e., choosing the same action creates a benefit rather than a cost.
- Anti-coordination games. They capture the notion of negative externalities; i.e., choosing the same action creates a cost rather than a benefit. A typical example of this type is the frequency selection game. In the latter, each node can select one of the available frequencies. When players use the same frequency at the same time and same interference area, there is congestion or a collision and the data are lost. Table 2.1 represents a simplified example with two nodes and two frequencies. The frequencies are denoted by f_1 and f_2. Node (user) 1 chooses a row and node 2 chooses a column. A successful configuration gives a payoff of 1 to the corresponding node.

Table 2.1 Strategic form Representation of 2 Nodes – 2 Frequencies

1\2	f_1	f_2
f_1	(0,0): collision	(1,1): success
f_2	(1,1): success	(0,0): collision

- Minority games. A minority game is a game where the objective of each player is to be part of the smallest of two populations: each player wants to be in the minority. The example in Table 2.2 illustrates this scenario:

Table 2.2 Strategic Form Representation for 3 Nodes – 2 Frequencies

1\2	f_1	f_2
f_1	(0,0,0): collision	(0,1,0): success
f_2	(1,0,0): success	(0,0,1): collision
f_1	(0,0,1): collision	(1,0,0): success
f_2	(0,1,0): success	(0,0,0): collision

Minority games are exemplified by the so-called *El Farol* bar problem (Arthur, 1994): There is an odd number of people. Every Saturday night, all these people want to go to a prestigious bar. However, the bar is quite small, and it is no fun to go if it is too crowded. The preference of the population is to go to the bar if the total number of people there does not exceed some threshold which is proportional to the capacity of the bar.

- Majority games. A majority game is a game where the objective of each player is to be part of the bigger of two populations: each player wants to be in the majority.
- Crowding games (Milchtaich, 1998). A crowding game is a game where each player's utility is non-increasing with respect to the number of other players using the same action.
- Congestion games (Milchtaich, 1996a,b). A congestion game is a crowding game over networks.
- Weakly acyclic games (Milchtaich, 1996a). A better reply path is a sequence of action profiles $a(1), \ldots, a(T)$, where for every $1 \leq t \leq T - 1$, there is exactly one player i at iteration t such that $a_i(t) \neq a_i(t + 1)$ and $a_{-i}(t) = a_{-i}(t + 1)$, and $u_i(\underline{a}(t)) < u_i(\underline{a}(t + 1))$. This means that only one player moves at each iteration, and each time a player moves he/she increases their own utility. A finite game $\mathcal{G}$ is weakly acyclic if for any $\underline{a} \in \prod_i \mathcal{A}_i$ there exists a better reply path starting at $\underline{a}$ and ending at some pure Nash equilibria. Potential games and dominance solvable games are special cases of weakly acyclic games.
- Multipotential games (Monderer, 2007). They consist of a generalization of potential games. Let $\bar{\mathcal{P}}$ be a subset of real-valued functions defined on $\prod_i \mathcal{A}_i$. The subset is a cover for $\bar{\mathcal{P}}$ of the game $\mathcal{G}$ if for every player j there exists $P \in \bar{\mathcal{P}}$, which is a potential function for i. A game $\mathcal{G}$ is a q-potential game, if it has a cover with cardinality less than q. As a consequence, every K-player finite game is a multipotential game. Many games of this class have at least one pure equilibrium. A potential game corresponds to a 1-potential game.

2.3 EQUILIBRIUM UNIQUENESS

Once one is assured that an equilibrium exists, a natural question is to know whether it is unique. Depending on the context, uniqueness can be a desirable feature of the game or not. For instance, if the goal is to predict the performance of a system, or the outcome of a dynamic process (e.g., a game between learning automata), uniqueness is generally desired. On the other hand, if we take the example of a repeated game with a finite number of stages, it is well known (see Chapter 3 and Benoît and Krishna (1985, 1987)) that Nash equilibria of the repeated game cannot be subgame perfect if the constituent (static) game has only one Nash equilibrium. Unfortunately, there are few general studies on the equilibrium uniqueness. In this section, for the sake of clarity, we will distinguish between two types of situations in which the best responses of the players can be explicated or otherwise. The methodology for proving uniqueness is summarized on the chart in Figure 2.4.

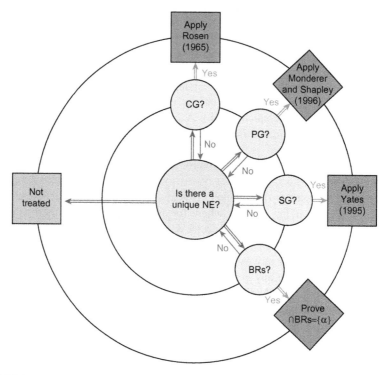

FIGURE 2.4

A (non-exhaustive) methodology for proving the uniqueness of a Nash equilibrium in strategic form games. The meaning of the used acronyms is as follows: CG? (Is the game concave?; Rosen, 1965), PG (potential games; Monderer and Shapley, 1996), SG (games with standard best responses; Yates, 1995), BRs? (Are the best-responses known?).

2.3.1 The Best Responses Do Not Need to be Explicated

A natural question would be to ask whether the Debreu-Fan-Glicksberg theorem has a counterpart for uniqueness, that is, is there a general uniqueness theorem for quasi-concave K-player games. To the best of the author's knowledge, the answer is no. However, there is a powerful tool for proving the uniqueness of a pure NE when the players' utilities are concave: this tool is the uniqueness theorem derived by Rosen (1965). This theorem states that if a certain condition, called diagonally strict concavity (DSC), is met, then uniqueness is guaranteed. This theorem is as follows:

Theorem 70: Rosen (1965)

Assume that: $\forall i \in \mathcal{K}$: $\mathcal{S}_i$ is a non-empty, compact, and convex set; $u_i(\underline{s})$ is a continuous function in $\underline{s} \in \mathcal{S}$ and concave in s_i. Let $\underline{r} = (r_1, \ldots, r_K)$ be an arbitrary vector of fixed positive parameters. Define the pseudogradient of the function $w_{\underline{r}} = \underline{r} \times \underline{u}^T$ by $\underline{\gamma}_{w_{\underline{r}}}(\underline{s}) =$

$\left[r_1 \frac{\partial u_1}{\partial s_1}(\underline{s}), \ldots, r_K \frac{\partial u_K}{\partial s_K}(\underline{s}) \right]^T$. If the following condition holds for some $\underline{r} > \underline{0}$:

$$\forall (\underline{s}, \underline{s}') \in \mathbb{S}^2, \underline{s} \neq \underline{s}' : (\underline{s} - \underline{s}') \left(\underline{\gamma}_{w_{\underline{r}}}(\underline{s}') - \underline{\gamma}_{w_{\underline{r}}}(\underline{s}) \right)^T > 0 \qquad (2.12)$$

then the game $\mathcal{G}$ has a unique NE.

The condition (2.12) is called the diagonally strict concavity (DSC) condition. To illustrate this theorem we provide an example of wireless game where it has been applied successfully.

Example 71

In Belmega et al. (2009b) the authors generalized the water-filling game of Lai and El Gamal (2008) to fast fading multiple access channels (MACs with one base station -BS- and K mobile stations -MS-) with multiple antennas both at the transmitters and receiver (uplink case). In this game, each user wants to maximize his ergodic transmission rate by choosing the best precoding strategy (namely a non-negative matrix). While showing that this game is concave is quite easy, proving its uniqueness is less trivial. It turns out that the DSC of Rosen translates into a trace inequality. Indeed, proving the DSC in the K-player power allocation game boils down to proving that:

$$\mathrm{Tr} \left\{ \sum_{k=1}^{K} (\mathbf{A}_k - \mathbf{B}_k) \left[\left(\sum_{\ell=1}^{k} \mathbf{B}_\ell \right)^{-1} - \left(\sum_{\ell=1}^{k} \mathbf{A}_\ell \right)^{-1} \right] \right\} > 0 \qquad (2.13)$$

for arbitrary $K \geq 1$ where $\mathrm{Tr}(\cdot)$ denotes the matrix trace operator, $\mathbf{A}_1$, $\mathbf{B}_1$ are any positive definite matrices and $\mathbf{A}_k$, $\mathbf{B}_k$, for all $k \in \{2, \ldots, K\}$, are any positive semidefinite matrices.

To conclude this section, we will mention a trivial but useful theorem for potential games which is a direct consequence of Monderer (1996) (e.g., see Scutari et al. (2006)).

Theorem 72: Uniqueness in Potential Games

If the potential function is strictly concave and the strategy spaces are non-empty, compact, and convex, then there is a unique pure Nash equilibrium.

A simple example of a potential game is the special case of the power allocation game of Example 71 when the channel transfer matrices are diagonal (fast fading parallel multiple access channels) and single-user decoding is used by the receiver. The game can be proven to have a concave potential but the latter is not strict; more elaborate reasoning is needed to prove that uniqueness is almost certain (Mertikopoulos et al., 2011). Now, in games where the potential is strictly concave, a useful result (Neyman, 1997) can be mentioned: if the game possesses a strictly concave potential, the set of correlated equilibria boils down to a singleton, which is the unique pure Nash equilibrium of the game.

2.3.2 When the Best Responses Can be Explicated

If the BRs of every player can be expressed, it is possible to analyze their properties, and for some classes of functions (or correspondences) to characterize the number of intersection points between them. A nice class of BRs is the class of standard BRs. Standard functions have been introduced by Yates (1995); they are defined as follows:

> **Definition 73: Standard Functions**
>
> A vector function $g : \mathbb{R}_+^K \to \mathbb{R}_+^K$ is said to be standard if it has the two following properties:
>
> - Monotonicity: $\forall (\underline{x}, \underline{x}') \in \mathbb{R}_+^{2K}, \underline{x} \leq \underline{x}' \Rightarrow g(\underline{x}) \leq g(\underline{x}')$;
> - Scalability: $\forall \alpha > 1, \forall \underline{x} \in \mathbb{R}_+^K, g(\alpha \underline{x}) < \alpha g(\underline{x})$.

In Yates (1995) it is shown that if a function is standard then it has a unique fixed-point. Applying this result to the best responses of a game gives the following:

> **Theorem 74: Yates (1995)**
>
> If the best responses of the strategic-form game $\mathcal{G}$ are standard, then the game has a unique NE.

A simple wireless game where the BRs possess these properties is the energy-efficient power control game introduced by Goodman and Mandayam (2000), which we have already mentioned. In the non-saturated regime in which no transmitter uses all his power, the BRs can be shown to be (e.g., see Lasaulce et al. (2009b) and Saraydar et al. (2002)):

$$\forall i \in \{1, \ldots, K\}, \ \mathrm{BR}_i(\underline{p}) = \frac{\beta^*}{|h_i|^2} \left(\sigma^2 + \sum_{j=0}^{i-1} p_j |h_j|^2 \right), \tag{2.14}$$

where β^* is a constant, h_i the instantaneous channel gain (which is assumed to be different from zero) of user i and σ^2 the reception noise variance. It can be checked that the BRs are monotonic and scalable.

As mentioned at the end of Section 2.2, if the BRs are available and reasonably simple to exploit, it might be possible to find their *intersection points*. The number of intersection points corresponds to the number of possible equilibria. A well-known game that exemplifies this kind of approach is the Cournot duopoly (Cournot, 1838), for which the BRs of the two players are affined and intersect in a single point. A counterpart of the Cournot duopoly in wireless networks is treated, for instance, in Belmega et al. (2009b).

2.4 EQUILIBRIUM SELECTION

There are important scenarios where the NE is not unique. This typically happens in routing games (Milchtaich, 1996a; Roughgarden and Tardos, 2004) and coordination games (Cooper, 1998). Another important scenario where such a problem arises, corresponds to games where the choice of actions of different players is not independent, for instance, non-cooperative games with correlated constraints and generalized NE (Altman and Shwartz, 2000; Debreu, 1952). A central feature in these constrained games is that they often possess a large number of equilibria. The selection of an appropriate equilibrium is the natural concern. What can be done when one has to deal with a game having multiple equilibria? Are there some dominant equilibria? Are some equilibria fairer and more stable than others? Obviously, the selection rule is strongly related to the used fairness criteria.

Equilibrium selection is a theory in itself (Harsanyi and Selten, 2003). The goal of the theory of equilibrium selection is to provide a convincing theory of rationality which selects a unique outcome in every game and this theory has to be common knowledge. While the theory developed by Harsanyi and Selten (2003) is recognized as an important contribution, it takes some positions which are still being discussed and with which some authors disagree (Hillas and Kohlberg, 2002). We will just mention one point about this theory and then focus on more pragmatic ways of solving the problem of equilibrium selection. Some equilibria can be utility/payoff dominant or risk dominant in the sense of Harsanyi and Selten (2003) and Damme (2002). An example of utility dominance occurs when one equilibrium Pareto-dominates the others. With respect to risk dominance, this notion translates into the degree of certainty a player has about an outcome (e.g., see Damme (2002) where it is shown how to measure the riskiness of an outcome). Depending on what the players believe about the others (e.g., Can I trust the other players?), a utility-dominant or risk-dominant equilibrium may appear in games (either in static games with well-informed players or as a result of dynamic/evolutionary process). As already mentioned, these issues will not be mentioned here. We will discuss only a very partial view of the general problem and give only partial answers to these questions. In fact, we will mention only two issues related to the equilibrium selection problem: how to select an equilibrium in concave games; and the role of the game dynamics in the selection. These issues have been chosen because they are closely connected to the content of the rest of this chapter.

2.4.1 Equilibrium Selection in Concave Games

In the case of concave games, Rosen (1965) again gives a very neat way to tackle the problem. In Rosen (1965), the author introduced the notion of *normalized equilibria*, which gives a way of selecting an equilibrium.

....

Definition 75: Normalized Equilibrium

Let $\mathcal{G}$ be a concave strategic form game and $\underline{r}$ a vector of positive parameters; $\mathcal{C} = \{\underline{s} \in \mathcal{S}, \ h(\underline{s}) \geq 0\}$ is a set of constraints. An equilibrium $\underline{s}^*$ of this game is said to be a normalized NE associated with $\underline{r}$ if there exists a constant λ such that $\lambda_i = \frac{\lambda}{r_i}$ where λ_i are the multipliers corresponding to the Kuhn-Tucker conditions $\lambda_i h(\underline{s}) = 0$.

This concept has been applied by Altman et al. (2009a) to study decentralized MACs with constraints. The impact of these constraints is to correlate the players' actions. As mentioned at the beginning of this section, there can be multiple NE in this type of game. This is precisely what happens in decentralized MACs. One of the problems that arises in such contexts is to know how a player values the fact that the constraints on another player are satisfied or violated. Some extreme cases are as follows:

i. A player is indifferent to satisfaction of constraints of other players.
ii. Common constraints: if a constraint is violated for one player then it is violated for all players.

The concept of normalized equilibrium, applicable to concave games, is one possible way of predicting the outcome of such a game and/or selecting one of the possible equilibria. Specifically, the authors of Altman et al. (2009a) have shown that, in the context of multiple access game with multiuser detection, the normalized equilibrium achieves maxmin fairness and is also proportionally fair (for these notions see for example Mo and Walrand (1998)).

2.4.2 The Role of Dynamics in Equilibrium Selection

As pointed out in Section 2.1, Nash equilibria of a certain game can be observed in a degraded version of this game where players have less information; or more generally, a different knowledge. Typically, a certain static game with complete information can have multiple equilibria, but if it is played several times under the assumption of partial information, it might happen that only one equilibrium is effectively observed. For example, if the players play sequentially such that each player observes the actions played by the others, and reacts to them by playing his best response, and then the others update their strategy accordingly, and so on, it can happen that this game converges to the NE that would be obtained if the players knew the game completely and played it in one shot. Figure 2.5 shows the possible NE in the power allocation game of Belmega et al. (2009a) in two-band two-user interference relay channels. In this figure, θ_i represents the power fraction user i allocates to a frequency band, $1 - \theta_1$ being the fraction allocated to the other band. It can be shown that the sequence $\left\{ \theta_i^{(0)}; \theta_{-i}^{(1)} = BR_{-i}\left(\theta_i^{(0)}\right); \theta_i^{(2)} = BR_i\left(\theta_{-i}^{(1)}\right); \theta_{-i}^{(3)} = BR_{-i}\left(\theta_i^{(2)}\right), \ldots \right\}$ will converge to one of the three possible NE, depending on the *game starting point*, i.e., on the value of $i \in \{1, 2\}$ and the value of $\theta_i^{(0)} \in [0, 1]$. As a more general conclusion, we see that the initial operating state of a network can determine the equilibrium state in decentralized networks with certain convergence properties. We will not expand on this issue here, but it is important to know that games with standard best-responses, potential games, and S-modular games have attractive convergence properties. For example, the authors of Sastry et al. (1994b) have shown how simple learning procedures, based on mild information assumptions, converge to the NE predicted in the associated game with complete information.

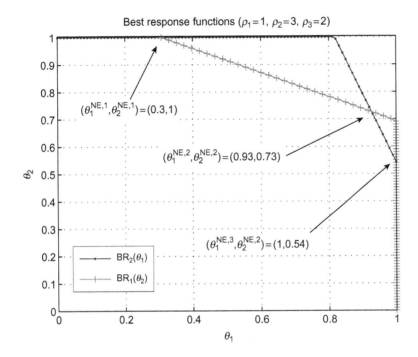

FIGURE 2.5

In two-band two-user interference relay channels, power allocation games have multiple Nash equilibria (Belmega et al., 2009a). If the users play and observe each other alternately, the game converges to an NE depending on the starting point of the game.

2.4.3 Additional Comments

In this section, we have seen how some equilibria can be selected based on fairness or as a result of a dynamic process. There are many other ways of selecting an equilibrium or equilibria. For instance, it can happen that some equilibria are never effectively observed because they have no properties which make them relevant in practice (subgame perfection, robustness to observation noise, stability, etc.): the fact that some equilibria verify these properties can play a role in the natural selection of an equilibrium. In some games, there can be a super-player (an operator, a base station, etc.) who is able to select a given equilibrium or induce a mechanism that leads to a given equilibrium. As far as wireless communications are concerned, equilibrium efficiency is a way of selecting an equilibrium. In the scenario of Belmega et al. (2009b) where two point-to-point communications compete with each other (interference channels), the network owner chooses the best location for the added relay in order to maximize the network sum-rate at equilibrium. In the next section, different ways of measuring the efficiency of an equilibrium are presented.

2.5 EQUILIBRIUM EFFICIENCY

We have seen that decentralized wireless networks represent simple scenario where game theory is a natural paradigm. The power control problem in the uplink of cellular systems (multiple access channels) is a much-studied example. But there is, a priori, no reason why a decentralized network of partially or totally autonomous and selfish terminals should perform as well as its centralized counterpart, for which the power control problem can be optimized at the base station. This poses the problem of efficiency of the network. More specifically, for a decentralized network having a unique NE, it is important to characterize the equilibrium efficiency since it is the state at which the network will spontaneously operate. As mentioned above, if there are multiple equilibria, the equilibrium efficiency can be used as a discriminant factor to select one of them. Whatever the equilibrium is – unique or resulting from selection – two critical issues arise:

- how to measure the network equilibrium efficiency (Sec. 2.5.1);
- how to improve the efficiency (Sec. 2.5.2) when the overall network performance at equilibrium is found to be unsatisfactory.

2.5.1 Measuring Equilibrium Efficiency

A well-known way of characterizing the efficiency of an equilibrium is to determine whether or not it is Pareto-optimal (PO). Pareto-optimality is a requirement which can be used for selecting equilibria, and was defined in Chapter 1. A strategy profile $\underline{s}^{PO}$ is Pareto-optimal if there exists no other profile of strategies for which one or more players can improve their utility without reducing the utilities of the others. A simple example of Pareto-optimal profile of strategies is as follows.

Example 76

Assume a canonical 2-user MAC (Cover, 1975; Wyner, 1974), $Y = X_1 + X_2 + Z$ with a receiver implementing successive interference cancellation (SIC). Consider the very simple game where, knowing the decoding order used by the receiver, the users want to maximize their Shannon transmission rate $u_i(p) = \log_2(1 + \text{SINR}_i)$ by choosing the best value for their transmit power $p_i \in [0, P_i^{\max}]$. It can be shown that all the profiles of strategies corresponding to the full cooperation segment of the capacity region frontier are Pareto-optimal. In fact, along this segment, we even have a zero-sum game $\forall \underline{p} \in [0, P_1^{\max}] \times [0, P_2^{\max}]$, $u_1(\underline{p}) + u_2(\underline{p}) =$

R_{sum}, where $R_{\text{sum}} = \log_2\left(1 + \frac{P_1^{\max} + P_2^{\max}}{\sigma^2}\right)$ is the MAC sum-rate.

In the example we have used to illustrate Pareto-optimality, we see that the Pareto-optimal profiles of strategies are such that the sum of the utilities is maximized. This observation corresponds to the following result[4] (e.g., see Arora (2004) for more

---

[4]The result typically holds when the utility region (or its complementary in $\mathbb{R}^K$) is convex.

details): every strategy profile $\underline{s}$ which maximizes the weighted sum $\sum_{i \in \mathcal{K}} \alpha_i u_i(\underline{s})$ is Pareto-optimal, with $\alpha_i > 0$. This result is useful for determining Pareto-optimal profiles of a game. In particular, the results holds when $\alpha_i = 1$. The corresponding function corresponds to a very well-known quantity: social welfare. The social welfare of a game is defined as follows (Arrow, 1963):

Definition 77: Social Welfare

The social welfare of a game is defined as the sum of the utilities of all players:

$$w = \sum_{i=1}^{K} u_i. \tag{2.15}$$

Social welfare, which corresponds to the average utility of the players (up to a scaling factor), is a well-known absolute measure of efficiency of a society, especially in economics. Is this quantity relevant in wireless communications? In theory, and more specifically in terms of the ultimate performance limits of a network (Shannon theory), social welfare coincides with the network sum-rate. In contrast with many economic studies, there is in communications, thanks to Shannon theory, a fundamental limit to the social welfare. For example, if we have K terminals, each of them implementing a selfish power control algorithm to optimize his Shannon transmission rate, and communicating with two base stations connected with each other, we know that w cannot exceed the transmission rate of the equivalent virtual $K \times 2$ multiple input multiple output (MIMO) system. In practice, social welfare can be a good measure if the players experience quite similar propagation conditions, in which case their utilities after averaging (e.g., over fading gains) can be similar. If the users experience markedly different propagation conditions, then the use of social welfare can be sometimes arguable, and can even lead to very unfair solutions.

There are at least three reasons why social welfare must be replaced with other measures of global network performance. First, it is an absolute measure, and therefore does not indicate the size of the gap between the performance of the decentralized network and its centralized counterpart. Second, as mentioned, it can be unfair. Third, while it has a very good physical interpretation when the users' utilities are Shannon transmission rates, its meaning is much less clear in contexts where other utilities are considered (e.g., energy-efficiency). Before providing a way to deal with the first drawback, let us illustrate the second and third ones by using the example of energy-efficient power control games.

Example 78

In Lasaulce et al. (2009b) the authors studied a non-cooperative, energy-efficient power control game (see Example 51 for more details) when the BS implements SIC. When the BS can optimize the decoding order associated with SIC, the authors have shown that after optimization of $w = \sum_{i=1}^{K} u_i = \sum_{i=1}^{K} \frac{f(\mathrm{SINR}_i)}{p_i}$, the users who were "rich" in terms of link quality are now even "richer". This shows that w can be an unfair measure of energy-efficiency of the network. On other hand, the other performance metric considered in Lasaulce et al. (2009b), the equivalent virtual MIMO system, $v = \frac{\sum_{i=1}^{K} f(\text{SINR}_i)}{\sum_{i=1}^{K} p_i}$, is fairer in terms of energy-efficiency and has a better physical interpretation than w.

Let us go back to the first drawback of w. To deal with it, Papadimitriou (e.g., see Papadimitriou (2001)) introduced the concept of the price of anarchy (PoA). The PoA was initially used to measure the inefficiency of equilibria in non-atomic selfish routing games, with a large number of players. In the context of games with a finite number of players, it is defined as follows:

Definition 79: Price of Anarchy

The PoA of the game $\mathcal{G}$ is equal to the ratio of the highest value of the social welfare (joint optimization) to the least optimal NE of the game:

$$\text{PoA} = \frac{\max\limits_{\underline{s} \in \mathcal{S}} w(\underline{s})}{\min\limits_{\underline{s}^* \in \mathcal{S}^{\text{NE}}} w(\underline{s}^*)} \tag{2.16}$$

where $\mathcal{S}^{\text{SE}}$ is set of Nash equilibria of the game.

The price of stability (PoS) is defined similarly, by replacing the denominator of the PoA with the best NE of the game:

$$\text{PoS} = \frac{\max\limits_{\underline{s} \in \mathcal{S}} w(\underline{s})}{\max\limits_{\underline{s}^* \in \mathcal{S}^{\text{NE}}} w(\underline{s}^*)}. \tag{2.17}$$

The term "price of stability" indicates the price for having a stable solution. Indeed, Nash equilibria are profiles which are stable to single deviations. Obviously, the PoA and PoS coincide if there exists a unique NE. Interestingly, these quantities can be bounded in some cases. For instance, in the case of non-atomic routing games the PoA can be upper bounded by using variational inequalities. For polynomial costs of maximum degree d, the PoA for the Nash equilibrium is bounded by Correa et al. (2004):

degree	1	2	3	4	...	d
PoA	$\frac{4}{3}$	1.626	1.896	2.151	...	$\Omega\left(\frac{d}{\ln(d)}\right)$

2.5.2 Improving Equilibrium Efficiency

When the performance of the network at the (considered) NE is found to be insufficient, the network or the game can be modified. There are many ways of doing this and we will just mention a few of them. What is important to have in mind is that the corresponding changes generally require allocating some resources (time-slots, band, ...) in order for the nodes to exchange some information and implement the new strategies.

A possible way of improving the equilibrium efficiency is to transform the non-cooperative game/network into a cooperative game/network. Note that we distinguish between cooperative networks and cooperative games. Often, cooperative networks are those in which some nodes can act as relays for other nodes. Implementing a cooperative solution based on game theory in a wireless network does not necessarily imply that relaying nodes have to be involved. Cooperation in game theory typically means that some groups of players have common objectives; they do not necessarily exchange signals for this purpose. A team formulation of the power control game introduced in Goodman and Mandayam (2000) will however require full channel state information at all the transmitters. Indeed, as mentioned in Lasaulce et al. (2009b), if all the transmitters want to maximize w, every transmitter will need to know $h_1, \ldots, h_K$, in contrast with the non-cooperative game where only h_i is required at transmitter i.

Of course, it is also possible and even required in a large network that smaller groups of players are formed: this is the principle of coalitional games (e.g., see Aumann and Peleg (1960) and von Neumann and Morgenstern (1944) for coalitional games with and without transferable utility respectively). If each member of a group or coalition has enough information, the coalition can even form a virtual antenna (Saad et al., 2008) and the gain brought by cooperation is shared between the players of the group. In such games, we can have both locally cooperative networks, and a non-cooperative game between the coalitions. In cooperative games, one of the counterparts of the NE solution concept for non-cooperative games is the Shapley value (Shapley, 1953a) (as mentioned in Chapter 1, other types of values and cooperative solution concepts exist).

Note that there are other forms of cooperation, sometimes more implicit. This is the case in repeated games (e.g., see Aumann and Maschler (1966) for infinitely repeated games), which are a special case of dynamic games in which the same static game is repeated at each step and the utilities result from averaging the utilities of the static game over time. In such games, certain agreements between players on a common plan can be implemented, and a punishment policy can also be implemented to identify and punish deviators. A simple wireless game that applies the concept of repeated games is the water-filling game by Lai and El Gamal (2008) used in the fast fading MAC with single-antenna transmitters and a multi-antenna receiver. The authors show that the capacity region frontier is reached by repeating the non-cooperative game where selfish users maximize their transmission rate. In Treust and Lasaulce (2010a) the authors show that it is possible to induce a cooperative solution in the repeated version of the energy-efficient power control game of Goodman and Mandayam (2000). Interestingly, in contrast with the cooperative multiple access channel (Sendonaris et al., 2003; Willems, 1983) no cooperation channels between the transmitters are required to obtain an efficient solution.

Finally we will mention another form of cooperation: bargaining. Very interestingly, Nash has proved that cooperative games (von Neumann and Morgenstern, 1944) can be studied using the same concepts as those used for non-cooperative games: a cooperative profile of strategies can be obtained as a subgame perfect equilibrium (which is an equilibrium having a certain robustness against changes in the main equilibrium plan; see Selten (1975) for more details) resulting from a bargaining procedure between the players. An example of the application of this concept is Larsson and Jorswieck (2008), in which two, multi-antenna base stations cooperate by implementing a Nash bargaining solution.

The ways of improving the efficiency of the network equilibrium that have been mentioned so far are generally very demanding in terms of signaling,[5] and can possibly need new physical links between some nodes. Since cooperation can be costly in terms of additional resources, a more reasonable solution can be to merely coordinate the players. Here, we do not mean creating coordination games (Cooper, 1998), which corresponds to a certain type of games that we will not study here, but adding a certain degree of *coordination* in a non-cooperative game. Coordination between users can be stimulated, for instance, by using existing broadcasting signals like DVB or FM signals (case of public information), or by introducing dedicated signals sent by a BS (which can send both private and public information). The fact that all players of the game have access to certain (public or/and private) signals will generally modify the players' behaviors. This can lead to a more efficient equilibrium, which can be a new NE or even a correlated equilibrium (Aumann, 1974). The authors of Bonneau et al. (2006) have shown that, in a MAC where a slotted ALOHA protocol is assumed, a public coordination signal induces correlated equilibria that are more efficient in terms of reducing the frequency of collisions between users. More generally, it can be shown (Aumann, 1974) that the set of achievable equilibria is enlarged by using common and private messages in the context of correlated equilibria.

Finally we will just mention two other techniques for improving the performance of a network at equilibrium: (1) implementing a *pricing technique* or (2) introducing a certain degree of *hierarchy* in the game. Technique (1) has been used by Saraydar et al. (2002) for energy-efficient power control games, and by Lai and El Gamal (2008) for transmission rate-efficient power control games. Technique (2) has been used by Lasaulce et al. (2009b) for energy-efficient power control games and by Basar and Srikant (2002) for transmission rate-efficient power allocation games.

2.6 CONCLUSION

As mentioned in the introduction to this chapter, the Nash equilibrium concept is very important due to the fact it is the solution for numerous types of interactive

---

[5]Even if the amount of signaling can be high, the cooperative solutions mentioned in this chapter have the property that the decision is left to the transmitter.

scenarios. Nash equilibria are not only relevant in highly structured games where a lot of information is assumed (e.g., repeated games with complete information), but also in situations like a game between learning automata (no structure is assumed for the environment seen from player i; rationality does not need to be common knowledge; only the value of the instantaneous utility at each stage of the game is used by a player). Of course, the Nash equilibrium concept has also some limitations. For instance, in dynamic games there can be a large number of equilibria, which makes the effective solution difficult to predict. In differential games (studied in Chapter 3), there are many scenarios in which the existence of Nash equilibria is still an open problem. Sometimes, the complexity of calculating a Nash equilibrium can be very high and even exceed the available computational capacity of existing machines. In such situations, approximating the solution can be very useful. Wardrop equilibria (Wardrop, 1952), the fluid approximation (Silva et al., 2008), random matrix theory based approximates (Belmega et al., 2009b), mean field games (Tembine et al., 2010), etc., are approaches which attempt to simplify the problem at hand and obtain approximate Nash equilibria, or solutions close to them. When the Nash equilibrium is found to be an unsuitable concept, other solutions such as those listed in Chapter 1 can be used. If the latter are still not relevant, this probably means that game theory needs to be enriched with new solution concepts. If game theory is found to be a strong and durable paradigm for wireless networks over the coming decade, this is likely to take place.

Moving from Static to Dynamic Games

3.1 INTRODUCTION

The preceding chapter emphasized one-shot/simultaneous/static games (both terms will be used interchangeably). This model is generally insufficient for addressing many important scenarios often encountered in wireless communications. For example, the outcome of a situation where two operators have to deploy a network of hot spots depends on the way the deployments occur. Assume that each operator wants to maximize the total number of connections to his network. If the two operators deploy their network in a rational way, and are fully informed about the options of their competitor, the one-shot game model may be appropriate. However, if one operator deploys his entire network first and the other starts deploying once he/she has finished, the strategies played (i.e. the set of location points for the hot spots) will in general be different from those in the one-shot deployment game. In fact, in the second situation there are two stages to the game and the corresponding "two-shot" game model is a dynamic game – a Stackelberg game to be precise.

3.1.1 Informal Definition of a Dynamic Game

Although a static game is generally considered to be a special case of a dynamic game, the difference between the two is not always obvious. For example, if a machine implements a sophisticated program which has been programmed once and for all but, once it is run, reacts dynamically to its environment and the actions played by the other players, is this a static or a dynamic game? In fact, in dynamic games it is generally assumed that players can extract some information from past moves and strategies, and take this into account to adjust their current and future moves. In static games, players have given knowledge (information assumptions, behavior assumptions) and have to make their decision once and for all. On the other hand, a game is dynamic if at least one player is allowed to use a strategy which depends on past actions, moves, or strategies. Informally, dynamic games are characterized by three problems:

- How to model the environment in which the players act. To obtain a mathematical model of the players' environment, a set of differential or difference equations is usually specified. These equations are assumed to capture the main dynamic features of the environment.

- How to model the players' objectives. Usually these are formalized as cost/payoff/reward/utility functions which have to be optimized. The underlying optimization problems have to be performed subject to the specified dynamic model of the environment.
- How to specify the order of play in the decision process, the information available to the players at the time of their decision, and the evolution of the game.

In this book, mathematical definitions resulting from these modeling problems will be given for several important special cases of dynamic games. These special cases are outlined in the next section.

3.1.2 Classifying Dynamic Games

Five important classes of dynamic games will be studied in this book. They are introduced in this chapter, and illustrated by case studies in Chapters 7, 8, and 9. These classes are:

- Stackelberg games (interpreted as multi-stage games);
- standard repeated games;
- stochastic games;
- difference games and differential games;
- evolutionary games.

Multi-stage Stackelberg games can be analyzed as dynamic games. They will not be detailed in this chapter, but are illustrated in Chapter 8. In their original formulation, Stackelberg games, named after their inventor H. Stackelberg (1934), comprise two players: one leader and one follower. The leader chooses his action first, knowing that the follower is going to observe the played action and react to it in a rational way. In this situation, the dynamic game comprises two stages – each player plays once, and only one player observes past moves. Standard repeated games with complete information are analyzed in Sec. 3.2. A standard repeated game consists of the repetition of the same static game a certain (finite of infinite) number of times. Although simple, repeated games allow one to introduce new concepts that are not present in static games (such as cooperation and punishment), and highlight important differences with respect to static games. One of these differences is that there is generally a large number of Nash equilibria (when they exist), which are not characterized by their action or strategy profiles but by their utility profiles, which is the purpose of "Folk theorems".

Stochastic games generalize repeated games. In stochastic games, discussed in Sec. 3.3, the instantaneous or current utility of a player depends not only on the played actions but also on the realization of a state. Sec. 3.4 describes difference and differential games. Here, in contrast with repeated and stochastic games, time is continuous. This assumption is generally made in evolutionary games to describe the evolution of populations but it is not necessary. Typically, differential games are characterized by a state which follows a certain evolutionary law (e.g., a deterministic or stochastic differential equation). Difference games are the discrete counterpart of differential games; this type of game is considered to bridge the gap between stochastic games and differential games. They allow the reader to become familiar with one of the most general definitions of dynamic games having complete information which is in use, and for which there are some theoretical results.

To conclude on differential games, note that they can also be considered as a limiting case of difference games. Continuous-time dynamics have been much used to design a discrete-time strategy for power control (e.g., see Foschini and Milijanic (1993)). Finally, evolutionary games will be studied in Sec. 3.5. This type of game involves a large number of players, and studies the evolution of the population of players as they interact with each other (e.g., in a pairwise manner). The (potentially strong) rationality assumption is not necessary in evolutionary games, and the solution concepts (e.g., the evolutionarily stable state) which are used ensure stability with respect to a fraction of the population of players. Usually, time is assumed to be continuous in evolutionary games but this assumption is not necessary.

3.2 REPEATED GAMES

3.2.1 Introduction

The formalization of repeated games dates back to the 1960s. Among the pioneering works we find Aumann (1960), Aumann and Maschler (1966), Luce and Raiffa (1957) and Friedman (1971). As defined in Aumann and Maschler (1966), a repeated game means repeating the same static game a certain (possibly infinite) number of times. Standard repeated games (Aumann, 1981; Mertens et al., 1994; Sobel, 1992) correspond to a very special structure of dynamic games which allow one to derive a relatively large number of results compared to more general discrete-time games (e.g., stochastic or difference games). Clearly, repeated games try to model the fact that players can interact more than once and want to optimize a certain performance metric, resulting from averaging their utility over the whole duration of the game. In contrast with iterative or learning techniques (studied in Chapters 5 and 6), which are based on relatively mild information assumptions and different behavior assumptions (transmitters can be modeled by automata), repeated games generally require more demanding information and behavior assumptions. They also aim to optimize an averaged utility and reach more efficient points than those available to the associated one-shot game Nash equilibria. For instance, it can be shown that if the forwarders' dilemma (see Chapter 1) is repeated an infinite number of times, selfish and rational players who observe past actions perfectly (the so-called perfect monitoring assumption is defined a little further on) have an interest in cooperating by playing the action profile (F, F) instead of (D, D). This leads to the equilibrium utility profile $(1 - c, 1 - c)$; this means that in repeated games new types of behaviors appear (based on cooperation, punishment, etc.).

In fact, there is an infinite number of possible equilibrium utility profiles. Indeed, in dynamic games, where equilibria exist there are generally many of them.

"Folk theorems" are theorems which aim to characterize the complete set of equilibrium utilities in repeated games (and more generally in dynamic games, when they are relevant). One methodological difference between static and repeated games is that equilibrium utilities are characterized in the latter, while most often equilibrium action/strategy profiles are considered for the former. One of the reasons for this is the combinatorial aspect involved in repeated games – that is the number of possible action plans or strategies can be large, or even infinite.

3.2.2 Fundamental Notions of Repeated Games

In this section we define several fundamental concepts used to analyze repeated games. Some notions, like sequential equilibria or communication equilibria, will not be used in this book and are therefore not defined here. The interested reader is directed to Sobel (1992) and Gossner and Tomala (2009).

As mentioned in the preceding section, a repeated game is built from a one-stage game; the latter is often called the stage game or constituent game. We will assume a one-shot game with K players and $\mathcal{A}_i$ will stand for the action set of player $i \in \mathcal{K}$ with $\mathcal{K} = \{1,\ldots,K\}$. As in the previous chapters, the vector $\underline{a}$ will stand for the action profile of the one-shot game; that is $\underline{a} = (a_1,\ldots,a_K) \in \mathcal{A}$ with $\mathcal{A} = \mathcal{A}_1 \times \cdots \times \mathcal{A}_K$. In contrast with the previous chapters, we restrict our attention to **finite**[1] action sets, but the results can be applied to a large extent to the case of infinite action sets (e.g., compact sets). When the action profile $\underline{a}$ is played, player i has a utility which is equal to $u_i(\underline{a})$. The game $\mathcal{G} = (\mathcal{K}, \{\mathcal{A}_i\}_{i\in\mathcal{K}}, \{u_i\}_{i\in\mathcal{K}})$ is played a certain number of times, which leads to a repeated game. The points at which players can update their actions are called the stages of the game. The index $t \geq 1$ will be used to refer to a given stage. To define a standard repeated game, several preliminary definitions are in order.

···

Definition 80: Perfect Monitoring

A repeated game is said to have perfect monitoring if, at each stage $t \geq 1$, every player knows the actions played by all the players at the previous stage.

This definition indicates that all the players can observe each other. Even though at each stage of the game, the last action profile is perfectly observed it can happen that older profiles have been forgotten by some players. This would typically happen in a wireless terminal with limited storage capabilities. The following definition accounts for scenarios where every player remembers at each stage all of the past action profiles.

---

[1] The approach chosen here is reminiscent of what is done in Shannon theory. Many coding theorems are derived for discrete-input discrete-output channels but can be applied most of the time to continuous channels (e.g., Gaussian channels).

Definition 81: Perfect Recall

A repeated game is said to be with perfect recall if every player knows perfectly at each stage the actions played by all the players since the beginning of the game.

The perfect recall assumption is not inherent to repeated games, and it can be made in any sequential game, where it refers to the assumption that, at every opportunity to act, each player remembers what he did in prior moves, and each player remembers everything that he knew before: players never forget information once it is acquired. In the framework of repeated games, it is relevant to define a notion which is called the history of the game. It is defined below in the case of games with perfect monitoring. Having perfect recall means knowing the game history at every stage.

Definition 82: Game History

The vector $\underline{h}_t = (\underline{a}(1),\ldots,\underline{a}(t-1))$, where $\underline{a}(t) = (a_1(t),\ldots,a_K(t))$ is the action profile played at stage t, is called the history of the game at stage t and lies in the set:

$$\mathcal{H}_t = \mathcal{A}^{t-1}. \tag{3.1}$$

From this definition a pure strategy can be properly defined.

Definition 83: Pure Strategy

A pure strategy for player $i \in \mathcal{K}$ is a sequence of causal functions $(\tau_{i,t})_{t \geq 1}$ with:

$$\tau_{i,t} : \left| \begin{matrix} \mathcal{H}_t & \to & \mathcal{A}_i \\ \underline{h}_t & \mapsto & a_i(t) \end{matrix} \right. \tag{3.2}$$

By denoting $\mathcal{T}_i$ as the set of pure strategies for player i, a mixed strategy for player i is merely a probability distribution $\widehat{\tau}_i$ over $\mathcal{T}_i$. It turns out that in repeated games with perfect recall, simpler mathematical constructs, namely behavior strategies, can be used.

Definition 84: Behavior Strategy

A behavior strategy for player $i \in \mathcal{K}$ is a sequence of causal functions, $(\widetilde{\tau}_{i,t})_{t \geq 1}$ with:

$$\widetilde{\tau}_{i,t} : \left| \begin{matrix} \mathcal{H}_t & \to & \Delta(\mathcal{A}_i) \\ \underline{h}_t & \mapsto & \pi_i(t) \end{matrix} \right. \tag{3.3}$$

where $\Delta(\mathcal{A}_i)$ is always (see Chapter 1) the unit simplex associated with $\mathcal{A}_i$ namely, the set of possible probability distribution over $\mathcal{A}_i$.

After Kuhn's theorem was formulated (Kuhn, 1953), it was realized that, in finite extensive-form games with perfect recall, there is an equivalence in terms of game outcome between mixed and behavior strategies. More specifically, a mixed strategy $\widehat{\tau}_i$ is said to be (realization) equivalent to a behavior strategy $\widetilde{\tau}_i$ when, for given mixed strategies chosen by the other players $(-i)$, both strategies lead to the same outcome (every player obtains the same utility). Kuhn's result has been generalized by Aumann (1961) to infinitely repeated games with a finite number of players and action sets. This is the reason why mixed strategies do not need to be considered in standard repeated games with perfect recall. When the perfect recall assumption does not hold, this result needs to be reconsidered. It can turn out that mixed strategies and even general strategies[2] are needed. These aspects will not be here considered and the reader is referred to papers like Kaneko and Kline (1995) and Wichardt (2008) for further details about perfect recall refinements or imperfect recall.

In repeated games, a player does not care about what he gets at a given stage, but about what his gains over the whole duration of the game. This is why utility functions resulting from averaging over the instantaneous utility have to be considered. There are three dominant models used in the literature. The corresponding expressions for the utility are provided by the following three definitions; pure strategies are considered but the expressions for the case of mixed or general strategies readily follow.

Definition 85: Utility in Finitely Repeated Games (FRG)

Let $\underline{\tau} = (\tau_1, \ldots, \tau_K)$ be a joint strategy and T the number of stages of the game. The utility for player $i \in \mathcal{K}$ is defined by:

$$v_i^T(\underline{\tau}) = \frac{1}{T} \sum_{t=1}^{T} u_i(\underline{a}(t)) \tag{3.4}$$

where $\underline{a}(t)$ is the action profile of the action plan induced by the joint strategy $\underline{\tau}$ at time t.

Definition 86: Utility in Infinitely Repeated Games (IRG)

$$v_i^\infty(\underline{\tau}) = \lim_{T \to +\infty} \frac{1}{T} \sum_{t=1}^{T} u_i(\underline{a}(t)). \tag{3.5}$$

when the limit exists.

Definition 87: Utility in Discounted Repeated Games (DRG)

Let $0 < \lambda < 1$ be the discount factor. The utility for player $i \in \mathcal{K}$ is defined by:

$$v_i^\lambda(\underline{\tau}) = \sum_{t=1}^{+\infty} \lambda(1-\lambda)^{t-1} u_i(\underline{a}(t)). \tag{3.6}$$

---

[2]By denoting $\widetilde{\mathcal{T}}_i$ the set of behavior strategies for player i, a general strategy for player i is a probability distribution $\overline{\tau}_i$ over $\widetilde{\mathcal{T}}_i$.

The FRG model indicates that the players interact over a finite number of stages and the game stops at stage T. In this model, the average utility $v_i^T(\underline{\tau})$, which is the Cesaro mean of the instantaneous utilities, is always well defined. For the IRG and DRG models, it first must be verified, for the scenario under study, that these limits effectively exist. The main question which arises is as follows: Which model is the most suitable? The answer to this question is strongly scenario-dependent. For example, in wireless scenarios where the duration over which the transmitters interact is known (this is typically the case when transmit power levels can only be updated during the training phase of the block) or/and transmitters value the different instantaneous utilities uniformly, the first model seems to be preferable. On the other hand, the third model can be used as follows: the discount factor is used in Etkin et al. (2007) as a way of accounting for the delay sensitivity of the network; the discount factor is used in Wu et al. (2009) to give the transmitters the possibility of valuing short-term and long-term gains differently. Interestingly, the authors of Neyman and Sorin (2010), Shapley (1953b) and Sobel (1992) offer another interpretation of this model. They see the discounted repeated game model as a finitely repeated game where T would be unknown to the players, and is considered as an integer-valued random variable, finite almost surely, whose law is known by the players. In other words, λ can be seen as the stopping probability at each stage: the probability that the game stops at stage t is thus $\lambda(1-\lambda)^{t-1}$. The function v_i^λ would correspond to an expected utility given the law of T. This shows that the discount factor is also useful for studying wireless games where a player enters/leaves the game (Treust and Lasaulce, 2010a). This approach can also model a heterogeneous game where players have different discount factors, in which case λ represents $\min_i \lambda_i$ (as pointed out in Shapley (1953b)).

Notation: in the sequel, the repeated game $\left(\mathcal{K}, \{\mathcal{T}_i\}_{i\in\mathcal{K}}, \{v_i\}_{i\in\mathcal{K}}\right)$ will be referred to as $\mathcal{G}^T$, $\mathcal{G}^\infty$, or $\mathcal{G}^\lambda$ depending whether the FRG, IRG, or DRG model is considered.

3.2.3 Equilibria in Repeated Games with Perfect Monitoring

In the preceding chapter the importance of Nash equilibria in static games was highlighted. In the framework of repeated games, a pure Nash equilibrium becomes a profile of action plans (instead of a single action profile) which is stable to single deviations. Formally, a pure Nash equilibrium is defined as follows.

....

Definition 88: Pure Equilibrium Strategies

A joint strategy $\underline{\tau}$ supports an equilibrium of the repeated game $\left(\mathcal{K}, \{\mathcal{T}_i\}_{i\in\mathcal{K}}, \{v_i\}_{i\in\mathcal{K}}\right)$ if $\forall i \in \mathcal{K}, \forall \tau_i', v_i(\underline{\tau}) \geq v_i(\tau_i', \underline{\tau}_{-i})$ where v_i is defined by (3.4), (3.5), or (3.6).

An important question is whether such strategies exist in a given repeated game. The answer is simple in the case of the finitely repeated game and discounted repeated game. Indeed, the Nash existence theorem gives the answer for any finite game (see Chapter 2) and the Debreu existence theorem (see also Chapter 2) guarantees the existence of mixed equilibria in any game with compact action spaces and continuous utility functions. Clearly, the sets $\mathcal{T}_i$ are compact and the functions

v_i^T are continuous for the product topology. The continuity of v_i^λ is achieved by assuming the values of the one-shot game utility functions u_i to be bounded. The answer is also simple and even more general when there exists an equilibrium in the constituent one-shot game: in this case, playing at one of these equilibria corresponds to an equilibrium of the repeated game. It can be noticed that the equilibrium analysis of the repeated game conducted here follows from the features of the constituent game. This observation is in fact more general, and is at the heart of the so-called "Folk theorems". These aim to characterize the set of equilibrium utilities of repeated games for different types of repeated games (finite/infinite/discounted repeated games), equilibria (Nash equilibria, correlated equilibria, communication equilibria, etc.), information assumptions (complete/incomplete information), and observation structures (perfect/imperfect monitoring).

The authorship of the first Folk theorem is obscure: the corresponding theorem was used independently by several people in the field without being published; hence its name. The first Folk theorem, which is given below, concerns infinitely repeated games with complete information, perfect monitoring, and perfect recall. Note that other existing theorems fully characterizing the equilibrium utilities in repeated games (e.g., for the case of public signals) are also called Folk theorems, although their authorship is known. Respectively denoting by E^T, E^∞, and E^λ the set of possible equilibrium utilities in FRG, IRG, and DRG the following three theorems can be proved. What is called the first Folk theorem is the characterization of the equilibrium utilities in IRG. For this purpose, the following auxiliary set will be used here and throughout the chapter:

$$E \triangleq \mathrm{IR}(\mathcal{G}) \cap \mathrm{co}\,(\mathcal{U}(\mathcal{G})). \tag{3.7}$$

where[3] $\mathrm{IR}(\mathcal{G}) = \{\underline{u} \in \mathbb{R}^K, \forall i \in \mathcal{K}, u_i \geq \min\limits_{\underline{s}_{-i} \in \prod_{j \neq i} \Delta(\mathcal{A}_j)} \max\limits_{a_i} u_i(a_i, \underline{s}_{-i})\}$ and $\mathcal{U}(\mathcal{G}) = \{\underline{u}' \in \mathbb{R}^K, \exists \underline{a}, \underline{u}(\underline{a}) = \underline{u}'\}$.

Theorem 89: Folk Theorem for IRG with Perfect Monitoring

$$E^\infty = E. \tag{3.8}$$

The set $\mathrm{IR}(\mathcal{G})$ is the set of individually rational utilities. As a player can always guarantee that his instantaneous utility be greater than or equal to his minmax level in the one-shot game, a vector of equilibrium utilities of the IRG has necessarily to be in this set; the notation $\Delta(\mathcal{A}_j)$ has been introduced in the preceding chapter and stands

---

[3] A slight abuse of notations has to be mentioned here. The arguments of the function u_i comprise both pure and mixed strategies. Rigorously, a_i should be replaced with e_i, which is the Dirac distribution having a non-zero weight for action a_i.

for the unit simplex associated with the discrete set $\mathcal{A}_j$. The set $\mathcal{U}(\mathcal{G})$ corresponds to the set of vector utilities which are feasible or reachable in the one-shot game $\mathcal{G}$. The notation co(.) stands for the convex hull of a set. The proof of Theorem 89 will not be provided here but can be found for example, in Aumann (1981). The proof is essentially based on what is called grim-trigger: before starting the game, players agree on operating at a certain profile ((F,F) in the forwarder's dilemma) and punish the deviator by playing their minmax level ((D,D) in the forwarder's dilemma) forever if someone deviates. A natural question is whether the above theorem applies to finitely repeated games and discounted repeated games. It turns out that sufficient conditions can to be added to make this extension possible. To illustrate the need for additional conditions in FRG, consider the case of the forwarders' dilemma (which is a wireless counterpart of the prisoner's dilemma we will consider often in this book). In a T-stage repeated forwarder's dilemma, there is no incentive to cooperate at the final stage, since no punishment mechanism can be implemented once the game is over. Therefore the action profile played at the last stage $t = T$ will be (D,D). At stage $T - 1$, the players know what is about to be played at stage T, and there is again no way of punishing a player who deviates from the cooperation plan. By induction, both players will always play the action D (drop a packet or defect). This observation holds for any value of the number of stages. It corresponds, in fact, to a special case of a result derived by Neyman and Sorin (2010).

Proposition 90: Equilibria in FRG

If there is a unique equilibrium in the one-shot game $\mathcal{G}$ which corresponds to playing at the minmax levels, then for all $T \geq 1$ the set of equilibrium utilities of the associated FRG boils down to a singleton; namely, the players will always play at their minmax level.

This explains the following condition in the following theorem, due to Benoît and Krishna (1987).

Theorem 91: Folk Theorem for FRG with Perfect Monitoring

If there is a Nash equilibrium in the one-shot game $\mathcal{G}$ such that for all $i \in \mathcal{K}$, the equilibrium utility u_i^* verifies $u_i^* > \min_{\underline{s}_{-i} \in \prod_{j \neq i} \Delta(\mathcal{A}_j)} \max_{a_i} u_i(a_i, \underline{s}_{-i})$ then we have that:

$$\lim_{T \to +\infty} E^T = E = \mathrm{IR}(\mathcal{G}) \cap \mathrm{co}(\mathcal{U}(\mathcal{G})). \tag{3.9}$$

When using the limit for the set E^T in this theorem, the convergence in the sense of Hausdorff is implicitly assumed; that is, the Hausdorff distance between E^T and E goes to zero as $T \to +\infty$. The Hausdorff distance, which quantifies the distance between two sets, is built from the distance between a point and a set (the distance between a point x and a set $\mathcal{Y}$ denoted by $d(x, \mathcal{Y}) = \min_{y \in \mathcal{Y}} d(x, y)$):

> **Definition 92: Hausdorff Distance**
>
> Let $\mathcal{X}$ and $\mathcal{Y}$ be two compact sets.
>
> $$d(\mathcal{X}, \mathcal{Y}) = \max\left\{\max_{x \in \mathcal{X}} d(x, \mathcal{Y}), \max_{y \in \mathcal{Y}} d(y, \mathcal{X})\right\}. \tag{3.10}$$

Example (the repeated prisoner's dilemma). If one applies this theorem to the finitely repeated prisoner's dilemma, one can easily find that the set of equilibrium utilities associated with the Folk theorem boils down to a singleton: the only possible equilibrium is when both players play "Defect". Although we have mentioned some success stories in game theory (like the one about the Talmud bankruptcy problem), here one of its weaknesses appears: many experiments show that real players, who play the prisoner's dilemma for a finite number of times, generally do not act according to the forecasts of the FRG model. Indeed, in real life, people tend to cooperate, e.g., by playing tit-for-tat, meaning that "non-rational" players get better utilities than rational ones!

Finally, a Folk theorem for repeated games with discounting can be stated, provided a simple sufficient condition is assumed (Neyman and Sorin, 2010).

> **Theorem 93: Folk Theorem for DRG with Perfect Monitoring**
>
> If the set E comprises a point $\underline{u}$ such that for all $i \in \mathcal{K}$, u_i verifies $u_i > \min_{\underline{s}_{-i} \in \prod_{j \neq i} \Delta(\mathcal{A}_j)} \max_{a_i} u_i(a_i, \underline{s}_{-i})$ then we have:
>
> $$\lim_{\lambda \to 0} E^\lambda = E = \mathrm{IR}(\mathcal{G}) \cap \mathrm{co}(\mathcal{U}(\mathcal{G})). \tag{3.11}$$

As already mentioned in this chapter, Folk theorems aim to provide all possible equilibrium utilities which can be sustained in a repeated game. The asymptotic regimes $T \to +\infty$ or $\lambda \to 0$ allow one to reach all such points. In practice, the value of the game duration or the discount factor is usually given. In such a situation, a more pragmatic way of interpreting the Folk theorem (say for the DRG) is to answer questions like the following. For a given point in the achievable region defined by the Folk theorem, what is the minimum discount factor to be able to operate at this equilibrium point? In Chapter 8, a concrete example will be provided to show how to calculate the value of the discount factor.

Another issue of practical importance is to make sure that, when a single deviation occurs, the other players effectively implement the punishment procedure. Otherwise, the threat corresponding to the punishment is not credible, and the cooperation-punishment plan used in the proofs of the Folk theorems to reach any point of the equilibrium utility region cannot always be implemented. An additional requirement for equilibria in repeated games to avoid this situation is that they be subgame perfect. A strategy profile $\underline{\tau}^*$ is a subgame perfect equilibrium, if, for any history $\underline{h}$, $\underline{\tau}^*$ induces an equilibrium in the game played after the history $\underline{h}$ has been played. This can be translated more formally as follows.

∙∙∙

Definition 94: Subgame Perfect Equilibria

A (behavior) strategy profile $\underline{\tau}^*$ is a subgame perfect equilibrium if:

$$\forall i \in \mathcal{K}, \ \forall \tau_i' \in \mathcal{T}_i, \ \forall t \geq 1, \ \forall \underline{h}_t \in \mathcal{H}_t, \ v_i\left(\underline{\tau}^*_{|\underline{h}_t}\right) \geq v_i\left(\tau_{i|\underline{h}_t}', \tau^*_{-i|\underline{h}_t}\right) \tag{3.12}$$

where the notation $_{|\underline{h}}$ indicates the history $\underline{h}$ is given.

This definition applies perfectly to IRG and DRG by setting $v_i = v_i^\infty$ or $v_i = v_i^\lambda$. For FRG, $v_i = v_i^T$ and the stage index t has to verify $t \leq T$. When it comes to verifying whether a behavior strategy is a subgame perfect equilibrium or not, the following characterization can be useful.

∙∙∙

Theorem 95: Characterization of Subgame Perfect Equilibria

A (behavior) strategy profile $\underline{\tau}^*$ is a subgame perfect equilibrium if and only if there exist no player i, no deviation τ_i', and no game history $\underline{h}_t'$ such that:

- $\forall \underline{h}_t \neq \underline{h}_t', \ \tau_{i,t}^* = \tau_{i,t}'$;
- $v_i\left(\tau_{i|\underline{h}_t'}, \underline{\tau}^*_{-i|\underline{h}_t'}\right) > v_i\left(\underline{\tau}^*_{|\underline{h}_t}\right).$

Interestingly, the three Folk theorems stated so far hold for subgame perfect equilibria provided simple conditions are met (Sobel, 1992), such as the following.

- Theorem 89: it holds for subgame perfect equilibria, which means that imposing this property does not modify the region of possible equilibrium utilities (Aumann and Shapley, 1976).
- Theorem 91: as proved by Benoît and Krishna (1985, 1987), a sufficient condition for making Theorem 91 valid for subgame perfect equilibria is to have two Nash equilibria in the one-shot game $\underline{u}^*$ and $\underline{u}^{\text{diag}}$ such that $\underline{u}^* > \underline{u}^{\text{diag}}$ componentwise.
- Theorem 93: if the interior of $E = \text{IR}(\mathcal{G}) \cap \text{co}(\mathcal{U}(\mathcal{G}))$ is non-empty, Theorem 93 holds for subgame perfect equilibria (Maskin and Fudenberg, 1986).

What about subgame perfection in the forwarder's (or prisoner's) dilemma? We know that in IRG, the cooperation-punishment plan consists of always playing Forward (or Cooperate) as long as no deviation is observed, and Drop (or Defect) forever as soon as a deviation is observed, and this corresponds to a subgame perfect equilibrium. In FRG, Theorem 91 cannot be applied since the unique one-shot game Nash equilibrium coincides with the profiles of minmax strategies. Finally, in DRGs, the interior of the set E is clearly non-empty, and hence having subgame perfect equilibria in the repeated forwarder's dilemma with discounting is possible.

3.2.4 **Beyond the Perfect Monitoring Assumption**

In the previous section, the equilibrium analysis was conducted under the assumption that every player observes the actions of all the other players, which might be not be true in some wireless games. It turns out that Theorem 89 holds under milder assumptions. Indeed, Renault and Tomala (1998) provide a necessary and sufficient condition on the observation graph of the game in order for Theorem 89 to be valid. To express this condition, the authors of Renault and Tomala (1998) define a directed graph (V, E) whose vertices $i \in V$ are the players, and for which there is an edge from player i to player j if player j sees the actions chosen by player i. (See Fig. 3.1.) A graph is said to be strongly connected if, for all pair of vertices (i, j), there exists a directed path from i to j that is 2-connected, and if for every vertex i, the subgraph obtained by removing this vertex from the original graph is still strongly connected. The following Folk theorem has been proved for IRG (Renault and Tomala, 1998).

Theorem 96: Folk Theorem for IRG

The following two assertions are equivalent:

i. the observation graph (V, E) is 2-connected;
ii. $E^\infty = E$.

In particular, this theorem shows that it is not necessary to observe all players. This theorem has been exploited in a wireless scenario in Treust et al. (2010b) to analyze a coverage game where base stations have to tune their transmit power in an autonomous manner by observing a reduced number of neighboring base stations.

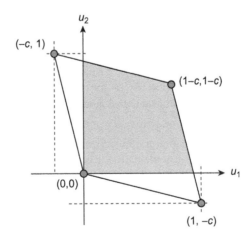

FIGURE 3.1

Illustration of the Folk theorem in the forwarders' dilemma.

Unfortunately, there is still no general Folk theorem that fully characterizes the region of possible equilibrium utilities for a game with an arbitrary observation structure (imperfect monitoring). In Chapter 4, special cases of observation structures for which the problem is solved will be provided.

Remark. The usefulness of the repeated game models will be made more obvious for the reader in Chapter 8 where they are applied to the distributed power control problem. We will make just one comment here to to give an idea of the impact of using a repeated game model instead of a one-shot game model for modeling this problem. Conventionally, for i.i.d. channels, power control schemes are designed such that the power levels are chosen in an independent manner from block to block. In distributed networks with selfish transmitters, the point of view has to be re-considered even if the channels are i.i.d., due to the fact that long-term interaction may change the behavior of selfish transmitters. In order to take into account this effect and the fact that channel gains may vary from block to block, dynamic models such as repeated games are considered in Chapter 8, leading to more efficient power control schemes.

3.3 STOCHASTIC GAMES

Stochastic games were introduced in Shapley (1953b) and generalize repeated games. In repeated games, the same game is played at each stage, whereas in stochastic games the game played at each stage can vary over time. A stochastic game is characterized by a state which can evolve over time; if the set of possible states reduces to a singleton, the stochastic game boils down to a repeated game. The instantaneous utility of a given player depends not only on the actions played, but also on the current state of the game. Additionally, the state evolves in a manner depending on previous states and played actions. A simple example is the fishery game, where two countries share the same area for fishing. The players are the two countries involved, their action corresponds to the quantity of fish they take, and the density of fish in the sea/ocean represents the state of the game. The density of fish at year t depends on its value at year $t - 1$ but also on the actions taken by the fishers of the two countries. Also, the action space of the players depends on the state of the game; for example, if there is no fish in the sea/ocean anymore, the action space of the fishers reduces to a singleton.

In wireless applications, many systems possess some states which evolve over time e.g., the battery level of a terminal (which generally depends on its transmit power), the state of a server, the channel gains. Typical examples are as follows. Atomic, unconstrained, stochastic games with individual states were proposed by Haykin (2005) to analyze dynamic spectrum sharing in cognitive radio networks, in which the state of the resource and messages are not taken in consideration. Atomic, stochastic games with individual states have been studied by Altman et al. (2009a,b), Altman and Solan (2009) in which the state of the resource is not considered (there is a single resource state) and there is neither signal nor message.

Zhu et al. (2010c) proposed a class of non-zero-sum stochastic games with individual states with two groups of players (attackers and defenders) to study intrusion detection in network security.

In this section, several models of stochastic games are provided. As already mentioned, the corresponding models generalize those studied in the section on repeated games. They are more general because the game possesses a state. For more generality, notions like ϵ-equilibria, and tools like the liminf in games with discounting, will be used in some places. The price to be paid for considering more general models is that there are fewer results available (e.g., in terms of Folk theorems). This section describes the case of stochastic games with a common state. The case of stochastic games with individual states is merely mentioned.

3.3.1 Stochastic Games with a Common State

3.3.1.1 Definition of a Stochastic Game with a Common State

Similarly to repeated games, stochastic games can be expressed under a normal form $\Gamma^m = \left(\mathcal{K}, \{\Sigma_i\}_{i \in \mathcal{K}}, \{v_i^m\}_{i \in \mathcal{K}}\right)$ where $m \in \{T, \infty, \lambda\}$ depending on the model of stochastic games under consideration. $\mathcal{K}$ is always the set of players, Σ_i the set of strategies of player i, and v_i the long-term utility of player i. The purpose of this subsection is to define all of these quantities. A stochastic game with a common state is defined from a 6-uplet $\left(\mathcal{K}, \{\overline{\mathcal{A}_i}\}_i, \Omega, \{\mathcal{A}_i\}_i, \{\alpha_i\}_i, q, \{u_i\}_i\right)$ where:

- $\{\overline{\mathcal{A}_i}\}_i$ is the set of all possible actions during the course of the game;
- Ω is the set of states of the game;
- $\mathcal{A}_i(\omega)$ is the set of possible actions for the state $\omega \in \Omega$;
- $\alpha_i : \Omega \to 2^{\overline{\mathcal{A}_i}}$ is the correspondence determining the possible actions at a given state of the game;
- under the Markov game assumption, the transition probability of the states is given by the conditional distribution q_t, which is defined a little further on.

The stochastic game starts at an initial state $\omega \in \Omega$ and is played as follows. The current state ω_t is known by all the players. At each stage $t \geq 1$, each player $i \in \mathbb{K}$ chooses an action $a_i(t) \in \mathcal{A}_i(\omega(t))$ and receives an instantaneous utility $u_i(\underline{a}(t), \omega(t))$, where $\underline{a}(t) = (a_1(t), \ldots, a_K(t)) \in \mathcal{A}(\omega(t))$ (always with the usual notation $\mathcal{A} = \mathcal{A}_1 \times \mathcal{A}_2 \times \cdots \times \mathcal{A}_K$). The game then moves to a new state according to a probability distribution given by $q_t(\omega(t+1) \mid \omega(t), \underline{a}(t), \omega(t-1), \underline{a}(t-1), \ldots, \omega(1)) \in \Delta(\Omega)$. The stochastic game is said to be Markovian or a Markov game if $q(\cdot \mid \omega(t), \underline{a}(t), \omega(t-1), \underline{a}(t-1), \ldots, \omega(1)) = q_t(. \mid \omega(t), \underline{a}(t))$. More formally, q_t can be written as:

$$q_t : \begin{vmatrix} \Omega \times \bigotimes_{i=1}^{K} 2^{\mathcal{A}_i} & \to & \Delta(\Omega) \\ (\omega, \underline{a}) & \mapsto & q_t(\omega' \mid \omega, \underline{a}). \end{vmatrix} \tag{3.13}$$

This class of stochastic game is also referred to as a competitive Markov decision process (Filar and Vrieze, 1997). Indeed, Markov decision processes (Feinberg and Shwartz, 2002) and Markov decision teams are stochastic games with a single player (or a team decision maker). In the sequel the transition probabilities q_t will be assumed to be independent of t and therefore denoted by q.

In stochastic games, the counterpart of the constituent game in repeated games is the 6-uplet defined previously. From this 6-uplet the game history, strategies, and players' utility can be defined. Assuming a stochastic game with perfect recall and perfect monitoring (see Sec. 3.2 for more details) the latter quantities can be defined as follows:

Definition 97: Game History

The vector $\underline{h}_t = (\omega(1), \underline{a}(1), \ldots, \omega(t-1), \underline{a}(t-1), \omega(t))$ is called the history of the game at stage t and lies in the set:

$$\mathcal{H}_t = \left(\bigotimes_{t'=1}^{t-1} \Omega \times \mathcal{A}(\omega(t')) \right) \times \Omega. \tag{3.14}$$

Two comments are in order. First, note that the knowledge of the state is not strictly causal since the realization of the state at stage t is assumed to be known to every player, to allow him to take his action at stage t. Second, the notation $\omega(t')$ has to be understood as the realization of the random state of the game at stage t', which means that the set of histories $\mathcal{H}_t$ is not a deterministic object, in contrast with repeated games. From this definition a pure strategy can be defined.

Definition 98: Pure Strategy

A pure strategy for player $i \in \mathcal{K}$, denoted by σ_i, is a sequence of functions $(\sigma_{i,t})_{t \geq 1}$ with:

$$\sigma_{i,t} : \begin{vmatrix} \mathcal{H}_t & \rightarrow & \mathcal{A}_i(\omega(t)) \\ \underline{h}_t & \mapsto & a_i(t) \end{vmatrix} \tag{3.15}$$

Definition 99: Behavior Strategy

A behavior strategy for player $i \in \mathcal{K}$, denoted by $\widetilde{\sigma}_i$, is a sequence of causal functions $(\widetilde{\sigma}_{i,t})_{t \geq 1}$ with:

$$\widetilde{\sigma}_{i,t} : \begin{vmatrix} \mathcal{H}_t & \rightarrow & \Delta(\mathcal{A}_i(\omega(t))) \\ \underline{h}_t & \mapsto & \rho_i(t). \end{vmatrix} \tag{3.16}$$

A joint behavior strategy (or a profile of behavior strategies) will be denoted by $\underline{\widetilde{\sigma}} = (\widetilde{\sigma}_1, \ldots, \widetilde{\sigma}_K)$.

Definition 100: Stationary Behavior Strategies

A behavior strategy is stationary if, for any pair of histories $\underline{h}_t$ and $\underline{h}'_{t'}$, $t \geq 1$, $t' \geq 1$:

$$\omega(t) = \omega(t') \Rightarrow \underline{\widetilde{\sigma}}(\underline{h}_t) = \underline{\widetilde{\sigma}}(\underline{h}'_{t'}). \tag{3.17}$$

In other words, a stationary behavior strategy only depends on the realization of the current state.

The long-term utilities in stochastic games can be defined in the same way as for repeated games. Exploiting the existence of a realization equivalent strategy for any given mixed strategy in stochastic games played (calling for direct extensions of Kuhn's theorem), the utilities of the mixed extension of the games Γ^m, $m \in \{T, \infty, \lambda\}$ are defined as follows.

Definition 101: Utility in T-Stage Stochastic Games

The utility for player $i \in \mathcal{K}$ is defined by:

$$v_i^T(\widetilde{\sigma}, \omega_1) = \mathbb{E}_{\widetilde{\sigma}, \omega_1}\left[\frac{1}{T}\sum_{t=1}^{T}u_i(\underline{a}(t), \omega(t))\right] \qquad (3.18)$$

where the expectation is taken with respect to the joint probability distribution $P_{\widetilde{\sigma}, \omega_1}$ and ω_1 corresponds to the notation $\omega(1)$; for the sake of clarity the notations $\omega(t)$ and ω_t will be used interchangeably.

In stochastic games with an infinite number of stages, a more general definition of utility than the one provided for repeated games can also be given.

Definition 102: Utility in Infinitely Long Stochastic Games

The utility for player $i \in \mathcal{K}$ is defined by:

$$v_i^\infty(\widetilde{\sigma}, \omega_1) = \mathbb{E}_{\widetilde{\sigma}, \omega_1}\left[\lim_{T \to +\infty} \inf \frac{1}{T}\sum_{t=1}^{T}u_i(\underline{a}(t), \omega(t))\right]. \qquad (3.19)$$

Finally, in infinitely long stochastic games with discounting, the long-term utility is given by the following definition:

Definition 103: Utility in Discounted Stochastic Games

The utility for player $i \in \mathcal{K}$ is defined by:

$$v_i^\lambda(\widetilde{\sigma}, \omega_1) = \mathbb{E}_{\widetilde{\sigma}, \omega_1}\left[\sum_{t=1}^{+\infty}\lambda(1-\lambda)^{t-1}u_i(\underline{a}(t), \omega(t))\right]. \qquad (3.20)$$

3.3.1.2 *Equilibrium Analysis*

A Nash equilibrium in stochastic games can be defined in the same way as for repeated games: a strategy profile is an equilibrium if it is stable to single deviations. Again, note that behavior strategies are implicitly assumed without any loss of optimality or generality. In Γ^T and Γ^λ, there is a simple condition guaranteeing the existence of an equilibrium, which is as follows:

Theorem 104

If both state and action spaces are finite then:

- the stochastic game Γ^T has at least one equilibrium;
- the stochastic game Γ^λ has at least one equilibrium.

This existence result was initially established by Shapley (1953b) for two-player zero-sum stochastic games. Takahashi (1962) and Fink (1964) extended Shapley's result to the case of non-zero-sum stochastic games with discounting. The proof is based on the dynamic programming principle (Bellman, 1952) or Bellman's optimality principle (which is reviewed in Note 2) and fixed point theorems. The dynamic programming principle can also be used to characterize stationary equilibria and the associated equilibrium utilities in games with discounting. This is the purpose of the following lemma.

Lemma 1

A stationary strategy $\widetilde{\sigma}$ is a discounted equilibrium if and only if the corresponding utility v_i^λ satisfies the dynamic programming principle:

$$v_i^\lambda(\widetilde{\sigma},\omega) = \sup_{x_k \in \Delta(\mathcal{A}_i(\omega))} \left\{ \lambda u_i(x_i,\widetilde{\sigma}_{-i},\omega) + (1-\lambda) \sum_{\omega' \in \Omega} q(\omega'|\omega,x_i,\widetilde{\sigma}_{-i}) v_i^\lambda(\omega',\widetilde{\sigma}) \right\} \quad (3.21)$$

for all $\omega \in \Omega$ and $i \in \mathcal{K}$.

A proof of this lemma can be found in Blackwell (1968). This theorem provides a simple, necessary and sufficient condition for a strategy to be an equilibrium. In terms of existence, we have the following theorem.

Theorem 105: Fink (1964) and Takahashi (1962)

Every K-player stochastic game with finite state and action spaces has a discounted equilibrium in stationary strategies.

For T-stage stochastic games, the dynamic programming principle for the averaged utility is written:

$$Tv_{k,T}(\omega,\sigma) = \sup_{x_k \in \Delta(\mathcal{A}_k(\omega))} \left\{ u_k(\omega,x_k,\sigma_{-k}(\omega)) + (T-1) \sum_{\omega' \in \mathcal{S}} q(\omega'|\omega,x_k,\sigma_{-k}(\omega)) v_{k,T-1}(\omega',\sigma) \right\}$$

$$(3.22)$$

Little is known regarding the existence of the liminf equilibrium. One of the most significant result in this direction is available in Vieille (2000a,b, 2002), where it is proved that every two-player stochastic game with finite state and action spaces has a uniform ϵ-equilibrium for every $\epsilon > 0$. This existence has been proved in a specific class of stochastic games under ergodicity conditions and stochastic games on product state spaces (Flesch et al., 2008). It is known from a well-known counter-example by Gillette (1957) that even in the two-person-zero-sum case, a 0-equilibrium does not need to exist in the average utility case.

The existence of equilibria in Γ^∞ is known for the class of irreducible Markov decision processes.

Definition 106

A stochastic game is irreducible if, for each player i, and, for each pair of states $(\omega,\omega') \in \Omega^2$, there is a $T > 0$ and a sequence $(\omega_1,\dots,\omega_T)$ of states such that $\omega_1 = \omega$ and $\omega_T = \omega'$, and for each $t < T$, there is a profile $\underline{\alpha}$ such that $q(\omega_{t+1}|\omega_t,a_i,\underline{\alpha}_{-i}) > 0$ for all $a_i \in \mathcal{A}_i$.

Theorem 107: Sobel (1971, 1973)

Assume an irreducible stochastic game. Then, this game possesses an equilibrium for the average utility v_i^∞. Additionally, there exists a stationary strategy $\tilde{\sigma}$ and for each player $i \in \mathcal{K}$ a constant g_i and a bounded function $v_i(.)$ such that:

$$g_i + v_i^\infty(\omega) = \sup_{x_i \in \Delta(\mathcal{A}_i(\omega))} \left\{ u_i(\omega,x_i,\sigma_{-i}(\omega)) + \sum_{\omega' \in \mathcal{S}} q(\omega'|\omega,x_i,\sigma_{-i}(\omega))v_i^\infty(\omega') \right\} \quad (3.23)$$

for all $\omega \in \Omega$, where $\sigma_i(\omega)$ achieves the maximum of the right hand side of (3.23) for all $\omega \in \Omega$. Moreover, $v_i^\infty(\omega,\sigma) = g_i$.

As mentioned in the section dedicated to repeated games, subgame perfection is a desirable feature for an equilibrium in a dynamic game. In particular, it can be seen as being the requirement for the credibility of an equilibrium strategy. In the context of stochastic games, subgame perfect equilibria have been studied in Thuijsman and Raghavan (1997). In stochastic games with discounting the following result can be proven.

Theorem 108: Mertens and Parthasarathy (1991) and Sobel (1998)

Any K-player discounted stochastic game Γ^λ has a subgame perfect equilibrium.

3.3.1.3 Characterization of Equilibrium Utilities in Special Cases of Stochastic Game

The case of zero-sum stochastic games. Zero-sum stochastic games with complete information have been studied in Filar and Vrieze (1997) and Sorin (2002). Neyman and Sorin (2010) have studied zero-sum stochastic games with a public uncertain horizon.

The case of irreducible stochastic games. In Dutta (1991, 1995) Dutta identifies a class of stochastic games with an infinite horizon (infinitely repeated stochastic games) for which the folk theorem holds. First, it is assumed that the players have complete information, perfect monitoring, and know the current state. Second, the game is assumed to be irreducible; more details about this assumption are given a little further on. For this framework, Dutta derived a Folk theorem.

An additional difficulty appears in the corresponding proof compared to (single-state) repeated games, due to the fact that the players may want to deviate, not only in order to improve their current utilities but also to affect the probability distribution of the states. For this reason, asymptotic state invariance of feasible long-run utility and additional assumptions are needed. Dutta (1991) designs a punishment strategy such that for all individually rational utilities, there is a subgame perfect strategy whose utility approaches the former. The equilibrium punishment strategy consists in minmaxing the deviator for several periods, and then playing a strategy that rewards players who effectively applied the punishment procedure during the previous phase. Note that the strategy constructed by Dutta may not be subgame perfect in the case where the number of stages is finite.

An extension of this result is the Folk theorem for irreducible stochastic games with public monitoring made by Fudenberg and Yamamoto (2009) and by Hörner et al. (2009). However, the determination of folk theorems remains an open question even for finite stochastic games with complete information and perfect monitoring. It will be seen that the irreducibility conditions are not satisfied in the example of energy-constrained stochastic games illustrated in Chapter 4.

Here we briefly describe the results obtained by Fudenberg and Yamamoto (2009). Denote by $\mathcal{Y}$ the set of public signals. At each stage t, players observe the state ω_t, and then move simultaneously, with player i choosing an action a_i from a finite set $\mathcal{A}_i$. Given an action profile $\underline{a}$, players observe a public signal y from a finite set $\mathcal{Y}$ and the state at stage $t+1$ is chosen from the set Ω. Let $\pi(\omega', y \mid \omega, \underline{a})$ be the probability that players observe a signal y and the state for the next stage is ω' when today's state is ω and players play action profile $\underline{a}$. (Note that the distributions of y and ω' may be correlated.) Player i's utility is $r_i(\omega, a_i, y)$, so that his expected utility conditional on $\omega, \underline{a}$ is:

$$u_i(\omega, \underline{a}) = \sum_{\omega'} \sum_{y} \pi(\omega', y \mid \omega, \underline{a}) r_i(\omega, a_i, y). \qquad (3.24)$$

Denote by $E_{\omega_0}^{\lambda}$ the set of feasible utilities when the initial state is ω_1 and the discount factor is λ. Dutta (1991) has shown that for an *irreducible stochastic game* with a public signal, the set of feasible utilities becomes independent of the initial state when λ goes to zero. The minmax utility level for player i in a stochastic game with initial state ω_1 is defined to be $\inf_{\widetilde{\underline{\sigma}}_{-i}} \sup_{\widetilde{\sigma}_i} v_i^{\lambda}(\omega_1, \widetilde{\underline{\sigma}})$. Dutta has shown that for irreducible stochastic games, the limit of the minimax utility as delta goes to one is independent of the initial state ω_1.

We now define the notion of irreducibility for the case of public signals.

Definition 109

A stochastic game is irreducible if, for each player i, and for each pair of states $(\omega, \omega') \in \Omega^2$, there is a $T > 0$ and a sequence $(\omega_1, \ldots, \omega_T)$ of states such that $\omega_1 = \omega$ and $\omega_T = \omega'$, and for each $t < T$, there is a profile $\underline{\alpha}$ such that $\sum_{y \in \mathcal{Y}} \pi(\omega_{t+1}, y \mid \omega_t, a_i, \underline{\alpha}_{-i}) > 0$ for all $a_i \in \mathcal{A}_i$.

Define the matrix $\mathbf{M}_i^{\omega, \omega'}(\underline{\alpha})$ with rows $\pi(\omega', y \mid \omega, a_i, \underline{\alpha}_{-i})_{y \in \mathcal{Y}}$ for all $a_i \in \mathcal{A}_i$.

Definition 110

A profile α has full rank for the triplets (i,ω,ω') if the rank of $\mathbf{M}_i^{\omega,\omega'}(\alpha)$ is of rank $|\mathcal{A}_i|$.

Definition 111

For each $i \neq i'$ and (ω,ω'), the profile α a has pairwise full rank for (i,i',ω,ω') if the rank of the matrix obtained by stacking the two matrices $\mathbf{M}_i^{\omega,\omega'}(\alpha)$ and $\mathbf{M}_{i'}^{\omega,\omega'}(\alpha)$ is of rank $|\mathcal{A}_i| + |\mathcal{A}_{i'}| - 1$.

In order to state the Folk theorem proved by Fudenberg and Yamamoto (2009) three assumptions need to be introduced:

i. For each $(\omega,\omega') \in \Omega^2$, there is an integer $T > 0$ and a sequence $(\omega_1,\ldots,\omega_T)$ of states such that $\omega_1 = \omega, \omega_T = \omega'$, and for each $t < T$, every pure action profile has individual full rank for (ω_t,ω_{t+1}) and every player i.

ii. For each $(\omega,\omega') \in \Omega^2$, there is an integer $T > 0$ and a sequence $(\omega_1,\ldots,\omega_T)$ of states such that $\omega_1 = \omega, \omega_T = \omega'$, and for each (i,i') and $t < T$, there is a profile α_t which has pairwise full rank for $(\omega_t,\omega_{t+1},i,i')$ and player i.

iii. Let E be the set of feasible and individually rational utilities. The dimension of E is exactly the cardinality of $\mathcal{K}$.

Theorem 112

Assume (i), (ii), and (iii) are verified. Then,

$$\lim_{\lambda \to 0} E_{\omega_1}^\lambda = E. \tag{3.25}$$

Note that the full dimensionality assumption (iii) is not satisfied in important classes of games such as (anti)coordination games, and team problems and irreducibility conditions fail in the presence of absorbing states. When $\mathcal{Y} = \prod_i \mathcal{A}_i$, one obtains the perfect monitoring case and the Folk theorem holds under (i)–(ii). A similar result has been proved by Hörner et al. (2009)

3.3.2 Stochastic Games with Individual States

A stochastic game with individual states can be described in the same way as a stochastic game with common state. This is why the complete description will not be provided here. The main technical difference between the two types of game is that the environment of the players is now represented by a vector of states $\underline{\omega} = (\omega_1,\ldots,\omega_K)$ instead of a scalar. The set of possible actions for player i at stage t is now $\mathcal{A}_i(\omega_i(t))$. The value of the utility function of player i at stage t depends on the whole state $\omega(t)$. For a Markov game, the state transition probability always has the same form; namely, $q(\underline{\omega}'|\underline{\omega},\underline{a})$. Therefore the main change is in the instantaneous sets of actions for the players.

Here are a couple of results for this type of game.

- If both the action and state spaces are finite, there is at least one equilibrium in the T-stage game and the game with discounting.
- Players have time-dependent equilibrium strategies in product games for which the action of a given player does not affect the others (Flesch et al., 2008).
- Players have stationary equilibrium strategies in two-player zero-sum stochastic games (Flesch et al., 2008).

3.4 DIFFERENCE GAMES AND DIFFERENTIAL GAMES

Repeated games were presented in Sec. 3.2, in which the same one-shot game is repeated a certain (possibly infinite) number of times. It was seen that the special structure of these dynamic games allows results to be derived, such as the Folk theorem, which fully characterizes all possible equilibrium utilities. Stochastic games, which were presented in Sec. 3.3, generalize repeated games: the state of the game can vary from stage to stage and the instantaneous players' utility depends not only on the played actions but also on the current state. Fewer results are available for this class of game. Most often, sufficient existence conditions are assumed (e.g., the irreducibility assumption). In the present section, a more general game model is considered, although we remain in the framework of games with complete information. This game model is the one of difference games with a common state. Three important features are worth highlighting for such games:

1. The state of the game now comprises two components which are denoted by ω and x.
2. The evolution law of the state x is more general than the one assumed for stochastic games.
3. The observation structure for the players is more general than the one assumed for stochastic games.

Feature 1 allows one to analyze scenarios where the system has two states, e.g., a power control system where both the channel gains and battery levels are considered to be the state of the game. The case of a single common state is a special case of this scenario. Feature 2 is integrated via a difference equation, which is given below. Feature 3 corresponds the assumption of an arbitrary observation structure for the players in terms of states. The players can not necessarily observe the state (ω, x) perfectly. Rather, they generally observe a signal which is an image of the composite state. The price to be paid for having such a general model is the lack of general results. In fact, the purpose of this section is to give the reader an idea of what a general model of dynamic games can be, and not to gather the results available in the literature. For further information, the reader is referred to Basar and Olsder (1982).

Note that time is always discrete, as was the case for repeated and stochastic games. The continuous-time counterpart of the difference game model is given by differential games, which are presented in Sec. 3.4.2. Basar and Olsder (1982) presents many results and open problems for differential games.

3.4.1 Difference Games with a Common State

In this section, a generic formulation of difference games with a finite number of stages is provided. Discrete-time dynamic games with K players and complete information are considered. The game is described by the collection:

$$\Gamma = \left(\mathcal{K}, \mathcal{T}, \Omega, \mathcal{X}, (\mathcal{A}_{i,t}, f_t, \Theta_{i,t}, o_{i,t}, \mathcal{N}_{i,t}, \Sigma_{i,t})_{i \in \mathcal{K}, t \in \mathcal{T}}, (v_i)_{i \in \mathcal{K}} \right) \qquad (3.26)$$

which is itself defined by:

- a set of players $\mathcal{K} = \{1, 2, \ldots, K\}$;
- a finite set of stages $\mathcal{T} = \{1, 2, \ldots, T\}$;
- a finite or infinite set Ω, where $\omega_t \in \Omega$ is the state of the environment at stage t;
- a state space $\mathcal{X}$;
- a set of possible actions for player i at stage t, which is denoted by $\mathcal{A}_{i,t}$;
- a mapping:

$$f_t : \begin{vmatrix} \Omega \times \mathcal{X} \times \left(\prod_i \mathcal{A}_{i,t} \right) & \rightarrow & \mathcal{X} \\ (\omega_t, x_t, \underline{a}_t) & \mapsto & x_{t+1} = f_t(\omega_t, x_t, \underline{a}_t) \end{vmatrix} \qquad (3.27)$$

which represents the evolution law of the state x. This law is initialized with a given state denoted by (ω_1, x_1). Equation (3.27) is called a difference equation. For instance, the random process ω_t can represent a noise (which is typically distributed as a Wiener process). When ω_t has a constant value, the state equation is said to be *deterministic*;

- an observation set $\Theta_{i,t}$ for player i at stage t, $\theta_{i,t} \in \Theta_{i,t}$;
- an observation function:

$$o_{i,t} : \begin{vmatrix} \Omega \times \mathcal{X} & \rightarrow & \Theta_{i,t} \\ (\omega_t, x_t) & \mapsto & \theta_{i,t} = o_{i,t}(\omega_t, x_t) \end{vmatrix} \qquad (3.28)$$

- K information sets $\eta_{i,t}$, each of them being a subset of the set of histories of the game at stage t:

$$\mathcal{H}_{i,t} = \left(\prod_{1 \leq t' \leq t-1} \prod_i \Theta_{i,t'} \times \mathcal{A}_{i,t'} \right) \times \prod_i \Theta_{i,t}. \qquad (3.29)$$

These sets allow one to characterize the information structure of the players;
- the information space $\mathcal{N}_{i,t}$ induced by $\eta_{i,t}$. This set contains the action observation structure of player i;

- a class $\Sigma_{i,t}$ of mappings $\sigma_{i,t} : \mathcal{N}_{i,t} \to \mathcal{A}_{i,t}$ corresponding to the possible decisions at stage t. The collection $\sigma_i = (\sigma_{i,t})_{t \in \mathcal{T}}$ is the strategy of player i and Σ_i is the strategy set of player i;
- a utility function $v_i : \prod_{t \in \mathcal{T}} \left(\Omega \times \mathcal{X} \times \prod_{i'} \mathcal{A}_{i',t} \right) \to \mathbb{R}$.

Different approaches can be used to solve a difference game: (1) the Pontryagin Maximum Principle to derive open-loop equilibria; or (2) the (stochastic) dynamic programming principle to derive feedback equilibria. These two principles are reviewed in the continuous-time case at the end of this section. Open-loop and feedback equilibria are defined from the type of information structures. Player i's information structure is said to be with:

- open-loop information pattern if $\eta_{i,t} = \{x_1\}, t \in \mathcal{T}$
- closed-loop perfect state information if $\eta_{i,t} = \{x_1, \ldots, x_t\}, t \in \mathcal{T}$
- closed-loop imperfect state information if $\eta_{i,t} = \{\theta_{i,1}, \ldots, \theta_{i,t}\}, t \in \mathcal{T}$
- feedback perfect state information if $\eta_{i,t} = \{x_1, x_t\}, t \in \mathcal{T}$.
- feedback imperfect state information if $\eta_{i,t} = \{\theta_{i,t}\}, t \in \mathcal{T}$.

As an example of existence result, let us consider deterministic difference games; that is, ω is fixed at a certain value. It is therefore removed from the notations. In Basar and Olsder (1982) the authors show how to exploit the Pontryagin Maximum Principle to prove the existence of open-loop equilibria in deterministic difference games. This is the purpose of the following theorem. For the sake of clarity, the theorem is stated in the case where the state x is a real but the generalization to vector states follows (e.g., see Basar and Olsder (1982)).

Theorem 113

Consider a K-player deterministic difference game with finite horizon $T < \infty$. Assume that:

- the marginal function $x \in \mathcal{X} \mapsto f_t(x, a_{1,t}, \ldots, a_{K,t})$ is continuously differentiable on $\mathcal{X} \subseteq \mathbb{R}$;
- the marginal functions $(x, x') \in \mathcal{X}^2 \mapsto u_{i,t}(x, x', a_{1,t}, \ldots, a_{K,t})$ are continuously differentiable on $\mathbb{R}$;
- the function $(x, a_1, \ldots, a_K) \mapsto f_t(x, a_1, \ldots, a_K)$ is concave.

Then, $(\sigma_i^*(x_1) = \gamma_i^*)_{i \in \mathcal{K}}$ provides an open-loop Nash equilibrium and $(\underline{x}_t^*(x_0))_{i \in \mathcal{K}}$ is the corresponding state trajectory, and there exists a finite sequence $p_{i,1}, \ldots, p_{i,T}$ for each player $i \in \mathcal{K}$, such that the following relations are satisfied:

$$
\begin{cases}
x_{t+1}^* = f_t\left(x_t^*, \gamma_t^*\right), x_1^* = x_1 \\
\sigma_i^*(x_0) = \gamma_i^* = \arg\max_{a_{i,t} \in \mathcal{A}_{i,t}} H_{i,t}\left(p_{i,t+1}, x_t^*, a_{i,t}, \gamma_{-i,t}^*\right) \\
H_{i,t}\left(p_{i,t+1}, x_t^*, a_{i,t}, \gamma_{-i,t}^*\right) = u_{i,t}\left(x_t^*, f_t\left(x_t^*, a_{i,t}, \gamma_{-i,t}^*\right), a_{i,t}, \gamma_{-i,t}^*\right) + p_{i,t+1} f_t\left(x_t^*, a_{i,t}, \gamma_{-i,t}^*\right) \\
p_{i,T+1} = 0 \\
p_{i,t} = D_{x_t} f_t\left(x_t^*, \gamma_t^*\right) p_{i,t+1} + D_{x_{t+1}} u_{i,t}\left(x_t^*, x_{t+1}^* \gamma_t^*\right) + D_{x_t} u_{i,t}\left(x_t^*, x_{t+1}^*, \gamma_t^*\right)
\end{cases}
$$

where the notation D_{x_t} stands for the partial derivative operator with respect to x_t: $D_{x_t} \equiv \frac{\partial}{\partial x_t}$.

Note 1 (Review of Pontryagin's Maximum Principle). Consider a classical optimization problem; that is, $K = 1$. The canonical optimization problem to be solved consists in maximizing the utility functional:

$$v_{(x_0, a)} = q(x_T) + \int_0^T u(t, x_t, a_t)\mathrm{d}t \tag{3.30}$$

with respect to the control function $a(.)$ (or strategy) where:

- both the control policy a and state variable x are assumed to be scalar (the principle can be easily formulated in the vector case, simply by replacing scalar with vector quantities);
- the initial state $x(0) = x_0$ is given;
- the state variable satisfies the evolution law:

$$\dot{x}(t) = f(t, x(t), a(t)); \tag{3.31}$$

- q is the terminal utility function;
- both f and u are continuous with respect to all their variables;
- both f and u are partially differentiable with respect to x and a and have continuous partial derivatives;
- the admissible set of control policies for $a(.)$ is the set of continuous functions on $[0, T]$, which is denoted by $\mathcal{A}$.

Pontryagin's theorem is as follows. If $a^*(t)$ corresponds to a local maximum of v with a corresponding state trajectory $x^*(t)$ then there exists a costate function p^* : $[0, T] \rightarrow \mathbb{R}$ satisfying:

$$\begin{cases} \dot{x}^*(t) &= & f(t, x^*, a^*) \\ x^*(0) &= & x_0 \\ \dot{p}^*(t) &= & -\frac{\partial H}{\partial x}(t, x^*, a^*, p^*) \\ p^*(T) &= & \frac{\partial q}{\partial x}(x_T) \\ H &= & u + pf \end{cases} \tag{3.32}$$

Pontryagin's theorem therefore provides a necessary condition for optimality of a solution to the optimization problem described here. In the sequel, the function H will be referred to as the Hamiltonian, as it is kennedi in the literature of control theory.

Note 2 (Review of the dynamic programming principle). The purpose of this principle is to provide sufficient conditions for the existence of a solution for the optimization problem described in Note 1. Bellman and Isaacs derived a very useful result, which is also called Bellman's Optimality Principle. The idea is that if a control policy is optimal on a trajectory, then it must also be optimal on a portion of that trajectory. To illustrate this principle, consider the example used by Engwerda (2005), which is represented in Figure 3.2. If one wants to travel from A to D by taking the shortest path, ABCD is the optimal solution. If the starting point becomes B, the path BCD, which is a restriction of the solution ABCD, is still the optimal solution.

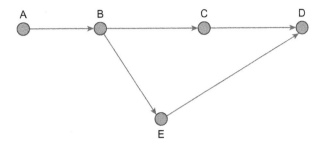

FIGURE 3.2

Illustration of the dynamic programming principle (Bellman's Optimality Principle).

For readers who are familiar with the Viterbi algorithm, this is exactly what happens when a survivor path is selected at a given node or state in the decoding trellis. To state the dynamic programming theorem, a few notations are in order. Assuming the optimization problem in Note 1 starts at $t = t_0$ instead of $t = 0$, the system state at this point is denoted by x_{t_0}, and the maximum utility v_i obtained for an optimal control policy $a^*(t)$ is denoted by $V(t_0, x_{t_0})$. The latter function is an auxiliary function which is usually called the value function. The dynamic programming theorem is as follows. If the value function verifies the Hamilton–Jacobi–Bellman (HJB) equation:

$$-\frac{\partial V}{\partial t}(t,x) = \max_{a \in \mathcal{A}} \left\{ u(t,x,a) + \frac{\partial V}{\partial x}(t,x) f(t,x,a) \right\} \tag{3.33}$$

with $V(T,x) = q(x_T)$, then the solution:

$$a^*(t,x) = \arg\max_{a \in \mathcal{A}} \left\{ u(t,x,a) + \frac{\partial V}{\partial x}(t,x) f(t,x,a) \right\} \tag{3.34}$$

is optimal. The proof of this theorem can be found, for example, in Engwerda (2005) and relies on Bellman's Optimality Principle.

3.4.2 Differential Games with a Common State

The branch of game theory which addresses continuous-time dynamic games is called differential game theory. The origin of differential games dates back to the 1940s with the work of Rufus Issacs, who modeled missile versus enemy aircraft pursuit schemes in terms of descriptive system evolution. The work he did in 1951 appeared in 1965 (Isaacs, 1965). During the same period, *optimal control theory* was developed, which includes the well-known Bellman's dynamic programming (Bellman, 1952) and Pontryagin's Maximum Principle; these two important results will be reviewed a little further on. Differential games extend optimal control theory by incorporating strategic behavior by decision makers. Different approaches have been proposed during the last seventy years. Berkovitz (1964) developed a variational approach to differential games and Leitman and Mon (1967) proposed a geometric point of view for differential games, Pontryagin (1967) proposing a maximum principle characterization of open-loop Nash equilibria in differential games. Case (1967) and Starr and

Ho (1969) studied feedback solutions for Nash equilibria in non-zero-sum determin-istic differential games, and observed that they avoid time inconsistency. Based on the stochastic control developed by Fleming (1969), stochastic differential games ana-lyze differential games under uncertainty. Closed-form solutions for linear-quadratic stochastic differential games can be found in the papers by Başar (1974–1977), see Basar (1974, 1975, 1977). Additionally, Aubin (2002) and Yeung and Petrosyan (2010) have studied dynamic cooperative games (with and without noise). More recent results on differential games can be found, e.g., Cardaliaguet (2007) and Bardi (2009).

In this section, we provide some results for the Nash equilibrium characteriza-tion in non-cooperative deterministic differential games with complete information. Below, a generic description of differential games with finite horizon is provided. Such a game is described by the following object:

$$\Gamma = \left(\mathcal{X}, \mathcal{T}, \mathcal{X}, (\mathcal{A}_{i,t}, f_t, \Theta_{i,t}, h_{i,t}, \mathcal{N}_{i,t}, \Sigma_{i,t})_{i\in\mathcal{K}, t\in\mathcal{T}}, (v_i)_{i\in\mathcal{K}}\right) \qquad (3.35)$$

which comprises:

- a set of players $\mathcal{K} = \{1, 2, \ldots, K\}$;
- a game interval $\mathcal{T} = [0, T]$, $T > 0$ – the game duration;
- an infinite set $\mathcal{X}_\infty$ which contains all the states. A state x_t is said to be admissible at time t if $x_t \in \mathcal{X}$. The set $\mathcal{X}$ is a subset of some vector topological space;
- an infinite set $\mathcal{A}_i$ representing the action set (control set) of player i at time t. Its elements are the possible actions $a_{i,t}$ for player i at time t;
- a mapping $f_t : \mathcal{X} \times \left(\prod_i \mathcal{A}_i\right) \to \mathcal{X}$, defined at each time $t \in \mathcal{T}$, such that:

$$\dot{x}_t = f_t(x_t, a_{1,t}, \ldots, a_{K,t}) \qquad (3.36)$$

where x_0 is called the initial state of the game. This equation is a (deterministic) differential equation, whose solution gives the state trajectory of the differential game;
- an observation set $\Theta_{i,t}$ for each player i at time t, $\theta_{i,t} \in \Theta_{i,t}$;
- an observation function for each player $o_{i,t} : \mathcal{X} \to \Theta_{i,t}$ defined by

$$\theta_{i,t} = o_{i,t}(x_{i,t}) \qquad (3.37)$$

- a set-valued function $\eta_{i,t} = \{x_t, 0 \leq t' \leq \epsilon_{i,t}\}$ defined for each player $k \in \mathcal{K}$ where $\epsilon_{i,t}$ is nondecreasing in t, and $\eta_{i,t}$ determines the state information gained and recalled by player i at time $t \in \mathcal{T}$. The set of $\epsilon_i = \{\epsilon_{i,t'}\}_{t'\in\mathcal{T}}$ characterizes the infor-mation structure of player i, and the collection of information structures of all the players, i.e., $\{\eta_i\}_i$ determines the information structure of the game;
- a sigma-field $\mathcal{N}_{i,t}$, in the set $\mathcal{X}$, generated for each player $i \in \mathcal{K}$ by the cylinder sets $\{(x, x_{t'}) \in \mathcal{X}_\infty \times \mathcal{B}\}$ where $\mathcal{B}$ is a Borel set in $\mathcal{X}$ and $0 < t' < \epsilon_{k,t}$. The sigma-field $\mathcal{N}_{i,t}$, $t \geq 0$, is called the information σ-field of player i;
- a class Σ_i of mappings $\sigma_i : [0, T] \times \mathcal{X}_\infty \to \mathcal{X}$, with the property $a_{i,t} = \sigma_{i,t}(x)$ being $\mathcal{N}_{i,t}$-measurable. The set Σ_i is the strategy space of player i;
- a function $u_i : \mathcal{T} \times \mathcal{X} \times \prod_i \mathcal{A}_i \to \mathbb{R}$ called intermediate utility function of player i in the dynamic game. The function $q_i : \mathcal{X} \to \mathbb{R}$ is called terminal utility function.

The long-term utility is the cumulative utility defined by:

$$v_i(\underline{a}, x_0) = q_i(x_T) + \int_{t \in \mathcal{T}} u_i(t, x_t, a_t)\, \mathrm{d}t. \tag{3.38}$$

The information structure of player i said to be with:

- open-loop information pattern if $\eta_{i,t} = \{x_0\}, t \in \mathcal{T}$;
- closed-loop perfect state information if $\eta_{i,t} = \{x_{t'}, 0 \leq t' \leq t\}, t \in \mathcal{T}$
- closed-loop imperfect state information if $\eta_{i,t} = \{\theta_{i,1}, \ldots, \theta_{i,t}\}, t \in \mathcal{T}$;
- feedback perfect state information if $\eta_{i,t} = \{x_t\}, t \in \mathcal{T}$.

Definition 114

A strategy profile $\left(\sigma_{i,t}^*\right)_{i \in \mathcal{K}}$ is said to be a Nash equilibrium for the K-player differential game if the following inequalities are satisfied for all $a_{i,t} \in \mathcal{A}_i,\ i \in \mathcal{K}, t \in \mathcal{T}$:

$$q_i\left(x_T^*\right) + \int_{t \in \mathcal{T}} u_{i,t}\left(x_t^*, \sigma_{i,t}^*, \sigma_{-i,t}^*\right)\, \mathrm{d}t \geq q_i\left(x_T^{*,(i)}\right) + \int_{t \in \mathcal{T}} u_{i,t}\left(x_t^{*,(i)}, a_{i,t}, \sigma_{-i,t}^*\right)\, \mathrm{d}t, \tag{3.39}$$

where:

$$\begin{cases} & \forall\, t \in \mathcal{T} \\ \frac{\mathrm{d}}{\mathrm{d}t} x_t^* = f_t\left(x_t^*, \sigma_{i,t}^*, \sigma_{-i,t}^*\right),\ x_0^* = x_0 \\ \frac{\mathrm{d}}{\mathrm{d}t} x_t^{*,(i)} = f_t\left(x_t^{*,(i)}, a_{i,t}, \sigma_{-i,t}^*\right),\ x_0^{*,(i)} = x_0,\ i \in \mathcal{K}. \end{cases}$$

Theorem 115

Consider a K-player deterministic differential game with finite horizon $T < \infty$. Assume that:

- the marginal function $x \in \mathcal{X} \mapsto f_t(x, a_{1,t}, \ldots, a_{K,t})$ is continuously differentiable on $\mathcal{X} \subseteq \mathbb{R}$ for all $t \in \mathcal{T}$;
- the marginal functions $x \in \mathcal{X} \mapsto u_{i,t}(x, a_{1,t}, \ldots, a_{K,t})$ and $x \in \mathcal{X} \mapsto q_i(x)$ are continuously differentiable on $\mathbb{R}$ for all $t \in \mathcal{T}$.

Then, $(\sigma_{i,t}^*(x_0) = \gamma_{i,t}^*)_{i \in \mathcal{K}}$ provides an open-loop Nash equilibrium and $(x_{i,t}^*(x_0))_{i \in \mathcal{K}}, t \in \mathcal{T}$ is the corresponding state trajectory. There exist K functions $p_i : \mathcal{T} \to \mathbb{R}$ for each player $i \in \mathcal{K}$ such that the following relations are satisfied:

$$\begin{cases} \dot{x}_t^* = f_t(x_t^*, \gamma_t^*), x_0^* = x_0 \\ \sigma_{i,t}^*(x_0) = \gamma_{i,t}^* = \arg\max_{a_{i,t} \in \mathcal{A}_{i,t}} H_{i,t}(p_{i,t}, x_t^*, a_{i,t}, \gamma_{-i,t}^*) \\ H_{i,t}(p_i, x, a_i, a_{-i}) = u_{i,t}(x, a) + p_i f_t(x, a_i, a_{-i}) \\ \dot{p}_{i,T} = \mathrm{D}_x q(x_T^*),\ i \in \mathcal{K} \\ \dot{p}_{i,t} = -\mathrm{D}_{x_t} H_{i,t}(p_{i,t}, x_t^*, a_{i,t}, \gamma_{-i,t}^*) \end{cases}$$

The function H_i is the Hamiltonian of player i. A detailed proof can be found in Basar and Olsder (1982, Theorem 6.11).

The information structure is said to be *closed-loop no-memory* if $\eta_{i,t} = \{x_0, x_t\}$; i.e., the strategies become functions of the initial state x_0, the current state x_t and current time t, and can be expressed as $(\gamma_{i,t} = \sigma_i(t, x_t, x_0))_{i \in \mathcal{K}}$. The following theorem provides a set of necessary conditions for any closed-loop no-memory Nash equilibrium to satisfy.

Theorem 116

A strategy profile $(\sigma^*_{i,t}(x_0, x^*_t) = \gamma^*_{i,t})_{i \in \mathcal{K}}$ constitutes a closed-loop no-memory Nash equilibrium and $(x^*_{i,t}(x_0))_{i \in \mathcal{K}}, t \in \mathcal{T}$ is the corresponding state trajectory, if there exist K costate functions $p_i : \mathcal{T} \to \mathbb{R}$ for each player $i \in \mathcal{K}$ such that the following relations are satisfied:

$$
\begin{cases}
\dot{x}^*_t = f_t(x^*_t, \gamma^*_t), x^*_0 = x_0 \\
\sigma^*_{i,t}(x_0, x^*_t) = \gamma^*_{i,t} = \arg\max_{a_{i,t} \in \mathcal{A}_i} H_{i,t}(p_{i,t}, x^*_t, a_{i,t}, \gamma^*_{-i,t}) \\
H_{i,t}(p_i, x, a_i, \underline{a}_{-i}) = u_{i,t}(x, \underline{a}) + p_i f_t(x, a_i, \underline{a}_{-i}) \\
p_{i,T} = D_x q(x^*_T), \ i \in \mathcal{K} \\
\dot{p}_{i,t} = -D_{x^*} H_{i,t}(p_{i,t}, f_t(x^*_t, \gamma^*_{i,t}, \sigma^*_{-i}(t, x^*_t, x_0)), \gamma_{i,t}, \sigma^*_{-i}(t, x^*_t, x_0))
\end{cases}
\tag{3.40}
$$

A detailed proof of this theorem can be found in Yeung and Petrosyan (2010, Theorem 2.2.2).

Definition 117

A strategy profile in the form $\gamma_{i,t} = \sigma^*_{i,t}(t, \eta_t)$ is a feedback Nash equilibrium solution if there exist functions $V_i(t, x)$ defined on $\mathcal{T} \times \mathbb{R}$ and satisfying the following relations for each $i \in \mathcal{K}$:

$$
\begin{cases}
V_i(T, x) = q_i(x), \\
V_i(t, x) = q_i(x^*_T) + \int_{t' \in [t,T]} u_{i,t'}(x^*_t, \sigma^*_{i,t'}(\eta_{t'}), \sigma^*_{-i,t'}(\eta_{t'})) \, dt' \\
\geq q_i(x^{*,(i)}_T) + \int_{t' \in [t,T]} u_{i,t'}(x^{*,(i)}_{t'}, a_{i,t'}(\eta_{t'}), \sigma^*_{-i,t'}(\eta_{t'})) \, dt',
\end{cases}
\tag{3.41}
$$

for all $a_{i,.}(.), x \in \mathbb{R}$, where on the interval $[t, T]$,

$$
\begin{cases}
\frac{d}{dt'} x^*_{t'} = f_{t'}(x^*_{t'}, \sigma^*_{i,t'}(\eta_{t'}), \sigma^*_{-i,t'}(\eta_{t'})), \ x^*_t = x \\
\frac{d}{dt'} x^{*,(i)}_{t'} = f_{t'}(x^{*,(i)}_{t'}, a_{i,t'}(\eta_{t'}), \sigma^*_{-i,t'}(\eta_{t'})), \ x^{*,(i)}_t = x, \ i \in \mathcal{K}.
\end{cases}
\tag{3.42}
$$

Theorem 118

A strategy profile $\gamma_{i,t} = \sigma_{i,t}(x) \in \mathcal{X}, i \in \mathcal{K}, t \in \mathcal{T}$ provides a feedback Nash equilibrium if there exist continuously differentiable functions $V_i : \mathcal{T} \times \mathbb{R} \to \mathbb{R}, i \in \mathcal{K}$ satisfying the following set of partial differential equations, called the Hamilton–Jacobi–Bellman equations:

$$
\begin{cases}
V_i(T, x) = q_i(x) \\
0 = \frac{\partial}{\partial t} V_i(t, x) + \sup_{a_i} \left\{ u_{i,t}(x, a_{i,t}(x), \sigma_{-i,t}(x)) + \frac{\partial V_i(t,x)}{\partial x} f_t(x, a_{i,t}(x), \sigma_{-i,t}(x)) \right\} \\
\quad = u_{i,t}(x, \sigma_{i,t}(x), \sigma_{-i,t}(x)) + \frac{\partial V_i(t,x)}{\partial x} f_t(x, \sigma_{i,t}(x), \sigma_{-i,t}(x))
\end{cases}
\tag{3.43}
$$

The above theorem and other results provided in this section hold for the special case of deterministic differential games and a finite horizon. Indeed, few works have addressed equilibrium utility characterization in differential games in a general setting. The work in Gaitsgory and Nitzan (1994) studies infinite horizon dynamic games satisfying the monotonicity property: the objective function of every player is to increase (decrease) the strategy of the other players. These authors prove that, when closed-loop strategies are allowed, the set of Nash equilibrium utilities coincides with the set of individually rational feasible utilities in the game obtained when restricting the strategy sets to open-loop strategies. Despite the existence of such interesting results, much work remains to be done in order to derive the counterpart of Folk theorems for general differential games. Differential game theory contains many known open problems (Basar and Olsder, 1982) and the relevance of using differential games in wireless communications is not fully understood at the time of writing this book. One of the main reasons for this is that continuous time evolution laws appear naturally in physics (e.g., fundamental principles of dynamics) but not (yet) in wireless models (see Olama et al. (2006) for a scenario exploiting a continuous-time law for the evolution of the channel state).

Application Example 1 (Bandwidth Allocation with Dynamic Service Selection). Here, we summarize in a few lines the work in Zhu et al. (2010a). Bandwidth allocation for different service classes in heterogeneous wireless networks is an important issue for service providers, in terms of balancing service quality and profit. In Zhu et al. (2010a), the underlying dynamic service selection is modeled as an evolutionary game based on replicator dynamics. An upper bandwidth allocation differential game is formulated to model the competition among different service providers. The service selection distribution of the underlying evolutionary game describes the state of the upper differential game. An open loop Nash equilibrium is considered to be the solution of this linear state differential game.

Application Example 2 (Rate control). Here, we summarize in a few lines the work in Wu et al. (2010). Network throughput and energy efficiency are paramount for network performance in an energy-constrained wireless network. However, it is difficult to achieve optimal objectives simultaneously. Therefore, it is necessary to find a rate control solution based on a tradeoff between network throughput and energy efficiency. In Wu et al. (2010), the authors propose a cooperative differential game model and find an optimal rate control for each player to get the total minimal cost with tradeoff between network throughput and energy efficiency of the networks.

3.5 EVOLUTIONARY GAMES

Evolutionary game theory studies the evolution of populations of players interacting strategically and the evolution of coalition, network formations, and fuzzy behaviors. Their original applications were in biology and social sciences. Examples of situations that evolutionary game theory helps us to understand include animal conflicts and communication behaviors (e.g., in the Hawk and Dove game). The interactions may be with players from the same population (a mobile with other mobiles, males fighting other males, etc.), or with players from other populations (males interacting with females or buyers with traders). The players are typically considered to be animals in biology, firms in economics, and transmitters in wireless networks. The starting point for an evolutionary game model is to believe that the players are *not always rational*. Instead of Nash equilibria in classical game theory, evolutionary game theory uses the concepts of evolutionary stability: the unbeatable state concept (Hamilton, 1967), and the evolutionarily stable state (ESS) concept (Smith and Price, 1973). A strategy is *unbeatable* if it is resilient to any fraction of deviants of any size. A strategy is *evolutionarily stable* if a whole population using this strategy cannot be invaded by a small fraction of deviants (mutants). These notions will be defined formally a little further on. The evolutionary stability concept is particularly interesting, because it provides some information about the robustness and the dynamic properties of the population (equilibria refinements). In situations where the game has multiple equilibria, the evolutionary stability concept can contribute to the selection of certain equilibria. A dynamic behavior for the population is proposed using models of *evolutionary game dynamics* such as replicator dynamics (Taylor and Jonker, 1978), fictitious play dynamics (Brown, 1949, 1951; Gilboa and Matsui, 1991), Brown-von Neumann-Nash dynamics (Brown and von Neumann, 1950), Smith dynamics (Smith, 1984), projection dynamics, best response dynamics (Fudenberg and Tirole, 1991), logit dynamics (Fudenberg and Levine, 1998) etc. Quite complete analyses on evolutionary game dynamics can be found in Hofbauer and Sigmund (1998, 2003), Weibull (1997) and in Sandholm (2010).

Originally formulated to explain complex evolutionary biology and economic behaviors problems, evolutionary game theory is becoming a useful technique in the field of evolving networks (Tembine et al., 2009a). This section examines the main fundamental ideas of evolutionary games, and gives an example of wireless scenarios in which has been applied.

3.5.1 Dynamic Procedures vs. Perfect Anticipation

Evolutionary games in large systems provide a simple framework for describing strategic interactions among a large number of players. Traditionally, predictions of behavior and outcome in game theory are based on some notion of equilibrium, typically Cournot equilibrium (Cournot, 1838), Bertrand equilibrium (Bertrand, 1883), conjectural variation (Bowley, 1924), Stackelberg solution (Stackelberg, 1934), Nash equilibrium (Nash, 1951), Wardrop equilibrium (Wardrop, 1952) or some refinement and/or extension thereof. Most of these notions require the assumption of equilibrium knowledge, which means that each player correctly anticipates how the other players will react. This assumption can be too strong, and is difficult to justify, especially in dense networks. As an alternative the evolutionary game approach proposes an explicitly dynamic updating choice, a model in which players myopically update their behavior in response to their current strategic environment. This dynamic procedure does not assume the automatic coordination of players' actions and beliefs. These procedures are specified formally by defining a revision of strategies called a revision protocol. A revision protocol takes current utilities (expected performance) and the system state as arguments;

its outputs are conditional switch rates which describe how frequently players in some class playing a given strategy are considering switching strategies, given the current expected cost vector and subpopulation state. This revision of pure strategies is flexible enough to incorporate a wide variety of paradigms, including those based on learning, imitation, adaptation, optimization, etc. The revision of pure strategies describes the procedures that players follow in adapting their behavior to the evolving environment, such as in evolving networks (Internet traffic, flow control, etc.).

3.5.2 **Pairwise Interaction Model**
3.5.2.1 *The Single Population Case*

Consider a homogeneous large population of players in which the players are randomly matched: each player j plays a symmetric bi-matrix game against some randomly selected player j'. Every player can select an action from the finite set $\mathcal{A}$. If a player j with action a meets another player j' with the action a', player j gets the utility $U(a, a')$. Denote by $\underline{x}$ the frequencies of use of actions, i.e., the population profile. The vector $x = (x_a)_{a \in \mathcal{A}}$ is a probability distribution on $\mathcal{A}$. The number x_a can be interpreted as the probability with which a given player in the population uses action a. We define by $\sum_{(a,a') \in \mathcal{A}^2} y_{a'} x_a U(a, a') := \langle \underline{x}, \mathbf{U}\underline{y} \rangle$ the expected utility for a given player if he uses a mixed strategy $\underline{x}$ when meeting another individual who chooses the mixed strategy $\underline{y}$, where $\mathbf{U}$ is the matrix whose entries are $U(a, a')$ for all $(a, a') \in \mathcal{A}^2$. This utility is sometimes referred to as "fitness" or "cost". Denote by $f: \mathbb{R}^{|\mathcal{A}|} \to \mathbb{R}^{|\mathcal{A}|}$ the vector of expected utilities $(f_{a_1}, \dots, f_{a_{|\mathcal{A}|}})$ where $f_a : \mathbb{R}^{|\mathcal{A}|} \to \mathbb{R}, f_a(\underline{x}) \triangleq \sum_{a'} U(a, a') x_{a'}$. The pair $(\mathcal{A}, f(.))$ defines a single-population game with random pairing.

A population profile $\underline{x} \in \Delta(\mathcal{A})$ is an equilibrium state if:

$$\forall \underline{y} \in \Delta(\mathcal{A}), \ \langle \underline{y} - \underline{x}, f(\underline{x}) \rangle \leq 0. \tag{3.44}$$

Assume that the whole population uses $\underline{x} \in \Delta(\mathcal{A})$ and a fraction ϵ (called "mutants") adopts another strategy $\underline{y}$. The population profile $\underline{x} \in \Delta(\mathcal{A})$ is robust against $\underline{y} \in \Delta(\mathcal{A})$ if:

$$\sum_{a \in \mathcal{A}} y_a f_a(\epsilon \underline{y} + (1 - \epsilon)\underline{x}) < \sum_{a \in \mathcal{A}} x_a f_a(\epsilon \underline{y} + (1 - \epsilon)\underline{x}) \tag{3.45}$$

that is, $\langle \underline{y} - \underline{x}, f(\epsilon \underline{y} + (1 - \epsilon)\underline{x}) \rangle < 0$.

....

Definition 119: Unbeatable State

A strategy $\underline{x}$ is an unbeatable state if for every $\underline{y} \neq \underline{x}$, one has that $\langle \underline{y} - \underline{x}, f(\epsilon \underline{y} + (1 - \epsilon)\underline{x}) \rangle < 0$ for all $0 < \epsilon < 1$.

This notion of unbeatable was introduced by Hamilton (1967). This definition of unbeatable state is related to a robustness property against deviations by any fraction of the population. It is thus a generalization of the strong equilibrium (see Chapter 1) in the context of large population. The evolutionarily stable state or strategy (ESS) introduced by Smith and Price (1973) is a weaker concept than the unbeatable state concept.

Definition 120: Evolutionarily Stable Strategy – ESS

A strategy $\underline{x}$ is an evolutionarily stable state or strategy (ESS) if, for every $\underline{y} \neq \underline{x}$, there exists some threshold fraction of mutants $\overline{\epsilon}_y > 0$ such that $\langle \underline{y} - \underline{x}, f(\epsilon \underline{y} + (1 - \epsilon)\underline{x}) \rangle < 0$ holds for all $\epsilon \in]0, \overline{\epsilon}_y[$.

For a symmetric bi-matrix game, the ESS definition is equivalent to the following:

Theorem 121

A population profile $\underline{x}$ is an ESS if and only if $\forall \underline{y} \neq \underline{x}$ the following conditions hold:

$$\langle \underline{y} - \underline{x}, f(\underline{x}) \rangle \leq 0, \text{ and} \tag{3.46}$$

$$\langle \underline{y} - \underline{x}, f(\underline{x}) \rangle = 0 \implies \langle \underline{y} - \underline{x}, f(\underline{y}) \rangle < 0 \tag{3.47}$$

A proof can be found in Weibull (1997, Proposition 2.1), and in Hofbauer and Sigmund (1998, Theorem 6.4.1, page 63). In fact, if condition (3.46) is satisfied, then the fraction of mutations in the population will tend to decrease, as it has a lower utility, meaning a lower growth rate. Thus the population profile $\underline{x}$ is then immune to mutations. If this is not case, but condition (3.47) holds, then a population using $\underline{x} \in \Delta(\mathcal{A})$ is "weakly" immune to a mutation using $\underline{y}$. Indeed, if the mutant population grows, then there will be more frequently players with strategy $\underline{x}$ competing with mutants. In such cases, the condition $\langle \underline{x} - \underline{y}, f(\underline{y}) \rangle > 0$ ensures that the growth rate of the original population exceeds that of the mutants.

Recall that a mixed strategy $\underline{x} \in \Delta(\mathcal{A})$ that satisfies (3.48) for all $\underline{y} \neq \underline{x}$ is a Nash equilibrium state. A mixed strategy $\underline{x}$ which satisfies (3.46) for all $\underline{y} \neq \underline{x}$ is called a strict Nash equilibrium. In conclusion, a strict Nash equilibrium is an ESS, and an ESS implies a Nash equilibrium. Note that for finite bi-matrix games, there is at least one mixed Nash equilibrium, but an ESS may not exist. However, an ESS exists in generic symmetric bi-matrix two-by-two games (i.e., two actions for each player); see Weibull (1997).

Proposition 122

Any generic two-by-two evolutionary game has at least one ESS (in pure or mixed strategies).

3.5.2.2 *The Two-Population Case*

Consider two large populations of players which are randomly matched from different populations. Each player j plays an asymmetric bi-matrix game against some randomly selected player j'. Denote by $\Theta = \{1, 2\}$ the set of populations. Every player from population 1 (resp. 2) can select an action among the finite set $\mathcal{A}^1$ (resp. $\mathcal{A}^2$). If a player j from population 1 with action a meets another player j' from 2 with the action a', player j gets the utility $U^1(a,a')$ and player 2 gets $U^2(a,a')$. Denote by x^θ the frequencies of use of actions in population θ. The vector $\underline{x} = (x_a^\theta)_{\theta \in \Theta, a \in \mathcal{A}^\theta}$ is a probability distribution on $\mathcal{A}$. The number x_a can be interpreted as the probability that a given player in the population uses the action a. Define by $\sum_a \sum_{a'} x_{a'}^2 x_a^1 U^1(a,a') \triangleq \langle x^1, U^1 x^2 \rangle$ the expected utility for a given player from population 1, if he uses a mixed strategy x^1 when meeting another individual who adopts the mixed strategy x^2, where U^1 is the matrix whose entries are $U^1 = (U^1(a,a'))$. Similarly the utility of player j' from population 2 is $\langle x^1, U^2 x^2 \rangle$. Denote by $f : \mathbb{R}^{\sum_\theta |\mathcal{A}^\theta|} \to \mathbb{R}^{\sum_\theta |\mathcal{A}^\theta|}$ the expected utilities $(f_a(.))_a$ where $f_a^\theta : \mathbb{R}^{|\mathcal{A}^\theta|} \to \mathbb{R}$, $f_a^1(x^1, x^2) := \sum_{a'} U^1(a,a') x_{a'}^2$. The triplet $(\{\mathcal{A}^\theta\}_{\theta \in \{1,2\}}, f^1(.), f^2(.))$ defines a two-population game with random pairing. With this population game we associate the asymmetric bi-matrix game defined by $(\Theta = \{1,2\}, \mathcal{A}^1, \mathcal{A}^2, f^1, f^2)$.

A population profile $x = (x^\theta)_{\theta \in \Theta}$ is an equilibrium state if:

$$\forall \theta, \ \forall y^\theta \in \Delta(\mathcal{A}^\theta), \ \sum_\theta \langle y^\theta - x^\theta, f^\theta(\underline{x}) \rangle \leq 0. \tag{3.48}$$

Assume that the whole population uses $\underline{x}$ and a small fraction ϵ (called "mutants") deviates and adopts $\underline{y}$. The new population profile after mutation is then $\epsilon \underline{y} + (1 - \epsilon)\underline{x}$. The population profile $\underline{x}$ is robust against $\underline{y}$ if:

$$\sum_\theta \sum_{a \in \mathcal{A}^\theta} y_a^\theta f_a^\theta (\epsilon \underline{y} + (1 - \epsilon)\underline{x}) < \sum_\theta \sum_{a \in \mathcal{A}^\theta} x_a^\theta f_a^\theta (\epsilon \underline{y} + (1 - \epsilon)\underline{x}) \tag{3.49}$$

which is $\sum_\theta \langle \underline{y}^\theta - x^\theta, \underline{f}^\theta (\epsilon \underline{y} + (1 - \epsilon)\underline{x}) \rangle < 0$.

....

Definition 123: Unbeatable State

A strategy $\underline{x}$ is an unbeatable state if for every $\underline{y} \neq \underline{x}$, the condition $\sum_\theta \langle \underline{y}^\theta - x^\theta, f^\theta (\epsilon \underline{y} + (1 - \epsilon)\underline{x}) \rangle < 0$ holds for all $\epsilon \in]0, 1[$.

This notion of globally unbeatable state has been introduced in Taylor and Jonker (1978) and studied in Weibull (1997). The global evolutionarily stable state (GESS) introduced by Smith and Price (1973) is a weaker notion than the globally unbeatable state concept.

....

Definition 124: Global Evolutionarily Stable Strategy – GESS

A strategy $\underline{x} = (x^\theta)_\theta$ is a global evolutionarily stable state or strategy (ESS) if for every $\underline{y} \neq \underline{x}$, there exists some threshold of fraction of mutants $\bar{\epsilon}_y > 0$ such that $\sum_\theta \langle \underline{y}^\theta - x^\theta, \underline{f}^\theta (\epsilon \underline{y} + (1 - \epsilon)\underline{x}) \rangle < 0$ holds for all $\epsilon \in]0, \bar{\epsilon}_y[$.

Note that a strict Nash equilibrium is a GESS.

Definition 125: Strategically Equivalent Games

Let $\mathcal{G} = (\{\mathcal{A}^\theta\}_{\theta \in \{1,2\}}, U^1(.), U^2(.))$ and $\bar{\mathcal{G}} = (\{\mathcal{A}^\theta\}_{\theta \in \{1,2\}}, \bar{U}^1(.), \bar{U}^2(.))$ be two-population games with finite actions: $|\mathcal{A}^1 \times \mathcal{A}^2| < +\infty$. The two-population games $\mathcal{G}$ and $\bar{\mathcal{G}}$ are strategically equivalent if there exist positive constants α_1^θ, $\theta \in \Theta$ and scalars α_2^θ, $\theta \in \Theta$, such that:

$$U^1(a_1,a_2) = \alpha_1^1 \bar{U}^1(a_1,a_2) + \alpha_2^1 \quad U^2(a_1,a_2) = \alpha_1^2 \bar{U}^2(a_1,a_2) + \alpha_2^2 \bar{U}^1(a_1,a_2)$$

for all $(a_1,a_2) \in \mathcal{A}^1 \times \mathcal{A}^2$.

The reader can easily verify that a pair strategy equivalent game as defined in Definition 125 is indeed an equivalence relation since it is reflexive, symmetric, and transitive (see Appendix A.1 for more details).

Proposition 126

All strategically equivalent two-population games have the same set of equilibrium states.

Definition 127

The equilibrium states of $\mathcal{G}$, $(\underline{x}^1, \underline{x}^2)$ and $(\underline{x}_*^1, \underline{x}_*^2)$ are interchangeable if $(\underline{x}^1, \underline{x}_*^2)$ and $(\underline{x}_*^1, \underline{x}^2)$ are also equilibrium states.

Proposition 128

If $\mathcal{G}$ is strategically equivalent to the zero-sum population game $\bar{\mathcal{G}} = (\{\mathcal{A}^\theta\}_{\theta \in \{1,2\}}, \bar{U}^1(.), -\bar{U}^1(.))$ then all the equilibrium states of $\mathcal{G}$ are interchangeable.

3.5.2.3 *Computing Equilibrium States*

Let $U^i() = -C^i()$. A pair $(\underline{x}_*^1, \underline{x}_*^2)$ constitutes an equilibrium state for the two-population game $\mathcal{G} = (\{\mathcal{A}^\theta\}_{\theta \in \{1,2\}}, U^1(.), U^2(.))$ if and only if there exists a pair $(\lambda_*^1, \lambda_*^2)$ such that $(\underline{x}_*^1, \underline{x}_*^2, \lambda_*^1, \lambda_*^2)$ is a solution of the following bilinear programming problem:

$$\min_{\underline{x}^1, \underline{x}^2, \lambda^1, \lambda^2} \langle \underline{x}^1, \mathbf{C}^1 \underline{x}^2 \rangle + \langle \underline{x}^1, \mathbf{C}^2 \underline{x}^2 \rangle + \lambda^1 + \lambda^2 \tag{3.50}$$

subject to the constraints:

$$\mathbf{C}^1 \underline{x}^2 + \lambda^1 \, \mathbb{1}_{|\mathcal{A}^1|} \geq 0, \; \underline{x}^1 \mathbf{C}^2 + \lambda^2 \, \mathbb{1}_{|\mathcal{A}^2|} \geq 0 \tag{3.51}$$

$$\forall a_1 \in \mathcal{A}^1, x_{a_1}^1 \geq 0, \; \sum_{a_1 \in \mathcal{A}^1} x_{a_1}^1 = 1, \; x_{a_2}^2 \geq 0, \forall a_2 \in \mathcal{A}^2, \; \sum_{a_2 \in \mathcal{A}^2} x_{a_2}^2 = 1. \tag{3.52}$$

3.5.3 Generic Formulation of Population Games

Consider large populations (multi-classes) of players in which a frequent interaction occurs among a (finite or infinite) random number of players. Each player is thus involved in several interactions with a random number of players. Each interaction can be described as one stage of an evolving game. The set of players in each local interaction evolves over time. The actions of the players at each stage determine a utility for each player, as well as the transition rates of the actions. The transition rate is determined by the change of actions, the system state, and the fitnesses. This model extends the basic pairwise interaction model in evolutionary games by introducing a random number of interacting players, and a rate of transition for several players who are able to change their action. At each game instance, a one-shot game with an unknown number of players replaces the matrix game. Instead of choosing a possibly mixed action, a player is now faced with the choice of decision rules, and revisions of these strategies that determine what actions should be chosen at a given interaction, considering given present and past observations. In addition to the study of the equilibrium of the evolving game we also consider some population dynamics and study its convergence properties. We assume that players revise their strategies, and that actions with higher utilities are used with higher probabilities. The system evolves under some evolutionary game dynamic process, which describes the change in strategies (incoming flow and outgoing flow) within the population.

3.5.3.1 *Homogeneous Population*

Consider a large population of players with a finite set of actions $\mathcal{A}$. With each action a, we associate an expected utility function $f_a(\underline{x})$ expressed in terms of the population profile $\underline{x} \in \Delta(\mathcal{A})$ after taking into consideration the distribution of players.

Define a global optimum (GO) as the optimum value of the total expected utilities, i.e., the mapping $\Delta(\mathcal{A}) \mapsto \sum_{a \in \mathcal{A}} x_a f_a(\underline{x})$ is maximized. Define an equilibrium state in a more general setting.

Definition 129

A population profile $\underline{x}$ is an equilibrium state if the following variational inequality is satisfied:

$$\forall y \in \Delta(\mathcal{A}), \sum_{a \in \mathcal{A}} (y_a - x_a) f_a(\underline{x}) \leq 0. \tag{3.53}$$

At an equilibrium state, no player can unilaterally change actions without decreasing his expected utility. The variational inequality (3.53) is equivalent to the following characterization, also known as Wardrop's First Principle (Wardrop, 1952): *At the equilibrium, the expected utility in all actions actually used are equal to and greater than those which would be experienced by a single player on any unused action*, i.e., for any action a such that $x_a > 0$:

$$f_a(\underline{x}) = \max_{b \in \mathcal{A}} f_b(\underline{x}). \tag{3.54}$$

Note that the same relation is satisfied by Nash definition of the equilibrium point (Nash, 1950) and its mass interpretation of population behaviors.

Theorem 130: Equivalence to Wardrop's First Principle

A population profile $\underline{x}$ is an equilibrium state if and only if for all action $a \in \mathcal{A}$ such that $x_a > 0$, it holds that $f_a(\underline{x}) = \max_{b \in \mathcal{A}} f_b(\underline{x})$.

Proof. Let $\underline{x}$ be an equilibrium state. We first remark that the variational inequality implies for all $(a, a') \in \mathcal{A}^2$, if $f_a(\underline{x}) < f_{a'}(\underline{x})$ then $x_a = 0$. To prove this, assume by contradiction that $x_a > 0$ and $f_a(\underline{x}) < f_{a'}(\underline{x})$ then $\langle \underline{x}, f(\underline{x}) \rangle < f_{a'}(\underline{x}) = \langle \underline{e}_{a'}, f(\underline{x}) \rangle$ where $\underline{e}_{a'}$ is the population profile when all the players use a'. This implies $\langle \underline{e}_{a'} - \underline{x}, f(\underline{x}) \rangle > 0$, which is in contradiction with the equilibrium state definition of $\underline{x}$.

Conversely, if for all a such that $x_a > 0$, one has $a \in \arg\max_b f_b(\underline{x})$, then for any population $\underline{y} \in \Delta(\mathcal{A})$,

$$\langle \underline{y} - \underline{x}, f(\underline{x}) \rangle = \sum_{a,\, x_a>0} (y_a - x_a) f_a(\underline{x}) + \sum_{a,\, x_a=0} (y_a - x_a) f_a(\underline{x})$$

$$= \sum_{a,\, x_a>0} (y_a - x_a)[\max_b f_b(\underline{x})] + \sum_{a,\, x_a=0, f_a(\underline{x}) \leq \max_b f_b(\underline{x})} (y_a - 0) f_a(\underline{x})$$

$$= \sum_{a,\, f_a(\underline{x})=\max_b f_b(\underline{x})} (y_a - x_a)[\max_b f_b(\underline{x})]$$

$$+ \sum_{a,\, x_a=0, f_a(\underline{x})=\max_b f_b(\underline{x})} (y_a - 0)[\max_b f_b(\underline{x})]$$

$$+ \sum_{a,\, x_a=0, f_a(\underline{x})<\max_b f_b(\underline{x})} (y_a - 0) f_a(\underline{x})$$

$$\leq \sum_{a,\, x_a>0, f_a(\underline{x})=\max_b f_b(\underline{x})} (y_a - x_a)[\max_b f_b(\underline{x})]$$

$$+ \sum_{a,\, x_a=0, f_a(\underline{x})=\max_{b \in \mathcal{A}} f_b(\underline{x})} (y_a - x_a)[\max_b f_b(\underline{x})]$$

$$+ \sum_{a,\, x_a=0, f_a(\underline{x})<\max_b f_b(\underline{x})} (y_a - x_a)[\max_b f_b(\underline{x})]$$

$$= [\max_b f_b(\underline{x})] \sum_{a \in \mathcal{A}} (y_a - x_a) = 0$$

Similarly, a population profile $\underline{x}$ is an evolutionarily stable state (ESS) if for any other population profile $\underline{y} \neq \underline{x}$ there exists $\epsilon_y > 0$ such that, $\forall \epsilon \in]0, \epsilon_y[$

$$\sum_{a \in \mathcal{A}} (y_a - x_a) f_a((1 - \epsilon)\underline{x} + \epsilon\, \underline{y}) < 0. \tag{3.55}$$

When inequality (3.55) is non-strict, and $\epsilon = 0$, we find that the probability distribution $\underline{x}$ is a symmetric Nash equilibrium (NE). When inequality (3.55) is non-strict, the population profile $\underline{x}$ is a neutrally stable state (NSS). A NSS is in particular a NE. Denote by Δ_{ESS} (resp. Δ_{NE}, Δ_{NSS}) the set of ESS (resp. NE, NSS). Thus, the following inclusions hold:

$$\Delta_{ESS} \subset \Delta_{NSS} \subset \Delta_{NE}.$$

Theorem 131

Under continuity assumption of the utility functions f, the evolutionary game has at least one equilibrium.

Sketch of Proof We will show that the game has a symmetric Nash equilibrium. We first remark that the generating function of the random number of interacting players K is continuous in $(0,1)$. Thus, the vector of utility functions f is lower semi-continuous in Δ_{n-1} (which is a non-empty, convex and compact subset of Euclidean space $\mathbb{R}^n$). The existence of a symmetric Nash equilibrium in mixed strategies follows from the existence of solutions of the following variational inequalities:

$$\text{find } \underline{x}^* \in \Delta_{n-1} \text{ s.t } \langle (\underline{y} - \underline{x}^*), f(\underline{x}^*) \rangle \leq 0, \ \forall \ \underline{y}$$

where $\langle , \rangle$ is the inner product of $\mathbb{R}^n$. We prove that this is equivalent to the fixed point equation: $\underline{x}^* = \Pi_{\Delta_{n-1}} (\underline{x}^* - \varsigma C(\underline{x}^*))$, $\varsigma > 0$ where $\Pi_{\Delta_{n-1}}$ denotes the projection into the simplex Δ_{n-1} and $C(.) = -f(.)$. Multiplying the inequality $\langle (\underline{x}^* - \underline{y}), -C(\underline{x}^*) \rangle \geq 0$ by $\varsigma > 0$, and adding $\langle \underline{x}^*, \underline{y} - \underline{x}^* \rangle$ to both sides of the resulting inequality, one obtains $\langle \underline{y} - \underline{x}^*, \underline{x}^* - [\underline{x}^* - \varsigma C(\underline{x}^*)] \rangle \geq 0$. Recall that the projection map on Δ_{n-1} which is a convex and closed set is characterized by:

$$\underline{z} \in \mathbb{R}^n, \ \underline{z}' = \Pi_{\Delta_{n-1}} \underline{z} \iff \langle \underline{z}' - \underline{z}, \underline{x} - \underline{z}' \rangle \geq 0, \ \forall \underline{x} \in \Delta_{n-1}$$

Thus,

$$\underline{x}^* = \Pi_{\Delta_{n-1}} (\underline{x}^* - \varsigma C(\underline{x}^*))$$

According to Brouwer-Schauder's fixed point theorem, given a map $\alpha: \Delta_{n-1} \to \Delta_{n-1}$, with α continuous, there is at least one $\underline{x}^* \in \Delta_{n-1}$, such that $\underline{x}^* = \alpha(\underline{x}^*)$. Observe that since the projection $\Pi_{\Delta_{n-1}}$ and $(I - \varsigma C)$ are both continuous, $\Pi_{\Delta_{n-1}}(I - \varsigma C)$ is also continuous by composition. It follows from convexity and compactness of Δ_{n-1} and the continuity of $\Pi_{\Delta_{n-1}}(I - \varsigma C)$ that the Brouwer-Schauder fixed point theorem can be applied to the map $\Pi_{\Delta_{n-1}}(I - \varsigma C)$. We conclude that at least one stationary equilibrium state exists in the evolving game with a random number of players. This completes the proof.

3.5.3.2 *Heterogeneous Population*

Similarly to the case of a homogeneous population, we define the following solution concepts for heterogeneous populations. Let Θ be a finite set of populations or classes.

Definition 132: Equilibrium State

$\forall \theta, \forall \underline{y}^{\theta}, \sum_{a \in \mathcal{A}^{\theta}} (y_a^{\theta} - x_a^{\theta}) f_a^{\theta}(\underline{x}) \leq 0.$

Theorem 133: Equivalence to Multi-Class Wardrop's First Principle

A vector $\underline{x} \in \Delta(\mathcal{A})$ is an equilibrium state if for all classes θ and actions $a \in \mathcal{A}$ such that $x_a^{\theta} > 0$, one has, $f_a^{\theta}(\underline{x}) = \max_{b \in \mathcal{A}^{\theta}} f_b^{\theta}(\underline{x})$.

Proof. The proof follows the same lines as in Theorem 130.

Definition 134: Global Equilibrium

$\forall \underline{y}, \sum_{\theta} \sum_{a \in \mathcal{A}^{\theta}} (y_a^{\theta} - x_a^{\theta}) f_a^{\theta}(\underline{x}) \leq 0.$

Definition 135: Global Evolutionary Stable State

$\forall \underline{y} \neq \underline{x}$ there exists a $\epsilon_y > 0$ such that, $\forall \epsilon \in (0, \epsilon_y), \sum_{\theta} \sum_{a \in \mathcal{A}^{\theta}} (y_a^{\theta} - x_a^{\theta}) f_a^{\theta}((1-\epsilon)\underline{x} + \epsilon \underline{y}) < 0.$

Theorem 136

Under continuity of the utilities, the evolutionary game has at least one global equilibrium.

Proof. The proof follows the same lines as in Theorem 131 using Kakutani's fixed point theorem.

3.5.4 Evolutionary Game Dynamics

In this section, several dynamics models are presented. The reason for this is twofold. First, they allow one to adopt a different approach to equilibria in evolutionary games since they can be the result of a dynamic process exploiting milder information and behavior assumptions. Second, the models will be re-used in Part B to study the convergence of the learning algorithms under consideration. As far as evolutionary games are concerned, evolutionary game dynamics propose an explicitly dynamic updating choice, a model in which players myopically update their behavior in response to their current strategic environment. These dynamic procedures do not assume the automatic coordination of players' actions and beliefs.

Evolutionary game dynamics are models of strategy change that are commonly used in evolutionary game theory and learning theory. A strategy which does better than the average, or than its opponent, increases in frequency at the expense of strategies that do worse than the average or the opposed action. Many evolutionary game dynamics models are used in the literature: replicator dynamics, best response dynamics, Brown–von Neumann–Nash dynamics, Smith dynamics, gradient dynamics, projection dynamics, etc.

Consider large populations of players. Denote by Θ the set of subpopulations or classes. Players from subpopulation θ select their actions from a finite action set $\mathcal{A}^\theta$. Denote by $\underline{x}(t) = (x_a^\theta(t))_{a \in \mathcal{A}^\theta, \theta \in \Theta}$ the population profile at time t and by $f_a^\theta(\underline{x}(t))$ the expected utility of players from subpopulation θ with a when the population profile $\underline{x}(t)$. Let $\beta_{ba}^\theta(\underline{x}(t), f)$ be the conditional switch rate from strategy b to strategy a in subpopulation θ. The flow of the population is specified in terms of the functions $\beta_{ba}^\theta(\underline{x}(t), f)$ which determine the rates at which a player who is considering a change in strategies chooses to switch to his various alternatives.

The *incoming flow* for the action a is:

$$\Phi_+ = \sum_{b \in \mathcal{A}^\theta} x_b^\theta(t) \beta_{ba}^\theta(\underline{x}(t), f) \tag{3.56}$$

and the *outgoing flow* is:

$$\Phi_- = x_a^\theta(t) \sum_{b \in \mathcal{A}^\theta} \beta_{ab}^\theta(\underline{x}(t), f) \tag{3.57}$$

where $x_a^\theta(t)$ represents the fraction of players of the population which uses action a at time t.

Let:

$$V_{a,f}^\theta(\underline{x}) = \sum_{b \in \mathcal{A}^\theta} x_b^\theta \beta_{ba}^\theta(x, f) - x_a^\theta \sum_{b \in \mathcal{A}^\theta} \beta_{ab}^\theta(x, f).$$

The evolutionary game dynamics is given by:

$$\dot{x}_a^\theta(t) = V_{a,f}^\theta(\underline{x}(t)), \ a \in \mathcal{A}^\theta, \ \theta \in \Theta. \tag{3.58}$$

Below we briefly mention some standard evolutionary game dynamics.

- Brown–von Neumann–Nash dynamics. Denote by g_a^θ the positive part of the excess utility of subpopulation a of class θ. There is a close relation between BNN dynamics and Nash's original proofs of his equilibrium existence theorem. BNN dynamics can be interpreted as follows. Assume that there are large populations in which there is steady incoming and outgoing flow. New players joining the system use only strategies that are better than the expected utility, and better strategies are more likely to be adopted. On the other hand, randomly chosen players leave the game. More precisely, the strategy x^θ is adopted with a probability that is proportional to the excess utility $g_a^\theta(x)$. In continuous-time, Brown–von Neumann–Nash dynamics is given by:

$$\frac{d}{dt} x_a^\theta(t) = \left[m^\theta g_a^\theta(\underline{x}(t)) - x_a^\theta(t) \sum_{b \in \mathcal{A}^\theta} g_b^\theta(\underline{x}(t)) \right] \tag{3.59}$$

where:

$$g_a^\theta(\underline{x}(t)) = \max \left[0, f_a^\theta(\underline{x}(t)) - \frac{1}{m^\theta} \sum_{b \in \mathcal{A}^\theta} x_b^\theta(t) f_b^\theta(\underline{x}(t)) \right]$$

The discrete time version is given by:

$$x_a^\theta(t+1) = \frac{\kappa x_a^\theta(t) + g_a^\theta(\underline{x}(t))}{m^\theta \kappa + \sum_{b\in A^\theta} g_b^\theta(\underline{x}(t))} \tag{3.60}$$

where κ is an appropriate constant.

The set of the rest points of the BNN dynamics is exactly the set of equilibrium states.

- Replicator dynamics. Among the most studied evolutionary game dynamics are *replicator dynamics*, such as are applied in several contexts of the evolution of road traffic congestion in which the fitness is determined by the strategies chosen by all drivers, base station assignment in hybrid systems, and the migration constraint problem in wireless networks. In replicator dynamics, a fraction of members of a subpopulation of class θ grows when its utility is greater than the expected average utility of all the subclass of θ in the population. In continuous time, the system of ordinary differential equations (ODE) is given by:

$$\frac{d}{dt}x_a^\theta(t) = x_a^\theta(t)\left[f_a^\theta(\underline{x}(t)) - \frac{1}{m^\theta}\sum_{b\in A^\theta} x_b^\theta(t)f_b^\theta(\underline{x}(t))\right] \tag{3.61}$$

$$x_a^\theta(t+1) = x_a^\theta(t)\frac{\kappa + f_a^\theta(\underline{x}(t))}{\kappa m^\theta + \sum_{b\in A^\theta}^{n^\theta} x_b^\theta(t)f_b^\theta(\underline{x}(t))}. \tag{3.62}$$

The set of the rest points in replicator dynamics contains the set of equilibrium states, but it can be bigger since these dynamics may not lead to equilibria. Typically, the corners are rest points, and the faces of a simplex are invariant but may not be an equilibrium.

An important result in the field of *evolutionary game dynamics* is the *Folk theorem* (evolutionary version). It states that, for the replicator dynamics of the expected two-player game, we have that:

- Every Nash equilibrium of the game is a rest point.
- Every strict Nash equilibrium of the game is asymptotically stable.
- Every stable rest point is a Nash equilibrium of the game.
- If an interior orbit converges, its limit is a Nash equilibrium of the game.

In the two-player case, there are many other interesting properties: If the time average $\frac{1}{T}\int_0^T x_{j,t}(s_j)\,dt$ of an interior orbit converges, the limit is a Nash equilibrium. The limit set of $P_{s_j s_k}^T := \frac{1}{T}\int_0^T x_{j,t}(s_j)x_{j,t}(s_k)\,dt$ leads to coarse correlated equilibria.

- Generalized Smith dynamics: $\gamma \geq 1$

$$\frac{d}{dt}x_a^\theta(t) = \sum_{b\in A^\theta} x_b^\theta(t)\max(0, f_a^\theta(x) - f_b^\theta(x))^\gamma - x_a^\theta(t)\sum_{b\in A^\theta}\max(0, f_b^\theta(x) - f_a^\theta(x))^\gamma. \tag{3.63}$$

The set of the rest points of the generalized Smith dynamics is precisely the set of equilibrium states.

- Best response dynamics. Members of each subpopulation revise their strategy and choose the best replies $BR(\underline{x}(t))$ at the current population state $\underline{x}(t)$.

$$\frac{d}{dt}x^\theta(t) \in m^\theta BR^\theta(\underline{x}^{-\theta}(t)) - x^\theta(t) \qquad (3.64)$$

where $BR^\theta(\underline{x}^{-\theta}(t)) = \arg\max_{y^\theta \in \Delta(A^\theta)} \left\{ \sum_{b\in A^\theta} y_b^\theta f_b^\theta(\underline{x}^{-\theta}(t)) \right\}$.

The set of rest points of BR dynamics is exactly the set of equilibrium states. More details on best reply and best response dynamics can be found in Gilboa and Matsui (1991).

- Fictitious play. One widely used model of learning is the process of fictitious play and its variants (stochastic fictitious play, fictitious play with inertia, etc.). In this process, the players behave as if they think they are facing an unknown distribution of strategies for the others players. In it, each player presumes that her/his opponents are playing stationary (possibly mixed) strategies. A player is said to use fictitious play if, at every time instant t, he chooses an action that is a best response to the empirical distribution of the opponents' play up to time $t-1$. At each round, each player thus best responds to the empirical frequency of play of his opponent. Such a method is of course adequate if the opponent indeed uses a stationary strategy, but it is flawed if the opponent's strategy is not stationary. The discrete time version of fictitious play is given by:

$$x_t^\theta = \left(1 - \frac{1}{t}\right)x_{t-1}^\theta + \frac{1}{t}BR^\theta\left(x_{t-1}^{-\theta}\right). \qquad (3.65)$$

As we can see, with an appropriate time normalization, the discrete-time fictitious play asymptotically approximates the continuous time best response dynamic. More precisely, the set of limit points of discrete time fictitious play is an invariant subset of the continuous time best-response dynamics.

In continuous time, fictitious play is given by:

$$\frac{d}{dt}x^\theta(t) \in \frac{1}{t}(m^\theta BR^\theta(\underline{x}^{-\theta}(t)) - x^\theta(t)) \qquad (3.66)$$

- Logit dynamics is based on an exponential weight of the utility function. This form of dynamics is sometimes referred to Boltzmann-Gibbs dynamics or smooth best response dynamics.

$$\frac{d}{dt}x_a^\theta(t) = \left[m^\theta \frac{e^{\frac{f_a^\theta(\underline{x}(t))}{\eta^\theta}}}{\sum_{b\in A^\theta} e^{\frac{f_b^\theta(\underline{x}(t))}{\eta^\theta}}} - x_a^\theta(t) \right] \qquad (3.67)$$

Using an explicit representation in terms of the logit map, Hofbauer et al. (2009) have shown that the time average of the replicator dynamics is a perturbed solution of best reply dynamics. The set of the rest points of Smooth dynamics is exactly the set of η-equilibrium states. More details on logit learning can be found in Fudenberg and Levine (1998).

- Orthogonal projection dynamics is a myopic adaptive dynamic in which a sub-population grows when its expected utility is greater than the arithmetic average utility of all the class.

$$\dot{x}_a^\theta(t) = \left[f_a^\theta(\underline{x}(t)) - \frac{1}{n^\theta} \sum_{b \in \mathcal{A}^\theta} f_b^\theta(\underline{x}(t)) \right] \tag{3.68}$$

- Ray-projection dynamics is a myopic adaptive dynamic in which a subpopulation grows when its expected utility is greater than the ray-projection utility of all the class. In this formulation, the transition rate from a to b is independent of the next strategy b.

$$\frac{d}{dt} x_a^\theta(t) = \left[m^\theta h_a^\theta(\underline{x}(t)) - x_a^\theta(t) \sum_{b \in \mathcal{A}^\theta} h_b^\theta(\underline{x}(t)) \right] \tag{3.69}$$

It is easy to see that the ray-projection dynamics is obtained for the revision protocol:

$$\beta_a^{\theta,b}(f,\underline{x}) = h_a^\theta(\underline{x}). \tag{3.70}$$

Notice that replicator dynamics, best response dynamics, and logit dynamics can be obtained as a special case of ray-projection dynamics.
- Target projection dynamics:

$$\frac{d}{dt} x^\theta(t) = \left[\Pi_{X^\theta} \left(x^\theta(t) - \varsigma C^\theta(\underline{x}(t)) \right) - x^\theta(t) \right] \tag{3.71}$$

From Proposition 131, $\underline{x}^*$ is an equilibrium equivalent to:

$$\underline{x}^* = \Pi_{\prod_\theta X^\theta} \left(\underline{x}^* - \varsigma C(\underline{x}^*) \right). \tag{3.72}$$

Thus, the set of rest points of target projection dynamics coincides with the set of equilibria of the evolving game.
- Dynamic equilibrium:
  - In many cases in evolving games, it is known that the trajectories of evolutionary game dynamics may not converge, can be chaotic, may have cycle limits, etc. Under several forms of evolutionary game dynamics, the expected utility obtained in the cycle is not so far from being a candidate for a "static" equilibrium utility. This suggests a *time dependent equilibrium* approach in evolving games. A trajectory is a *dynamic equilibrium* if the *time average utility under this trajectory leads to an equilibrium utility.*
  The trajectory $x: [t_0, \infty[\rightarrow \prod_\theta \Delta(\mathcal{A}^\theta)$ is a dynamic equilibrium if:

$$\sum_\theta \int_0^T \langle x^\theta(t) - \underline{y}^\theta(t), f^\theta(\underline{x}(t)) \rangle \, dt \geq 0,$$

for any other trajectory $y(.)$. The existence of such equilibrium is guaranteed under continuity of the utility function.

3.5.5 **Price of Anarchy in Evolutionary Games**

The so-called price of anarchy (PoA) is one of the approaches used to measure how much the performance of decentralized decision-making systems is affected by the selfish behavior of its components. For a definition of the PoA see Chapter 1 or Koutsoupias and Papadimitriou (1999). We will now present a similar concept for evolutionarily stable states. The concept of the price of anarchy can be seen as an *efficiency metric* that measures the *cost of selfishness* or decentralization, and it has been used extensively in the context of congestion games or routing games where players typically have to minimize a cost function. If the evolutionary game has an ESS, we can define the analog of the "price of anarchy", denoted by PoA_{ESS}, as the ratio between the cost of the worst evolutionary equilibrium and the global optimum value. Conversely, the "price of stability" (PoS_{ESS}) is the ratio between the cost of the "best" evolutionary equilibrium and the global optimum (GO) value defined as the minimum value of the expected global cost of $\underline{x} \in \Delta(\mathcal{A}) \mapsto \sum_{b \in \mathcal{A}} x_b C_b(x)$ which we assume to be strictly positive.

$$PoA_{ESS} = \frac{\max_x ESS \sum_{b \in \mathcal{A}} x_b C_b(x)}{GO} \geq 1, \qquad (3.73)$$

$$PoS_{ESS} = \frac{\min_x ESS \sum_{b \in \mathcal{A}} x_b C_b(x)}{GO} \geq 1. \qquad (3.74)$$

Using the inclusions of subsection (3.5.3.1), we obtain the following inequalities:

$$PoA_{NE} \geq PoA_{NESS} \geq PoA_{ESS} \geq PoS_{ESS} \geq PoS_{NESS} \geq PoS_{NE} \geq 1 \qquad (3.75)$$

3.5.6 **Applying Evolutionary Games to the Medium Access Control Problem**

Random Medium Access Control (MAC) algorithms have played an increasingly important role in the development of wired and wireless networks, and the performance and stability of these algorithms, such as slotted-Aloha and Carrier Sense Multiple Access (CSMA), is still an open problem. Distributed Random Medium Access Control, starting from the first version of Abramson's Aloha to the most recent algorithms used in IEEE802.11, have enabled the rapid growth of both wired and wireless networks. They aim at efficiently and fairly sharing a resource among users, even though each user must decide independently (eventually after receiving some messages or listening) when and how to attempt to use it. MAC algorithms have generated a lot of research interest, especially recently in attempts to use multi-hop wireless networks to provide high-speed access to the Internet with low-cost and low-energy consumption.

In this section, we consider a wireless communication network with distributed receivers, in which some mobiles contend for access via a common, wireless communication channel. We characterize this distributed multiple access problem in terms of many random access games at each time. Random multiple access games introduce the problem of medium access. We assume that with mobiles randomly placed over an area and receivers distributed in the corresponding area, the channels are ideal for transmission and all errors are due to collisions. A mobile decides to transmit a packet with some power level or not to transmit (null power level) to a receiver when they are within transmission range of each other. Interference occurs as in the Aloha protocol where power control is introduced: if more than one neighbor of the receiver transmits a packet with a power level greater than the corresponding power of the mobile at the same time there is a collision. The evolutionary random multiple access game is a non-zero-sum dynamic game, while the mobiles in the same neighborhood have to share a common resource, the wireless medium. To cover the non-saturated transmission case, the random selection of mobiles is done only on the set of mobiles which have a packet to transmit. Alternatively, those that do not have a packet to send at the current slot are automatically constrained to the state "0" (with null power).

We assume that for each packet, its source can choose the transmitted power from several power levels $A = \{p_0, p_1, p_2, \ldots, p_n\}$ with $p_0 < p_1 < \cdots < p_n$; i.e., a mobile's strategy corresponds to the choice of a power level in $\mathcal{P}$. $p_0 = 0$ means that the mobile does not transmit, and p_n is the maximum power level available. If mobile j transmits a packet using a power p_j, it incurs a transmission cost of $c(p_j) \geq 0$. The packet transmission is successful if the other users in the range of its receiver use power levels that are lower than p_j in that time slot, otherwise there is a collision. If there is no collision, mobile j gets a payoff 1 from the successful packet transmission. If $c(a) > 1$ for some power a, then a is dominated by 0 (not transmit). For the remainder, suppose that the cost of transmission $\max_{p_j} c(p_j)$ is strictly less than 1. All packets of a lower power level that are involved in a collision are assumed to be lost, and will have to be retransmitted later. In addition, if more than one packet of a higher power level is involved in a collision then all packets are lost. The power differentiation thus allows one packet of a higher power level to be successfully transmitted in collisions that do not involve other packets of the same higher power level. Then, a transmission of mobile j is successful if its transmission power is strictly greater than the power levels used by the others mobiles at the same slot. When the number of mobiles transmitting to the receiver is $k+1$, the payoff is given by $u_{j,k+1} : A^{k+1} \rightarrow \mathbb{R}$

$$u_{j,k+1}(a_j, a_{-j}) = -c(a_j) + \begin{cases} 1 & \text{if } a_j > \max_{l \neq j} a_l \\ 0 & \text{otherwise} \end{cases}$$

where a_{-j} denotes $(a_1, \ldots, a_{j-1}, a_{j+1}, \ldots, a_{k+1})$, $c : \mathbb{R}_+ \rightarrow \mathbb{R}_+$ is a pricing function. We assume that:

$$c(p_0) = c(0) = 0 \leq c(p_1) \leq c(p_2) \ldots \leq c(p_n).$$

3.5.6.1 *Control Without Pricing*

Consider the case where there is no cost for transmission power, i.e., $0 = c(p_0) = c(0) = c(p_1) = c(p_2) \ldots = c(p_n) = 0$. Table 3.1 illustrates an example of two power levels (to transmit or to wait).

Table 3.2 illustrates an example of three power levels.

Table 3.1 Random Access Game with Two Mobiles and Two Actions: Mobile 1 Chooses a Row, Mobile 2 Chooses a Column. The Symbol ♯ Represents a Pareto Optimal Solution, ⋆ Indicates a Pure Nash Equilibrium. The Game is a Unique Interior Equilibrium State. The Unique Interior Equilibrium State is an ESS

	p_1	p_0
p_1	$(0,0)^\star$	$(-1,0)^{\sharp,\star}$
p_0	$(0,-1)^{\sharp,\star}$	$(0,0)$

Table 3.2 Random Access Game with Two Mobiles and Three Actions: Mobile 1 Chooses a Row, Mobile 2 Chooses a Column. The Symbol ♯ Represents a Pareto Optimal Solution, ⋆ Indicates a Pure Nash Equilibrium

	p_2	p_1	p_0
p_2	$(0,0)^\star$	$(-1,0)^{\sharp,\star}$	$(-1,0)^{\sharp,\star}$
p_1	$(0,-1)^{\sharp,\star}$	$(0,0)$	$(-1,0)^{\sharp}$
p_0	$(0,-1)^{\sharp,\star}$	$(0,-1)^{\sharp}$	$(0,0)$

3.5.6.2 *Control With Pricing*

Consider the case where:

$$c(p_0) = c(0) = 0 < c(p_1) < c(p_2) \ldots < c(p_n).$$

The game is represented by Table 3.3.

Table 3.3 Multiple Access Game with Two Mobiles and Three Actions: Mobile 1 Chooses a Row, Mobile 2 Chooses a Column. $0 < c(p_1) < c(p_2) < 1$. There is no Pure Equilibrium

	p_2	p_1	p_0
p_2	$(c(p_2),c(p_2))$	$(-1+c(p_2),c(p_1))$	$(-1+c(p_2),0)$
p_1	$(c(p_1),-1+c(p_2))$	$(c(p_1),c(p_1))$	$(-1+c(p_1),0)^{\sharp}$
p_0	$(0,-1+c(p_2))$	$(0,-1+c(p_1))^{\sharp}$	$(0,0)$

Games with Incomplete Information and Learning

Bayesian Games

4.1 INTRODUCTION

As mentioned in Chapter 1, a game is said to be with complete information when every player knows all data of the game. For a strategic form game $\mathcal{G} = \left(\mathcal{K}, \{u_j\}_{j \in \mathcal{K}}, \{\mathcal{S}_j\}_{j \in \mathcal{K}} \right)$, this means that every player knows the number of players and the strategy sets of all the players and their utility functions. Most often, in practical situations, the assumption that every player knows all the data of the game is too demanding. For instance, in a wireless network it may happen that each transmitter has only individual channel state information, and therefore ignores the channels of the others. In this book, the focus is on learning algorithms because ultimately this is what matters to wireless engineers. In Chapters 5 and 6, we show an interactive situation where several decision-makers (e.g., automata) who implement learning algorithms can be modeled as a game with incomplete information. The present chapter, which is deliberately short, is provided for essentially three reasons.

- First, it allows the reader to see the difference between what is commonly called a game with incomplete information, or Bayesian game in the game-theoretic literature (e.g., see Zamir (2009) for an excellent survey), and what is called a game with incomplete information in the framework of learning.
- Second, the framework of learning may be inapplicable, e.g., when the game is played only once. In such a case, players will not receive samples or feedbacks from their environment, preventing them from learning from experience.
- Bayesian games possess some special features which may lead to surprising conclusions, not encountered in games with complete information. We mention some of them in this chapter.

As mentioned above, Bayesian games are characterized by the fact that the data of the game are not common knowledge, allowing concepts such as private information and secret information to appear. When a player has private information, he is the only player who knows this information but all the players are aware he possesses it. On the other hand, information is said to be secret from a given player if all the other players do not know he possesses it. Elaborating on the last item, it may happen that, in a Bayesian game, possession of more information may decrease the utility of a player (Aumann, 1976). We will not go into the details here but in essence, when a player has more information, he has more options, strategically speaking. And we already know from the Braess paradox (see Chapter 1) that having a larger action or strategy set may lead to a less beneficial outcome for some (and even all) players.

An important notion in Bayesian games is the notion of "type", introduced by Harsanyi (1967). We have mentioned that in Bayesian games, the data of the game are generally not known to the players. All the data which influence the utility of a player can be seen as a state of nature (the actions, the realization of a random variable, etc.). A type corresponds to a "state of mind" of a player – that is, the description of his belief about the state of nature. For example, in a wireless multiple access channel, the knowledge of a transmitter about the channel gains of the different links between the transmitters and receiver may be considered as the type of this transmitter. The Harsanyi model of Bayesian games is described in Sec. 4.2.1 and is based on the notion of type. Remarkably, this model is relatively simple to understand and apply, compared to more elaborate models (Mertens and Zamir, 1985). Indeed, Harsanyi's model has been shown to integrate the notion of hierarchy of beliefs in a simple manner (Mertens and Zamir, 1985).

Assume that each player chooses a prior distribution over the parameters he does not know (e.g., the overall channel state); this is his belief. But, a player would also have to assume what he knows about the belief of the other players. Going further, a player would need to have belief about the belief of the other players about his own belief. This leads to the quite complex notion of a hierarchy of beliefs. This approach seems to be inapplicable to wireless scenarios: Why should wireless terminals implement such an elaborate level of reasoning? The excellent news for wireless engineers is that it has been proved by Mertens and Zamir (1985) that the simple model of Harsanyi captures the whole hierarchy of beliefs. This is why we will not describe in this chapter concepts such as the universal belief space, and restrict our attention to the model of Harsanyi.

4.2 BAYESIAN GAMES IN A NUTSHELL

4.2.1 The Harsanyi Model

Following the model of Harsanyi (1967), a possible definition for a Bayesian game is as follows.

Definition 137: Bayesian Game

A Bayesian game can be described by a 5-tuplet:

$$\mathcal{G} = \left(\mathcal{K}, \{\mathcal{T}_i\}_{i \in \mathcal{K}}, q, \{\mathcal{S}_i(\tau_i)\}_{i \in \mathcal{K}}, \{u_i\}_{i \in \mathcal{K}} \right) \tag{4.1}$$

where:

- $\mathcal{K} = \{1, 2, \ldots, K\}$ is the set of players of the game;
- $\mathcal{T}_i$ is the set of types for player i;
- q is a (joint) probability distribution over the profiles of types $\underline{\tau} = (\tau_1, \ldots, \tau_K)$;
- $\mathcal{S}_i(\tau_i)$ is the set of possible actions or strategies for player i when his type is $\tau_i \in \mathcal{T}_i$;

- u_i is the utility function of player i:

$$u_i : \left| \begin{array}{ccc} \mathcal{T} \times \mathcal{S} & \rightarrow & \mathbb{R} \\ (\tau_1, \ldots, \tau_K, s_1(\tau_1), \ldots, s_K(\tau_K)) & \mapsto & u_i(\underline{\tau}, \underline{s}) \end{array} \right. \tag{4.2}$$

with $\mathcal{T} = \mathcal{T}_1 \times \mathcal{T}_2 \times \cdots \times \mathcal{T}_K$, $\mathcal{S} = \mathcal{S}_1 \times \mathcal{S}_2 \times \cdots \times \mathcal{S}_K$, $\underline{\tau} = (\tau_1, \ldots, \tau_K)$, $\underline{s} = (s_1, \ldots, s_K)$.

At least two comments are in order. First, in this model, the set of strategies depends on the type of the player. A useful special case of this model is when the set of strategies is fixed and is not type-dependent. For example, in a power control game where the type of a player is his individual channel gain, it is generally assumed that the admissible set of powers does not depend on the value of the channel gains. Second, the above definition implicitly assumes a consistent description. Indeed, with this description, each player can evaluate *ex ante* his expected utility by following a common procedure, namely marginalizing q given his type τ_i. However, it may happen that some players have a different belief on a same event (e.g., the global channel state information in power control). In this case, the provided model is no longer valid and one has to consider a more general model, namely the interim model (Fudenberg and Tirole, 1991; Zamir, 2009) in which each player has his own conditional probability. With such a model, the expected utility evaluated by a player may differ from what he actually gets.

4.2.2 Bayesian Equilibrium

In this section and the rest of this chapter, only the Harsanyi model is considered. Assume a Bayesian game $\mathcal{G} = (\mathcal{K}, \{\mathcal{T}_i\}_{i \in \mathcal{K}}, q, \{\mathcal{S}_i(\tau_i)\}_{i \in \mathcal{K}}, \{u_i\}_{i \in \mathcal{K}})$. In this setup, a Bayesian equilibrium of $\mathcal{G}$ is merely a Nash equilibrium in the expected game where players consider their expected utility:

$$u_i^B(\underline{s}(\underline{\tau}), \underline{\tau}) = \sum_{\underline{\tau}_{-i}} q_i(\underline{\tau}_{-i} | \tau_i) u_i(s_i(\tau_i), \underline{s}_{-i}(\underline{\tau}_{-i})) \tag{4.3}$$

where finite sets of types are assumed (note that the counterpart of the above equation for continuous sets of types can be easily expressed). More formally we have the following definition for a Bayesian equilibrium.

Definition 138: Bayesian Equilibrium

The strategy profile $\underline{s}^*$ is a Bayesian equilibrium of $\mathcal{G}$ if:

$$\forall i \in \mathcal{K}, \ \forall \underline{s}_{-i}(\cdot) \in \mathcal{S}_{-i}(\cdot), \ u_i^B\left(s_i^*(\tau_i), \underline{s}_{-i}^*(\underline{\tau}_{-i}), \underline{\tau}\right) \geq u_i^B\left(s_i(\tau_i), \underline{s}_{-i}^*(\underline{\tau}_{-i}), \underline{\tau}\right). \tag{4.4}$$

To know whether such an equilibrium exists, most of the tools presented in Chapter 2 can be applied. In particular, for finite Bayesian games we have the following theorem.

Theorem 139: Existence Theorem

If the set of players, the sets of players' actions, and the sets of players' types are finite, then there necessarily exists at least one mixed Bayesian equilibrium.

This theorem is due to Harsanyi (1967) and the proof directly follows from the Nash theorem (Nash, 1950). In fact, this theorem also holds for the interim model (which is not consistent).

Example 140

Assume a game with two players (e.g., mobile phone manufacturers): $\mathcal{K} = \{1,2\}$. Each player has two possible actions lying in the set $\mathcal{S}_1 = \mathcal{S}_2 = \{\texttt{WiMax}, \texttt{LTE}\}$. Additionally, each player has two possible types lying in $\mathcal{T}_1 = \mathcal{T}_2 = \{\texttt{cooperative}, \texttt{non-cooperative}\}$. The type $\texttt{cooperative}$ (C) means that the player in question wants to reach a consensus on the standard used whereas $\texttt{non-cooperative}$ (NC) corresponds to the oppositive behavior. Denote by $q_i(\underline{\tau}_{-i})$ the belief of player i about the type of player $-i$, that is, the probability player $-i$ chooses type $\underline{\tau}_{-i} \in \mathcal{T}_{-i}$. Assuming the beliefs to be independent, the game can be described by the following matrix form.

$q_2(\text{C}) = \frac{1}{2}$

$q_2(\text{NC}) = \frac{1}{2}$

$q_1(\text{C}) = \frac{2}{3}$

	LTE	WiMax
LTE	(2,1)	(0,0)
WiMax	(0,0)	(1,2)

	LTE	WiMax
LTE	(2,0)	(0,2)
WiMax	(0,1)	(1,0)

$q_1(\text{NC}) = \frac{1}{3}$

	LTE	WiMax
LTE	(0,1)	(2,0)
WiMax	(1,0)	(0,2)

	LTE	WiMax
LTE	(0,0)	(2,2)
WiMax	(1,1)	(0,0)

Depending on the four possible pairs of types, different games are played, as represented by four utility matrices. In the chosen example, player 1 believes player 2 is going to cooperate with probability $\frac{2}{3}$, player 1 believes player 2 is going to not cooperate with probability $\frac{1}{3}$, etc. As the beliefs are independent, one can express the players' utilities in the expected game. For instance, the expected utility player 1 gets when the action profile is (LTE, LTE) is given by:

$$u_1^B((\texttt{LTE}, \texttt{LTE})) = \frac{2}{3} \times \frac{1}{2} \times 2 + \frac{2}{3} \times \frac{1}{2} \times 2 + \frac{1}{3} \times \frac{1}{2} \times 0 + \frac{1}{3} \times \frac{1}{2} \times 0 = \frac{4}{3}. \quad (4.5)$$

By computing the expected utilities for all game outcomes, we obtain the following matrix form for the expected game.

	LTE	WiMax
LTE	$\left(\frac{4}{3}, \frac{1}{2}\right)$	$\left(\frac{2}{3}, 1\right)$
WiMax	$\left(\frac{1}{3}, \frac{1}{2}\right)$	$\left(\frac{2}{3}, 1\right)$

From this matrix, it is easily seen that the game has two pure Bayesian equilibria:

1. $\underline{s}^*(C) = (\text{LTE},\text{LTE})$, $\underline{s}^*(NC) = (\text{WiMax},\text{WiMax})$;
2. $\underline{s}^*(C) = (\text{LTE},\text{WiMax})$, $\underline{s}^*(NC) = (\text{LTE},\text{WiMax})$.

4.3 APPLICATION TO POWER CONTROL GAMES

4.3.1 Bayesian Energy-Efficient Power Control Games

We consider a multiple access channel with K transmitters, one receiver, and flat block fading links. The received signal can be written:

$$y = \sum_{j=1}^{K} h_j x_j + z \tag{4.6}$$

where h_j is is the channel gain for user j, x_j is the signal transmitted by user j, and $z\mathcal{N}(0,\sigma^2)$ is an additive white Gaussian noise. For each block, each transmitter tunes his power level to maximize his energy-efficiency:

$$u_i(p_i,\underline{p}_{-i}) = \frac{Rf(\text{SINR}_i)}{p_i} \tag{4.7}$$

where R is the transmission rate, $f(\cdot)$ the packet success rate, SINR_i the signal to interference plus noise ratio, and p_i is the transmit power for user i. By assuming single-user decoding at the receiver, the SINR has the following form:

$$\text{SINR}_i = \frac{p_i \eta_i}{\sum_{j\neq i} p_j \eta_j + \sigma^2} \tag{4.8}$$

where $\eta_i = |h_i|^2 \in \Gamma_i$ and σ^2 is always the noise variance.

For the sake of simplicity, for all $i \in \mathcal{K}$, each set Γ_i is assumed to be discrete. Additionally, the set $\mathcal{P}_i$ is assumed to be discrete. Here, we want to include the fact that the transmitter does not know perfectly the global channel state $\underline{\eta} = (\eta_1, \eta_2, \ldots, \eta_K)$. A suitable game model to take this into account is a Bayesian game model consisting of an extension of the one-shot power control game $(\mathcal{K}, \{\mathcal{P}_i\}_i, \{u_i\}_i)$ to which a certain observation structure is added:

$$\mathcal{G}^B = \left(\mathcal{K}, \{\theta_i\}_{i\in\mathcal{K}}, q, \{\mathcal{P}_i\}_{i\in\mathcal{K}}, \{\Sigma_i\}_{i\in\mathcal{K}} \left\{u_i^B\right\}_{i\in\mathcal{K}} \right) \tag{4.9}$$

where:

-

$$\theta_i : \begin{array}{c|ccc} & \Gamma_1 \times \cdots \times \Gamma_k & \rightarrow & \mathcal{S}_i \\ & \underline{\eta} & \mapsto & s_i = \theta_i(\underline{\eta}) \end{array} \tag{4.10}$$

is a measurable application from the set of channel states, $\Gamma_i = \{\eta_i^1, \ldots, \eta_i^{m_i}\}$, $m_i \in \mathbb{N}^*$, to the set of observation of transmitter i which is $\mathcal{S}_i$, with $|\mathcal{S}_i| < +\infty$;

- $q(\underline{s}_{-i}|s_i)$ is the probability that the transmitters other than i observe $\underline{s}_{-i}$ given that transmitter i observes s_i;
- a strategy for a player i is a function $\sigma_i : \mathcal{S}_i \to \mathcal{P}_i$ where $\sigma_i \in \Sigma_i$ and Σ_i is the set of possible mappings;
- u_i^B is the expected utility for transmitter i which is defined by:

$$u_i^B(\sigma_1, \ldots, \sigma_K, s_i) = \mathbb{E}_{\underline{\eta}, \underline{s}_{-i}} \left[u_i \left(\sigma_1(s_1), \ldots, \sigma(s_K), \underline{\eta} \right) | s_i \right]. \tag{4.11}$$

For each transmitter i, denote by $\Lambda_i(s_i)$ the ambiguity set associated with the observation s_i. For instance, in the case of perfect global CSI, this set boils down to a singleton: $\Lambda_i(s_i) = \{\theta^{-1}(s_i)\} = \{\underline{\eta}\}$. Using this notation the expected utility can be rewritten as:

$$u_i^B(\sigma_1, \ldots, \sigma_K, s_i) = \sum_{\underline{\eta} \in \Lambda_i(s_i)} \frac{P(\underline{\eta})}{\Pr[\underline{\eta} \in \Lambda_i(s_i)]} \sum_{\underline{s}_{-i}} q(\underline{s}_{-i}|s_i) u_i \left(\sigma_1(s_1), \ldots, \sigma(s_K), \underline{\eta} \right). \tag{4.12}$$

where $P(\underline{\eta})$ is the probability that the channel state $\underline{\eta}$ is drawn. Using the fact that channel gains of the different links are generally independent in wireless networks the expected utility can be expressed by:

$$u_i^B(\sigma_1, \ldots, \sigma_K, s_i) = \sum_{\underline{\eta} \in \Lambda_i(s_i)} \frac{P(\underline{\eta})}{\Pr[\underline{\eta} \in \Lambda_i(s_i)]} \sum_{\underline{s}_{-i}} \prod_{j \neq i} \Pr[\underline{\eta} \in \Lambda_j(s_j)] \frac{f\left(\frac{\eta_i \sigma_i(s_i)}{\sum_{j \neq i} \eta_j \sigma_j(s_j) + \sigma^2} \right)}{\sigma_i(s_i)}. \tag{4.13}$$

The next step is to find whether a Bayesian equilibrium exists, whether it is unique, etc. Therefore the classical machinery described in Chapter 2 has to be put to work, which is left to the reader as a problem to be solved. In the example below, it is shown how results from Chapter 2 can be directly applied.

4.3.2 Bayesian Rate Efficient Power Allocation Games

The previous section aimed to show how to model the problem, so here our goal is to solve the existence and uniqueness problem completely. For this, a more simple framework is considered (He et al., 2010a). We always assume a multiple access channel, but this time the player's utility is the transmission rate and the links are static, that is:

$$v_i(p_i, \underline{p}_{-i}) = \log \left(1 + \frac{p_i \eta_i}{\sum_{j \neq i} p_j \eta_j + \sigma^2} \right) \tag{4.14}$$

where $p_i \in \mathcal{P}_i = [0, P_i^{\max}]$ and $\eta_i = |h_i|^2$. For simplicity, η_i is assumed to be in a two-element set $\{\eta_{\min}, \eta_{\max}\}$. A suitable Bayesian game model to describe this problem comprises the following elements:

- the set of players $\mathcal{K} = \{1, 2, \ldots, K\}$;
- the sets of types: $\mathcal{T}_i = \{\eta_{\min}, \eta_{\max}\}$;
- the set of actions $p_i(\tau_i) \in [0, P_i^{\max}]$;
- joint probability distribution over the channel states $\underline{\eta} = (\eta_1, \eta_2, \ldots, \eta_K)$ which is assumed to be the product of its marginals: $Q(\underline{\eta}) = q(\eta_1) \times q(\eta_2) \times \cdots \times q(\eta_K)$;
- the expected utility for transmitter i which is defined by:

$$v_i^{\mathrm{B}}(p_1(\eta_1), \ldots, p_K(\eta_K)) = \mathbb{E}_{\underline{\eta}}[v_i(p_1(\eta_1), \ldots, p_K(\eta_K))]. \qquad (4.15)$$

Is there a Bayesian equilibrium in this game? Is it unique? Can it be easily determined? Let us answer these questions.

Existence of a Bayesian equilibrium. The strategic form game $\left(\mathcal{K}, \{\mathcal{P}_i\}_i, \left\{v_i^{\mathrm{B}}\right\}^i\right)$ can be shown to be a concave game in the sense of Rosen (see Chapter 2). Indeed:

- for any i, $\mathcal{P}_i = [0, P_i^{\max}]$ is a compact and convex set;
- for any i, the utility function v_i^{B} is continuous with respect to $\underline{p} = (p_1, \ldots, p_K)$;
- for any i, the utility function v_i^{B} is concave with respect to $p_i(\eta_i)$.

Therefore, there is at least one pure Bayesian equilibrium.

Uniqueness of the Bayesian equilibrium. In Chapter 2, a sufficient condition for having a unique pure Nash equilibrium in a concave game was provided. Indeed, it is

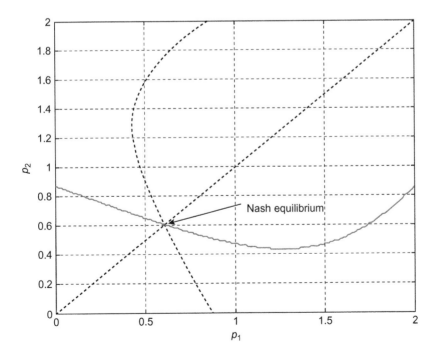

FIGURE 4.1

The figure depicts the best response of the players and their unique intersection, which is the unique Bayesian equilibrium of the game.

known from Rosen (1965) that the diagonally strict concavity (DSC) is a sufficient condition for having uniqueness. This condition is as follows. The DSC is met if there exists some vector of strictly positive parameters $\underline{r} = (r_1, r_2, \ldots, r_K)$ such that:

$$\forall \underline{p}' \neq \underline{p}, \ (\underline{p}' - \underline{p})\left(\gamma_{\underline{r}}(\underline{p}) - \gamma_{\underline{r}}(\underline{p}')^T\right) > 0 \tag{4.16}$$

where, by definition:

$$\gamma_{\underline{r}}(\underline{p}) = \left(r_1 \frac{\partial v_1^B}{\partial p_1}(\underline{p}), \ldots, r_K \frac{\partial v_K^B}{\partial p_K}(\underline{p})\right) \tag{4.17}$$

and $(\cdot)^T$ is the transposition operator. As proven in He et al. (2010a), the DSC is always met in the considered game, which proves that there exists a unique pure Bayesian equilibrium. Figure 4.1 represents the best responses of the players in a game where: $K = 2$, $P_i^{\max} = 1$, $\sigma^2 = 1$, $\eta^{\min} = 1$, $\eta^{\max} = 3$ and $q(\eta^{\min}) = q(\eta^{\min}) = 0.5$.

Partially Distributed Learning Algorithms

5

5.1 INTRODUCTION

Chapter 1 defined some game solution concepts. In particular, the Nash equilibrium was stressed as one of the central solution concepts of game theory. A possible setting where a Nash equilibrium may be observed is in games with complete information; players are assumed to be rational, rationality is assumed to be common knowledge, all the data of the game are known to every player, and this is also common knowledge. Although this modeling can be relevant for some interactive situations, game theory provides other important settings, where Nash equilibria (or other existing solution concepts) may be observed. Indeed, many game-theoretic results show that, when a game is played several times, the outcome after several sets of moves is often much closer to equilibrium predictions than initial moves, which supports the idea that equilibrium arises as a result of players learning from experience. The learning approach formalizes this idea, and studies how, which, and what kind of equilibrium might arise as a consequence of a long-run non-equilibrium process of learning, adaptation and/or imitation, and all of this by starting at an arbitrary starting point for the game[1]. Learning approaches are not procedures that are necessarily trying to reach a Nash equilibrium, but investigate ways in which players can optimize their own utility while simultaneously learning about the strategies of other players. An important and typical feature of the learning setting is that players are not necessarily assumed to be rational, and the structure of the game to be played is generally not assumed to be common knowledge. For instance, players can be modeled as automata which have partial information about their environment. In this chapter and in Chapter 6, both types of learning algorithms are described: some of them assume rationality (e.g., what we call game-theoretic learning algorithms) but most of them do not (e.g., when players are modeled by automata).

Classical game-theoretic learning procedures (Fudenberg and Kreps, 1988; Fudenberg and Levine, 1998; Young, 2004, 2009) assume that: the number of players is small and fixed; the rules of the game are known by every player; at least the distribution of possible utilities is known by all the players; and every player is rational. But even under these rigorous assumptions, and hyper-rationality of players, learning in games is difficult because it is interactive: each player's learning process correlates

---

[1] Indeed, while operating at an equilibrium point may be desirable, the question remains how to get there. The learning framework provides a possible answer to this question.

and affects what has to be learned by every other player over time. In many network-ing and communication scenarios, the above information and behavior assumptions regarding the other players may not hold or require a central authority or a coordina-tor. We will see now these assumptions can be relaxed. In particular, we will study the case where players can adjust their behavior only in response to their own realized utilities; they have no knowledge of the overall structure of the game, the number of players that interact can be random, they may not observe the actions or utilities of the other players, they occasionally make mistakes on their choices and readapt by experience. In situations such as wireless communications, it may be desirable to have a learning procedure that does not require any information about the other play-ers actions or utilities, and as little memory (small number of parameters in terms of past own-actions and past own-utilities) as possible. Such a rule is said to be uncou-pled or fully distributed. It has been shown in Hart and Mas-Colell (2003) that for a broad class of games, there is no general algorithm which would allow the play-ers' period-by-period behavior to converge to a Nash equilibrium[2]. Therefore, there is no guarantee for uncoupled dynamics that behaviors will approach Nash equilib-rium most of the time. The publications (Foster and Vohra, 1997; Hart and Mas-Colell, 2000) show that regret-minimizing procedures can cause the empirical frequency dis-tribution of play to converge to the set of correlated Nash equilibria. But there is no arbitrator, and no random device in their models. Note that the set of correlated equi-libria is convex, and includes the Nash equilibria; but is often much larger. There exist many ways of classifying learning algorithms (Fudenberg and Levine, 1998; Young, 2004). As far as this book is concerned we will only distinguish between two classes. A learning algorithm is said to be *fully distributed or uncoupled* if a player does not use information about the other players. Each player builds his strategies and updates them by using own-actions and own-utilities. On the other hand, a learning scheme is said to be *partially distributed* if each player implements his updating rule after receiv-ing a signal or some data about the other players. The latter is more sophisticated than fully distributed learning algorithms. Partially distributed learning is the purpose of this chapter and Chapter 6 focuses on fully distributed learning. In both chapters the following notation will be used to distinguish the situation where the utility function is known to the player from the one where only the value of the function at a given stage/time is known: $u_j(\underline{a}(t))$ will be used when the utility function is known and $u_j(t)$ will be used when only the instantaneous value of the function is known.

The chapter is organized as follows. Section 5.2 presents different algorithms based on the best response correspondence. Section 5.3 focuses on the fictitious play algorithm and its variants. The convergence time of logit response learning in potential games is discussed in Section 5.4. Section 5.5 focuses on continuous-kernel games with cost of learning, and Section 5.6 focuses on the fictitious play based algorithm for bargaining solutions. Section 5.7 provides a summary.

---

[2]If players cannot observe the strategies chosen by the other players, steady states of the learning process are not necessarily Nash equilibria because players can maintain incorrect beliefs about off-path play. Indeed, beliefs can be incorrect due to imperfect observations, errors, and anticipation mistakes.

5.2 BEST RESPONSE DYNAMICS

In 1838, Cournot introduced a dynamical process called the Cournot tâtonnement (Cournot, 1838). The setup was a duopoly (an economic competition between two firms) where each competitor has to decide the quantity of goods to produce. For the model used Cournot showed that the following dynamic procedure converges: firm 1 chooses a certain quantity of goods, firm 2 observes the quantity produced by firm 1 and plays his best response, that is the quantity which maximizes its profit, firm 1 re-adjusts its quantity to this reaction in order for its benefit to be maximal and so on. This is what best response dynamics (BRD) is about. More formally, if one considers the strategic form game $\mathcal{G} = \left(\mathcal{K}, \{u_j\}_{j \in \mathcal{K}}, \{\mathcal{A}_j\}_{j \in \mathcal{K}}\right)$, the asynchronous/sequential best response dynamics (ABRD) consists of the following sequence:

1. the game is assumed to have a certain starting state, say $\underline{a}(0) = (a_1(0), a_2(0), \ldots, a_K(0))$;
2. player $j \in \mathcal{K}$ updates his action by choosing his best response to $\underline{a}_{-j}(0)$: $a_j(1) \in$ BR$(\underline{a}_{-j}(0))$. If there is more than one best action, one of them is chosen randomly;
3. player $j' \in \mathcal{K}$ updates his action by choosing his best response to the new action profile for which only one action has been updated;
4. another player updates his action by choosing his best response to the new action profile for which two actions have been updated, and so on.

Another version of the BRD is the simultaneous BRD: all players update their actions synchronously. Both the sequential and simultaneous BRD have been used in the wireless literature. For instance, in Yu et al. (2004) and Scutari et al. (2009), it is applied to update the power allocation policy of the transmitters in multi-band interference channels where the utility function is given by an achievable Shannon transmission rate. In the mentioned references, the iterative dynamical procedure that was used is called iterative water-filling (IWF). In this particular setup, it is shown that the knowledge of the power of the received signal is sufficient (which is an aggregate function of the actions) and therefore observing the actions themselves is not necessary. Also, some conditions under which the IWF algorithms converge are provided in Scutari et al. (2009). Indeed, there are not many general results concerning the convergence of the BRD, which means that, often, an ad hoc proof has to be done. Below, we provide a couple of known convergence results for the ABRD. We also provide the definitions of the concepts used to derive these results.

5.2.1 An Acyclic Graph Game has a Pure Equilibrium

Let $\mathcal{G}^*$ be the graph (V, E) where $\phi = \prod_{j \in \mathcal{K}} \mathcal{A}_j$ is the set of action profiles and E is the set of edges defined by:

$$E = \{(\mathbf{a}, \mathbf{b}) \mid \exists j, \ \mathbf{b} = (\mathbf{b}_j, \mathbf{a}_{-j}), \tilde{u}_j(\mathbf{a}) < \tilde{u}_j(\mathbf{b})\} \qquad (5.1)$$

Lemma 2

If $\mathcal{G}^*$ is acyclic then the corresponding game $\mathcal{G}$ possesses a pure equilibrium.

Proof. Every acyclic graph has a sink (vertex without outgoing edges). Every sink on the graph $\mathcal{G}^*$ is a pure equilibrium (this follows immediately from the definition of the graph $\mathcal{G}^*$).

5.2.2 Weakly Acyclic Games and BRD

Consider any finite game with a set $\mathcal{A}$ of action profiles. A better reply path is a sequence of action profiles $\mathbf{a}_1, \mathbf{a}_2, \ldots, \mathbf{a}_T$ such that for each successive pair $(\mathbf{a}_t, \mathbf{a}_{t+1})$ there is exactly one player such that $a_{j,t} \neq a_{j,t+1}$ and for that player the utility $\tilde{u}_j(\mathbf{a}_t) < \tilde{u}_j(\mathbf{a}_{t+1})$. This states that only one player moves at a time, and each time a player moves he increases his own utility.

Definition 141

The game $\mathcal{G}$ is weakly acyclic if for any action profile $\mathbf{a} \in \mathcal{A}$, there exists a better reply path starting at a and ending at some pure Nash equilibrium of $\mathcal{G}$.

One of the nice properties of weakly acyclic games is that they have at least one pure Nash equilibrium. The following theorem gives convergence to pure Nash equilibria for the ABRD.

Theorem 142: Young (2004)

In weakly acyclic games, the asynchronous/sequential best response dynamics converge with probability one to a pure Nash equilibrium.

Since weakly acyclic games include potential games, the following result holds.

Corollary 143

In potential games, the asynchronous/sequential best response dynamics converge with probability one to a pure Nash equilibrium.

In addition to acyclic games, the ABRD also converges in pseudo-potential games, q-potential games, and super-modular games (see Chapter 2 for these concepts). When it converges, the convergence is generally fast (see for instance, the wireless example of Scutari et al. (2009)). The main problem of the BRD is that it is based on rigourous assumptions: in general, the played action profiles have to be perfectly observed by all players (although in some special cases like the IWF algorithm, less information is required). This motivates the use of other dynamical procedures such as reinforcement learning.

5.3 FICTITIOUS-PLAY-BASED ALGORITHMS

Aside from the Cournot tâtonnement, one of the oldest iterative algorithms for computing equilibria is given by the fictitious play (FP) procedure (Brown, 1949, 1951; Robinson, 1951). Fictitious play and its variants (stochastic fictitious play, smooth fictitious play, weakened generalized fictitious play, etc.) are simple models of learning which are extensively used in the literature of game theory. One of the reasons for considering FP is that its structure allows one to better understand reinforcement-type learning algorithms.

5.3.1 Brown's Fictitious Play Model

The standard *fictitious play* algorithm introduced by Brown (1949, 1951) and Robinson (1951) assumes that each player knows his actions and his own utility function, but also the last actions played by the other players. In it, each player presumes that the other players are playing stationary (possibly mixed) strategies. At each game instance/stage, each player thus best responds to the empirical frequency of play of his opponent (this presumes also that he/she observes the last actions chosen by the other players, and also that he/she is able to maximize and compute the best-response set).

Formally, the sequence of the fictitious play in discrete time is described as follows. To simplify the analysis, consider a finite number of players. The set of players is denoted by $\mathcal{K} = \{1, 2, \ldots, K\}$ where $K \geq 2$ is the cardinality of $\mathcal{K}$. Each player $j \in \mathcal{K}$ has $\mathcal{A}_j$ as a set of actions. We assume that the set $\mathcal{A}_j$ has at least two elements. Denote by $a_j \in \mathcal{A}_j$ a generic action of player j. The time space is the set $\{1, 2, 3, \ldots, \}$. Time index is denoted by t. Let $a_{j,t}$ be the action picked by player j at time t. Define the empirical frequency of use of action a_j up to t as:

$$\bar{f}_{j,t}(a_j) = \frac{1}{t} \sum_{t'=1}^{t} \mathbb{1}_{\{a_{j,t'}=a_j\}}. \tag{5.2}$$

where $\mathbb{1}_{\{a_{j,t'}=a_j\}}$ is equal to 1 if the action $a_{j,t}$ chosen by player j is exactly a_j, and 0 otherwise.

The next Lemma gives a recursive equation that is satisfied by the frequencies of actions and utilities. This equation is very useful for understanding the construction of the iterative learning algorithms presented in this chapter and in Chapter 6.

. . . .

Lemma 3

The following statements hold:

- The empirical process $\bar{f}_{j,t}(a_j) = \frac{1}{t} \sum_{t'=1}^{t} \mathbb{1}_{\{a_{j,t'}=a_j\}}$ satisfies the recursive equation:

$$\bar{f}_{j,t+1}(a_j) = \bar{f}_{j,t}(a_j) + \lambda_{t+1}\left(\mathbb{1}_{\{a_{j,t+1}=a_j\}} - \bar{f}_{j,t}(a_j)\right) \tag{5.3}$$

where $\lambda_{t+1} = \frac{1}{t+1}$.

- $\lambda_{t+1} = \frac{1}{t+1}$ satisfies:

$$\sum_{t\geq 0}\lambda_{t+1} = +\infty, \quad \sum_{t\geq 0}\lambda_{t+1}^2 = \sum_{t=1}^{\infty}\frac{1}{t^2} = \frac{\pi^2}{6} < +\infty.$$

Proof. We will first prove the first statement.

$$(t+1)\bar{f}_{j,t+1}(a_j) = \mathbb{1}_{\{a_{j,t+1}=a_j\}} + \sum_{t'=1}^{t}\mathbb{1}_{\{a_{j,t'}=a_j\}}$$

$$= \mathbb{1}_{\{a_{j,t+1}=a_j\}} + t\bar{f}_{j,t}(a_j)$$

By dividing both sides by $(t+1)$, one gets.

$$\bar{f}_{j,t+1}(a_j) = \frac{1}{t+1}\mathbb{1}_{\{a_{j,t+1}=a_j\}} + \frac{t}{t+1}\bar{f}_{j,t}(a_j)$$

$$\bar{f}_{j,t+1}(a_j) = \frac{1}{t+1}\mathbb{1}_{\{a_{j,t+1}=a_j\}} + \left(1 - \frac{1}{t+1}\right)\bar{f}_{j,t}(a_j)$$

$$= \bar{f}_{j,t}(a_j) + \frac{1}{t+1}\left(\mathbb{1}_{\{a_{j,t+1}=a_j\}} - \bar{f}_{j,t}(a_j)\right)$$

which is the announced recursive equation.

The second statement is obtained from the fact that the term $\frac{1}{t+1}$ is not summable and by Euler's identity one has $\sum_{t=1}^{\infty}\frac{1}{t^2} = \frac{\pi^2}{6}$. A simple proof of this relation follows from a Fourier series development of the polynomial $x \mapsto x^2$ and evaluating at the point $x = \pi$.

$$x^2 = \frac{\pi^2}{3} + \sum_{t\geq 1}\frac{4(-1)^t}{t^2}\cos(xt)$$

Using the relation $\cos(\pi t) = (-1)^t, t \in \mathbb{N}$, one gets Euler's identity. Note that one can use Parseval's identity or the ζ-function $\zeta(z) = \sum_{t=1}^{\infty}\frac{1}{t^z}$, $\Re(z) > 1$ to obtain the same result.

From Lemma 3, each player can compute this frequency iteratively and independently of the other players. The frequency $\bar{f}_{j,t} = (\bar{f}_{j,t}(a_j))_{a_j \in A_j}$ is a probability vector over A_j, i.e., $\bar{f}_{j,t} \in X_j = \Delta(A_j)$.

Consider a finite game $\mathcal{G} = (\mathcal{K}, (A_j, \tilde{u}_j)_{j\in\mathcal{K}})$ in strategic form. In fictitious play, each player j chooses his action at time $t+1$ to maximize his utility in response to the frequency of play $\bar{f}_{-j,t}$ of the other players. Then, Brown's fictitious play algorithm is determined by:

$$a_{j,t+1} \in \arg\max_{a_j \in A_j}\ u_j(e_{a_j}, \bar{f}_{-j,t}). \tag{5.4}$$

where e_{a_j} is the probability distribution consisting of choosing the action a_j in a deterministic way, $u_j : \prod_{j' \in \mathcal{K}} \mathcal{X}_{j'} \longrightarrow \mathbb{R}$ denotes the mixed extension utility of player j. This choice is myopic in the sense that the players are trying to maximize their current utility without considering the future. The players need to know their own utility function and observe the actions of the others at the end of each play. Thus, it belongs in the class of *partially distributed learning schemes*.

5.3.1.1 *How Can Fictitious Play Help in Solving the Game $\mathcal{G}$?*

Fictitious play can be used in solving simple games in strategic-form. In particular if the solution concept under consideration is the (Nash) equilibrium point, relation (5.4) defines a best response which is a key concept for Nash's theorem. Proposition 145 below answers how the fictitious play algorithm can be exploited when finding the equilibria of $\mathcal{G}$. In order to study the behavior of the fictitious play algorithm, we define the concept of convergence in terms of frequencies.

Definition 144

We say that the fictitious play algorithm converges if the sequence of empirical frequencies of play $\{\bar{f}_{j,t}\}_{j \in \mathcal{K}, \, t \in \mathbb{N}}$ converges to some point $\{\bar{f}_j^*\}_{j \in \mathcal{K}} \in \prod_j \mathcal{X}_j$ (componentwise).

5.3.1.2 *Convergence Issue*

Recall that the strategic form game $\mathcal{G}$ is an exact potential game (Chapter 2) if it admits a potential function $\phi : \prod_j \mathcal{A}_j \longrightarrow \mathbb{R}$ which is a function satisfying $\forall j$, $a_j \in \mathcal{A}_j$, $\underline{a} \in \prod_j \mathcal{A}_j$,

$$u_j(a_j, \underline{a}_{-j}) - u_j(\underline{a}) = \phi(a_j, \underline{a}_{-j}) - \phi(\underline{a}). \tag{5.5}$$

In terms of action frequencies, fictitious play is known to converge to Nash equilibria in two-player finite games with common utilities (also called common interest or team games) and in any potential game. However, fictitious play does not always converge. Shapley (1964), proved that the play can cycle indefinitely in the Rock-Paper-Scissors game (see Table 5.1). Note that, in the above formulation, the players update their beliefs simultaneously, but they can do it sequentially (alternating update of choices). Throughout this chapter and in Chapter 6, it will be seen that only partial convergence results are available, and a lot of work remains to be done

Table 5.1 The Rock-Paper-Scissors Game in Matrix Form. Fictitious Play does not Converge in Such a Game

	Rock	**Paper**	**Scissors**
Rock	(0, 0)	(−1, 1)	(1, −1)
Paper	(1, −1)	(0, 0)	(−1, 1)
Scissors	(−1, 1)	(1, −1)	(0, 0)

in this direction. As a rule of thumb for considered learning algorithms, it can be said that a sufficient condition for convergence is for the game to be potential, and a sufficient condition for non-convergence is the existence of cycles in the game.

5.3.1.3 *Connection with the Nash Equilibria of* $\mathcal{G}$

The next Proposition establishes a connection between the Nash equilibria (both pure and mixed) of the game $\mathcal{G}$ and the stationary distributions of the fictitious play process.

Proposition 145

Suppose a fictitious play sequence $\{a_{j,t}\}_{j,t}$ generates the sequence of empirical frequencies $\{\bar{f}_{j,t}\}_{j\in\mathcal{K},\ t\in\mathbb{N}}$ which converge to $\{\bar{f}_j^*\}_{j\in\mathcal{K}}$. Then $\{\bar{f}_j^*\}_{j\in\mathcal{K}}$ is an (Nash) equilibrium point of the one-shot game $\mathcal{G}$.

The result immediately follows from the definition in Equation (5.4) and the continuity of the mappings u_j.

Remark Note that for the continuous action space $\mathcal{A}_j$, the average: $\bar{f}_{j,t} = \frac{1}{t}\sum_{t'=1}^{t} a_{j,t'}$ also satisfies $\bar{f}_{j,t+1} = \bar{f}_{j,t} + \lambda_{t+1}\left(a_{j,t+1} - \bar{f}_{j,t}\right)$. Similarly, the average utility $\bar{U}_{j,t} = \frac{1}{t}\sum_{t'=1}^{t} u_{j,t'}$ satisfies:

$$\bar{U}_{j,t+1} = \bar{f}_{j,t} + \lambda_{t+1}\left(u_{j,t+1} - \bar{U}_{j,t}\right).$$

Bush and Mosteller (1955), Arthur (1993), Borgers and Sarin (1993), and Sastry et al. (1994a,b) studied variants of these iterative equations (in terms of reinforcement strategy, average utility learning, etc.) by re-scaling the utility function and/or by changing the step size λ_{t+1}. For constant step-size i.e., $\lambda_{t+1} = b$, $b \in \mathbb{R}_+$, one gets:

$$\mathbf{x}_{j,t+1} = \mathbf{x}_{j,t} + b\left(\mathbb{1}_{\{a_{j,t+1}=a_j\}} - \mathbf{x}_{j,t}\right) \tag{5.6}$$

If the step size is proportional to the current utility $\lambda_{t+1} = bu_{j,t+1}$, one gets:

$$\mathbf{x}_{j,t+1} = \mathbf{x}_{j,t} + bu_{j,t+1}\left(\mathbb{1}_{a_{j,t+1}=a_j} - \mathbf{x}_{j,t}\right) \tag{5.7}$$

5.3.2 Other Versions of Fictitious Play

The original version of fictitious play we presented above is not much used in the literature. We now describe one of the versions of fictitious play which is used in the game theory literature (see Fudenberg and Levine (1998) and Young (2009)). One of the motivations for considering this modified version is that each player takes into account the joint empirical frequency of the other players. As a consequence, better convergence results are obtained: coarsely correlated equilibria can be reached. Let us describe what the modifications consist of and show how fictitious play can be applied to simple wireless games. A learning scheme is said to be *adaptive* if the players consider past play as the best predictor of their opponents' future play, and they best-respond to their forecast. Fictitious play and its variants are *adaptive learning algorithms* used in standard repeated games (the same game is played many times). Assume that players play a standard repeated game with a fixed strategic form game $\mathcal{G}$ many times. Each player is assumed to know the joint strategy space, his own-utility function, and to observe the strategy of the other players at the end of each game stage. Players play as Bayesian players; they believe that the strategies of the other players correspond to some unknown mixed strategy and their beliefs are updated as follows:

Start with some initial references. For each player j, $\alpha_j(0) : \prod_{i \neq j} \mathcal{A}_i \longrightarrow \mathbb{R}_+$, which is updated by adding one if a strategy is played:

$$\alpha_{j,t+1}(s_{-j}) = \alpha_{j,t}(s_{-j}) + \mathbb{1}_{\{a_{-j,t}=s_{-j}\}}, \ s_{-j} \in \prod_{i \neq j} \mathcal{A}_i. \tag{5.8}$$

The probability that the player j assigns to the actions s_{-j} of the other players at stage $t+1$ is:

$$\tilde{x}_{j,t+1}(s_{-j}) = \frac{\alpha_{j,t+1}^{\gamma_j}(s_{-j})}{\sum_{s'_{-j}} \alpha_{j,t+1}^{\gamma_j}(s'_{-j})}, \ \gamma_j \geq 1 \tag{5.9}$$

where γ_j is a parameter to be tuned. Fictitious play assigns histories to actions by:

- computing the probabilities $\{\tilde{x}_{j,t}\}_t$ after computing the parameters $\{\alpha_{j,t}\}_t$
- choosing an action:

$$a_{j,t+1} \in \text{PBR}_j\left(\tilde{x}_{j,t+1}\right) := \arg\max_{a'_j \in \mathcal{A}_j} \sum_{s_{-j} \in \prod_{j' \neq j} \mathcal{A}_{j'}} u_j(a'_j, s_{-j})\tilde{x}_{j,t+1}(s_{-j})$$

at stage $t+1$.

Below we give an illustrative example which shows the limitation of the fictitious play learning scheme (Miyasawa, 1961).

Example 146: Channel Selection Game

Consider two receivers $R1, R2$ and two mobile terminals $MT1, MT2$. Each mobile terminal MT can transmit to one of the two receivers. If both MTs transmit at the same receiver at the same stage, there is collision, the packets are lost and the utilities are zero. If they choose different receivers, the probability $\mu > 0$ for a successful transmission is very high. One can represent this scenario by the following table: Mobile terminal 1 ($MT1$) chooses a row and $MT2$ chooses a column. The first component of the utility is for $MT1$ and the second component is for $MT2$.

$MT1\backslash MT2$	$R1$	$R2$
$R1$	$(0, 0)$	(μ, μ)
$R2$	(μ, μ)	$(0, 0)$

Suppose now that the players have implemented the fictitious play algorithm in their mobile terminal and the algorithm starts with the initial references $\alpha_j(0) = (1, \bar{\epsilon})$, $j \in \{1, 2\}$

with $0 < \bar{\epsilon} < 1$. At the first stage, the MTs compute:

$$x_{1,1}(R1) = x_{2,1}(R1) = \frac{1}{1+\bar{\epsilon}}, x_{1,1}(R2) = x_{2,1}(R2) = \frac{\bar{\epsilon}}{1+\bar{\epsilon}}.$$

The best response to $\left(\frac{1}{1+\bar{\epsilon}}, \frac{\bar{\epsilon}}{1+\bar{\epsilon}}\right)$ is reduced to the choice of receiver $R2$. Both MTs transmit at receiver $R2$ at stage 1. This leads to a collision and the utilities are $(0,0)$. At stage 2, the updated parameters become $\alpha_2^1(R1) = 1 = \alpha_2^2(R1)$, and $\alpha_2^1(R2) = 1 + \bar{\epsilon} = \alpha_2^2(R2)$. The new probabilities are:

$$x_{j,2}(R1) = \frac{1}{2+\bar{\epsilon}}, \ x_{j,2}(R2) = \frac{1+\bar{\epsilon}}{2+\bar{\epsilon}}.$$

So both mobile terminals transmit at receiver $R1$ at stage 2, and the outcome is again a collision. We then have an alternating sequence $(R2,R2),(R1,R1),$ $(R2,R2),(R1,R1),$ $(R2,R2),(R1,R1),\ldots$ and at each stage there is collision. This outcome cannot be a Nash equilibrium. Now, let us consider the frequencies generated by this cycling trajectory. It is not difficult to see that the empirical frequencies of use of each receiver $\bar{f}_{j,t}(R1) = \frac{1}{t}\sum_{t'=1}^{t} \mathbb{1}_{\{a_{j,t'}=R1\}}$ converge to $\frac{1}{2}$, and $\left(\frac{1}{2},\frac{1}{2}\right)$ is a mixed Nash equilibrium of the one-shot game. As can be seen, the realized utility of each MT via the fictitious play algorithm is zero (because there is a collision at each stage). Note that under the fictitious play algorithm $\frac{1}{t}\sum_{t'=1}^{t} \mathbb{1}_{\{a_{t'}=(R1,R1)\}} \neq \bar{f}_{1,t}(R1)\bar{f}_{2,t}(R1)$. In this channel selection example using the fictitious learning algorithm, even if the mobile terminals make their best choices at each stage, based on their best estimations and beliefs, it does not necessarily lead to a better outcome (the cumulated realized utility under the algorithm is worse than the worst Nash equilibrium utility of the one-shot game). We conclude that the equilibrium utility is not obtained by the players. This says that the equilibrium is not played.

The next example studies a weighted anti-coordination game with an aggregation structure.

Example 147: Base Station Selection Game

Consider two mobile stations (MS) and two base stations (BS). Each mobile station can transmit to one of the base stations. For simplicity, appropriate symmetry assumptions are made on the channel gains so that only two channel gains intervene. If mobile station 1 chooses a different base station to mobile station 2, then mobile station 1 gets the utility $\log_2\left(1 + \frac{p_1|h_1|^2}{N_0}\right)$ and mobile station 2 gets $\log_2\left(1 + \frac{p_2|h_2|^2}{N_0}\right)$, where p_1,p_2 are fixed positive powers, h_1,h_2 non-zero numbers representing the channel states, and N_0 is a variance noise parameter. If both mobile stations transmit at the same base station, mobile station 1 gets $\log_2\left(1 + \frac{p_1|h_1|^2}{N_0+p_2|h_2|^2}\right)$ and mobile station 2 gets $\log_2\left(1 + \frac{p_2|h_2|^2}{N_0+p_1|h_1|^2}\right)$. The following table summarizes the different configurations. Mobile station 1 ($MS1$) chooses a row and $MS2$ chooses a column. The first component of the utility is for $MS1$ and the second component is for $MS2$.

- It is easy to see that this game has the same equilibrium structure as the game defined by the utility matrix

$$\begin{pmatrix} (0,0) & (\mu_1,\mu_2) \\ (\mu_1,\mu_2) & (0,0) \end{pmatrix}$$

which is in the class of anti-coordination games.

$MS1 \backslash MS2$	$BS1$	$BS2$
$BS1$	$\log_2\left(1 + \frac{p_1\|h_1\|^2}{N_0 + p_2\|h_2\|^2}\right),$ $\log_2\left(1 + \frac{p_2\|h_2\|^2}{N_0 + p_1\|h_1\|^2}\right)$	$\log_2\left(1 + \frac{p_1\|h_1\|^2}{N_0}\right),$ $\log_2\left(1 + \frac{p_2\|h_2\|^2}{N_0}\right)$
$BS2$	$\log_2\left(1 + \frac{p_1\|h_1\|^2}{N_0}\right),$ $\log_2\left(1 + \frac{p_2\|h_2\|^2}{N_0}\right)$	$\log_2\left(1 + \frac{p_1\|h_1\|^2}{N_0 + p_2\|h_2\|^2}\right),$ $\log_2\left(1 + \frac{p_2\|h_2\|^2}{N_0 + p_1\|h_1\|^2}\right)$

- By applying the fictitious play algorithm with the initial belief point $\left(\frac{\bar{\epsilon}}{1+\bar{\epsilon}}, \frac{1}{1+\bar{\epsilon}}\right)$, $\bar{\epsilon} \in (0,1)$, one gets a cycling behavior between the two worst configurations. Since the two alternating cycling configurations lead to the worst utilities, the Cesaro-limit utility obtained under the fictitious play algorithm is worse than the worst Nash equilibrium utility. Now, if the initial belief point is fixed to $\left(\frac{1}{2}, \frac{1}{2}\right)$, then the outcome of the fictitious play algorithm is the completely mixed Nash equilibrium.

5.3.3 **On the Convergence of Frequencies**

As illustrated in the above example, even when the frequencies converge, the average utility under fictitious play algorithms can be worse than the worst Nash equilibrium utility. This leads to the following question: Is the convergence in terms of frequencies a natural notion of convergence in adaptive learning algorithms? In the classification of game theory, the game given in the example above is in the class of anti-coordination games. The game has a unique completely mixed (Nash) equilibrium and two pure Nash equilibria. As observed, if the standard fictitious play algorithm is used in this game, the players may never take opposite actions at the same stage (they may never anti-coordinate) but there is a convergence in terms of frequencies.

The next result is due to Fudenberg and Kreps (1988). It establishes that the strict Nash equilibria are absorbing for the fictitious play dynamics.

Proposition 148: Fudenberg and Kreps (1988)

For the fictitious play dynamics, the following results hold:

- strict Nash equilibria are absorbing for the fictitious play learning process;
- any pure strategy steady state of fictitious play learning must be a Nash equilibrium.

5.3.4 **Asynchronous Clocks in Stochastic Fictitious Play**

The model that we previously presented assumes that all the players play at every game instance (all the players are always active) so that every player knows that his current opponent has played the game exactly as often as he has. This is typically the case in some *atomic games* but in many settings, players play at varying frequencies, and do not know how many times their opponents have played the game. To take into account such situations, the authors in Fudenberg and Takahashi (2008) proposed an asynchronous stochastic fictitious play model, described below.

Consider a system with K players. Time is discrete. At each stage, two players are drawn randomly from a set of K players. Denote the random set of selected players by $\mathcal{B}^K(t)$. When a player j is in $\mathcal{B}^K(t)$, i.e., he/she is selected at time t, he/she chooses a smoothed best response in the stage-game, given his/her assessment of his/her current opponents' actions and beliefs. After the meeting, he/she updates his/her assessment based on the realized strategy of the other players. If the player j is not in $\mathcal{B}^K(t)$, j keeps his/her old assessment.

As in the stochastic fictitious play algorithm (with synchronous clocks) described above, at each stage, each player j has a parameter $\alpha_{j,t}$ and forms a probability $\mathbf{x}_{j,t}$ that he assigns to the other player. If player j is drawn from the system, one of the two active players at stage t (i.e., $j \in \mathcal{B}^K(t)$.), then he chooses an action $a_{j,t}$ best response to $\mathbf{x}_t$.

The algorithm is described as follows. Start with some initial references. For each player j, $\alpha_j(0) : \prod_{i \neq j} \mathcal{A}_i \longrightarrow \mathbb{R}_+$, is updated by adding one if a strategy is played and zero otherwise:

$$\alpha_{j,t+1}(s_{-j}) = \alpha_{j,t}(s_{-j}) + \mathbb{1}_{\{j \in \mathcal{B}^K(t), a_{-j,t}=s_{-j}\}}, \ s_{-j} \in \prod_{i \neq j} \mathcal{A}_i. \tag{5.10}$$

The difference between this model and stochastic fictitious play is that in this case, each player only updates his assessment when he can do so. The term $\alpha_{j,t+1}$ is equal to the initial weight on s_{-j}, $\alpha_{j,0}$ plus the number of stages in which player j was drawn and his opponents chooses s_{-j} before stage t. The number $\frac{1}{\sum_{s_{-j}} \alpha_{j,t}(s_{-j})}$ is player j's step size at time t which governs the weight of the influence of new observations.

Let $y_{j,t}$ be the ratio between step sizes of synchronous and asynchronous clocks, $y_{j,t} = \frac{1+\bar{\alpha}_{j,t}}{1+t}$ where $\bar{\alpha}_{j,t} = \sum_{s_{-j}} \alpha_{j,t}(s_{-j})$. The updating rule becomes:

$$\mathbf{x}_{j,t+1} - \mathbf{x}_{j,t} = \mathbb{1}_{\{j \in \mathcal{B}^K(t)\}} \frac{1}{t+1} \frac{1}{y_{j,t}} \left(\mathbb{1}_{\{a_{j,t+1}=a_j\}} - \mathbf{x}_{j,t} \right)$$

$$= \frac{1}{t+1} \left[\frac{2}{K} \frac{1}{y_{j,t}} \left(\frac{1}{K-1} \sum_{i \neq j} \beta_i(\mathbf{x}_{-i,t}) - \mathbf{x}_{j,t} \right) + M_{j,t+1} \right].$$

where $M_{j,t+1} = \frac{1}{y_{j,t}} \left[\mathbb{1}_{\{j \in \mathcal{B}^K(t)\}} \left(\mathbb{1}_{\{a_{j,t}=a_j\}} - \mathbf{x}_{j,t} \right) - \frac{2}{K} \left(\frac{1}{K-1} \sum_{i \neq j} \beta_i(\mathbf{x}_{-i,t}) - \mathbf{x}_{j,t} \right) \right]$.

The evolution of the y_j is given by:

$$y_{j,t+1} - y_{j,t} = \frac{1}{t+1} \left(\frac{2}{K} - y_{j,t} + \epsilon^1_{j,t+1} + \epsilon^2_{j,t+1} \right)$$

where $\epsilon^1_{j,t+1} = \mathbb{1}_{\{j \in \mathcal{B}^K(t)\}} - \frac{2}{K}$, $\epsilon^2_{j,t+1} = -\frac{1}{t+2} \left[\mathbb{1}_{\{j \in \mathcal{B}^K(t)\}} - y_{j,t} \right]$

The stochastic process $\{(\mathbf{x}_{j,t}, y_{j,t})\}_t$ satisfies the conditions in Benaïm (1999): these conditions are reported and explained in Appendix A.2. We then use it and approximate the stochastic process by the following system of ordinary differential equations:

$$\begin{cases} \dot{\mathbf{x}}_j(t) = \frac{2}{K}\frac{1}{y_j(t)}\left(\frac{1}{K-1}\sum_{i\neq j}\beta_i(\mathbf{x}_i(t)) - \mathbf{x}_j(t)\right) \\ \dot{y}_j(t) = \frac{2}{K} - y_j(t) \end{cases}$$

A rest point $\mathbf{x}^*$ is *linearly unstable* with respect to a dynamic process $\dot{\mathbf{x}} = \kappa(\mathbf{x})$ if its linearization $D\kappa(\mathbf{x})$ evaluated at $\mathbf{x}^*$ has at least one eigenvalue with a positive real part.

....

Theorem 149: Fudenberg and Takahashi (2008) and Pemantle (1990)

If $\mathbf{x}^*$ is a linearly unstable equilibrium for the system $\dot{\mathbf{x}} = \beta(\mathbf{x}) - \mathbf{x}$ then:

$$\lim_{t\longrightarrow\infty} \mathbf{x}_{1,t} = \ldots = \lim_{t\longrightarrow\infty} \mathbf{x}_{K,t} = \mathbf{x}^* \tag{5.11}$$

with probability zero.

....

Lemma 4

Let $\bar{f}_{j,t}(a_j) = \frac{\sum_{t'=1}^t w_{j,t'}\mathbb{1}_{\{a_{j,t'}=a_j\}}}{\sum_{t'=1}^t w_{j,t'}}$ be the weighted empirical frequency of player j evaluated at stage t. Then, $\bar{f}_{j,t}$ satisfies the recursive equation:

$$\bar{f}_{j,t+1}(a_j) = \left(1 - \frac{w_{j,t+1}}{\sum_{t'=1}^{t+1} w_{j,t'}}\right)\bar{f}_{j,t}(a_j) + \frac{w_{j,t+1}}{\sum_{t'=1}^{t+1} w_{j,t'}}\mathbb{1}_{\{a_{j,t+1}=a_j\}} \tag{5.12}$$

Assume that the coefficients $\lambda_{j,t+1} = \frac{w_{j,t+1}}{\sum_{t'=1}^{t+1} w_{j,t'}}$ satisfy $\lambda_{j,t} \geq 0$, $\sum_t \lambda_{j,t+1} = \infty$, $\sum_t \lambda_{j,t+1}^2 < \infty$.

The asymptotic pseudo trajectory is given by:

$$\frac{d}{dt}\bar{f}_{j,t}(a_j) \in \mathbb{E}\left(\mathbb{1}_{\{a_{j,t+1}=a_j\}} \mid \mathcal{F}_t\right) - \bar{f}_{j,t}(a_j) \tag{5.13}$$

where $\mathcal{F}_t$ is the filtration generated by the past play of actions and the frequencies.

The proof of this lemma follows the line of the Robbins-Monro algorithm given in Appendix A.2. Note that for w constant, i.e., λ uniform, one gets the standard recursive equations satisfied by the frequencies.

5.3.5 Is Fictitious Play Applicable in Wireless Games?

In the standard fictitious play algorithm and many of its variants, each player needs to know his own-strategy, own-utility, and the strategies of other players at the end of each stage. These assumptions are obviously not always valid in wireless networks. The observation of the actions of the other players is one of the critical points.

In many wireless scenarios, the players may not be able to observe the actions outside some range (neighborhood), hence observations about the actions of the others may not hold. Even when the fictitious play algorithm converges in specific games, there may be some cost for having the information about the actions played by the other players, or the players may need help from a local arbitrator (such as base station broadcasting, central coordinator, etc.). Thus, the applicability of this procedure is limited; it needs too much information for decentralized implementations. Chapter 6 shows how to relax the observation assumptions.

5.4 LEARNING LOGIT EQUILIBRIA

In this section, we present logit or Boltzmann-Gibbs learning. The logit equilibrium is a special case of the ϵ-Nash equilibrium (see Chapter 2) and will be defined after defining the Boltzmann-Gibbs distribution. The idea of logit learning is that the distribution over the possible actions or strategies for a given player is updated according to a formula (derived from physics) depending on the utility function of that player namely, $u_j(x_j(t), \underline{x}_{-j}(t))$ where $x_j(t)$ is the mixed distribution chosen by player j at stage t (sometimes the stage index t will be removed for the sake of clarity). Also some convergence results in potential games are provided. Indeed, most of the learning algorithms proposed in the literature fail to converge. Even when convergence occurs in some specific classes of games, the speed of convergence can be arbitrarily slow. For the standard logit distribution/rule, the convergence time is exponentially related to the size of the action set, which makes its implementation difficult. In this section, we present the convergence time of the log-linear, logit response learning, or Boltzmann-Gibbs learning algorithm, for potential games with bounded rationality. We describe the Boltzmann-Gibbs distribution as the way to update a strategy, and show that the convergence to stationary distributions can be reduced from exponential time to linear time (w.r.t. the number of players).

5.4.1 Information-Theoretic Interpretation of Boltzmann-Gibbs Rule

Information theory provides us with a principled way to modify strategic form games in order to accommodate bounded rationality. This is done by following information theory's proposal that, given only partial knowledge concerning the distributions the players are using, we should use the minimum information principle to infer those distributions. Doing so results in the principle that the outcome under bounded rationality is the minimizer of a certain set of coupled Lagrangian functions of the joint distribution. In the standard Boltzmann-Gibbs learning distribution, each player j chooses his action $a_j(t)$ with probability:

$$\tilde{\beta}_{j,\epsilon}(a_j(t)) = \frac{e^{\frac{1}{\epsilon_j} u_j(e_{a_j}(t), \underline{x}_{-j}(t))}}{\sum_{a_j'} e^{\frac{1}{\epsilon_j} u_j(e_{a_j'}(t), \underline{x}_{-j}(t))}} \qquad (5.14)$$

where the notation $e_{a_j}(t)$ stands for the special choice of probability distribution consisting in choosing the action/strategy a_j with probability 1 for stage t.

Note that to evaluate $\tilde{\beta}_{j,\epsilon}(a_j(t))$, player j is assumed to know his (expected) utility function, and the distributions chosen by the others. In Chapter 6, we will also provide a logit rule using less restrictive assumptions. The parameters $\{\epsilon_j\}_j$, are some non-negative constants. Let us interpret the meaning of these parameters. As $\bar{\epsilon} = \max_j \epsilon_j$ becomes smaller, the player chooses action with the highest utility with higher probability. The parameters $\{\frac{1}{\epsilon_j}\}_j$ can then be interpreted as the level of rationality of the players and the value $\tilde{\beta}_{j,\epsilon}(u)$ follows the distribution of the soft-max; in physics, $\frac{1}{\epsilon_j}$ corresponds to temperature. The following lemma corresponds to the information-theoretic interpretation of this distribution.

> ### Lemma 5
> The distribution $\tilde{\beta}_{j,\epsilon}$ is the maximizer of the modified game with utilities $\tilde{v}_j(\underline{x}) = u_j(\underline{x}) + \epsilon_j H(\underline{x})$ where H is the entropy function and $\underline{x} = (x_1, x_2, \ldots, x_K)$.

From this lemma, it can be seen that when choosing a high value for ϵ_j (low rationality levels or low temperatures) the optimal distribution tends to maximize the entropy; that is, it tends to a uniform distribution. This means that each player explores his action set. On the other hand, for high rationality levels, players try to maximize their individual utility without exploring (which may lead the procedure to cycle and therefore to never converge). At this point, we have all the necessary concepts to define a logit equilibrium, which corresponds to the limit of the Boltzmann-Gibbs learning procedure when the latter converges.

> ### Definition 150: Logit Equilibrium
> Consider the game $\mathcal{G} = (\mathcal{K}, \{u_j\}_{j \in \mathcal{K}}, \{\mathcal{A}_j\}_{j \in \mathcal{K}})$ and let the vector $\underline{x}^* = (x_1^*, \ldots, x_K^*) \in \Delta(\mathcal{A}_1) \times \cdots \times \Delta(\mathcal{A}_K)$. Then, $\underline{x}^*$ is a logit equilibrium with parameter $\underline{\epsilon} = (\epsilon_1, \ldots, \epsilon_K)$ if for all $j \in \mathcal{K}$, it holds that:
>
> $$x_j^* = \beta_{j,\epsilon_j}\left(u_j\left(e_1, \underline{x}_{-j}^*\right), \ldots, u_j\left(e_{|\mathcal{A}_j|}, \underline{x}_{-j}^*\right)\right). \tag{5.15}$$

As the case of potential games is a case where the Boltzmann-Gibbs distribution converges, we now provide the corresponding distribution and conduct the corresponding convergence analysis. In the special case of (exact) potential games the Boltzmann-Gibbs distribution is given by:

$$\tilde{\beta}_{j,\epsilon}(a_j) = \frac{e^{\frac{1}{\epsilon_j} u_j(\mathbf{e}_{a_j}, x_{-j})}}{\sum_{a_j} e^{\frac{1}{\epsilon_j} u_j(\mathbf{e}_{a_j}, x_{-j})}} = \frac{e^{\frac{1}{\epsilon_j} \phi^K(\mathbf{e}_{a_j}, x_{-j})}}{\sum_{a_j} e^{\frac{1}{\epsilon_j} \phi(\mathbf{e}_{a_j}, x_{-j})}}, \tag{5.16}$$

which is directly obtained from:

$$\tilde{\beta}_{j,\epsilon}(a_j) = \frac{e^{\frac{1}{\epsilon_j}u_j(\mathbf{e}_{a_j},x_{-j})}}{\sum_{s'_j} e^{\frac{1}{\epsilon_j}u_j(\mathbf{e}_{s'_j},x_{-j})}} = \frac{e^{\frac{1}{\epsilon_j}\left[u_j(\mathbf{e}_{a_j},x_{-j})-u_j(\mathbf{e}_{a_j^*},x_{-j})\right]}}{\sum_{s'_j} e^{\frac{1}{\epsilon_j}\left[u_j(\mathbf{e}_{s'_j},x_{-j})-u_j(\mathbf{e}_{a_j^*},x_{-j})\right]}}$$

$$= \frac{e^{\frac{1}{\epsilon_j}\phi(\mathbf{e}_{a_j},x_{-j})-\phi(\mathbf{e}_{a_j^*},x_{-j})}}{\sum_{s'_j} e^{\frac{1}{\epsilon_j}\phi(\mathbf{e}_{s'_j},x_{-j})-\phi(a_j^*,x_{-j})}} = \frac{e^{\frac{1}{\epsilon_j}\phi(\mathbf{e}_{a_j},x_{-j})}}{\sum_{a_j} e^{\frac{1}{\epsilon_j}\phi(\mathbf{e}_{a_j},x_{-j})}},$$

where ϕ^K is a potential function.

To establish a result on the speed of convergence of Boltzmann-Gibbs learning algorithms, we will need some reasonable bounds on the time it takes for the process to become close to its stationary distribution, and then we will quantify the distance between the stationary distribution and an ϵ-equilibrium. To this end, some definitions of stationary distributions of Markov processes are needed. For the reader who is not interested in the technicalities used to derive the convergence speed of the Boltzmann-Gibbs learning, please go directly to Proposition 152.

The full support of the probability of choosing action given the Boltzmann-Gibbs rule ensures that at each updating time, every action in $\mathcal{A}$ has a positive probability of being chosen by the revising player. Therefore, there is a positive probability that the Markov chain X_t^K will transit from any given current state x to any other given state y within a finite number of steps. A Markov process with this property is said to be irreducible (see Chapter 3). The constant jump rate Markov chain X_t^K is said to be reversible if it admits a reversible distribution: a probability distribution $\bar{\omega}^K$ on S^K satisfies the balance conditions:

$$\bar{\omega}_x^K P_{xy}^K = \bar{\omega}_y^K P_{yx}^K, \ \forall (x,y) \in S^K \times S^K. \tag{5.17}$$

By taking the sum of this equation over $x \in S^K$, we find that a reversible distribution is also a stationary distribution.

A continuous time Markov process over a finite state space S^K can be characterized using a discrete-time Markov chain with transition matrix P^K. The matrix $H_t^K = e^{t(P^K - I)}$ is the transition matrix of the continuous process where I is the identity matrix of size $|S^K|$.

Next, we provide conditions under which the distributions of a Markov chain or process converge to its stationary distribution, i.e., $H_t^{K,x} \longrightarrow \bar{\omega}^K$. In the continuous-time setting, irreducibility is sufficient for convergence in distribution; in the discrete-time setting, aperiodicity is also required.

Lemma 6

Suppose that X_t^K is either an irreducible aperiodic Markov chain or an irreducible Markov process, and that its stationary distribution is $\bar{\omega}^K$. Then, for any initial distribution μ_0, we have that:

$$\lim_{t \longrightarrow \infty} P_{\mu_0}\left(X_t^K = x\right) = \bar{\omega}_x^K, \quad \lim_{t \longrightarrow \infty} H_t^{K,x} = \bar{\omega}^K.$$

We now focus on the convergence time to the stationary distribution for normal-form potential games. We use logarithmic Sobolev inequalities for finite Markov chains, as established by Diaconis and Saloff-Coste (1995). These inequalities give bounds for the rate of convergence of Markov chains on finite state spaces to their stationary distributions. Now, suppose that the players in the game $\mathcal{G}$ have the same action set $\mathcal{A}_j = \mathcal{A}$, and each player's utilities are defined by the same function of the player's own action and the overall distribution of strategies $\xi(\underline{a}) \in S^K$, defined by $\xi_b(a_1,\ldots,a_K) = \frac{1}{K}\sum_{j=1}^K \mathbb{1}_{a_j=b}$. There is a population game $F^K : S^K \longrightarrow \mathbb{R}^K$ that represents G in the sense that $u_j(\underline{a}) = F_{j,a_j}^K(\xi(\underline{a}))$. Suppose that $\mathcal{G}$ is a strategic form potential game with the same action set and with potential function ϕ^K. If $\xi(\underline{a}) = \xi(\underline{a}')$ then $\phi^K(\underline{a}) = \phi^K(\underline{a}')$. Furthermore, the population game defined by F^K is a potential game, and its potential function is given by $\psi^K(\underline{x}) = \phi^K(\underline{a})$, where $\underline{a} \in \xi^{-1}\underline{x} = \{\underline{a}', \xi(\underline{a}') = x\}$. The Boltzmann-Gibbs learning rule induces an irreducible and reversible Markov chain X_t on the finite state space:

$$S^K = \left\{ \left(\frac{x_1}{K},\ldots,\frac{x_m}{K}\right) \mid x_i \in \mathbb{N}, \sum_a x_a = n \right\}.$$

The cardinality of S^K is:

$$\binom{K+m-1}{m-1}.$$

If the players use a Boltzmann-Gibbs learning rule, the Markov process X_t^K is reversible. The stationary distribution weight $\bar{\omega}_x^K$ is proportional to the product of two terms. The first term is a multinomial coefficient, and represents the number of ways of assigning the K agents to actions in $\mathcal{A}$ so that each action a is chosen by precisely Kx_a players. The second term is an exponential function of the value of the potential at state x. Thus, the value of $\bar{\omega}_x^K$ balances the value of potential at state x with the likelihood that profile x would arise were players assigned to strategies at random.

Proposition 151

Let F^K be a finite potential game with potential function ϕ^K, and suppose that the players follow a Boltzmann-Gibbs learning rule, with the same rationality level $\frac{1}{\epsilon}$, and revise their strategies by removing their own-influence, i.e., x is replaced by $\frac{Kx-e_a}{K-1}$;

then the Markov process X_t^K is reversible with stationary distribution:

$$\bar{\omega}_x^K = \frac{1}{\kappa^K} \binom{K}{Kx_1, Kx_2, \ldots, Kx_m} e^{\frac{1}{\epsilon} \phi^K(x)}$$

for $x \in S^K$, the number κ^K is chosen such that $\sum_{x \in S^K} \bar{\omega}_x^K = 1$.

For the proof, we need to verify the reversibility condition. It is enough to check that the equality:

$$\bar{\omega}_x^K P_{xy}^K = \bar{\omega}_y^K P_{yx}^K, \ \forall x, y \in S^K$$

holds for pairs of states x, y such that $y = x + \frac{1}{K}(\mathbf{e}_b - \mathbf{e}_a)$, $\bar{\omega}_x^K P_{xy}^K = \bar{\omega}_y^K P_{yx}^K$

Proof. For pairs of states x, y such that $y = x + \frac{1}{K}(\mathbf{e}_b - \mathbf{e}_a)$, denoted by $z = y - \frac{1}{K}\mathbf{e}_a = x - \frac{1}{K}\mathbf{e}_a$, z represents both the distribution of action of the other a-player when the Markov state is x, and the distribution of opponents of a b-player at state y. Thus, in both cases, a player who is revising his utility by omitting himself will consider the utility vector $\bar{F}_a^K(z) = F_a^K(z + \frac{1}{K}\mathbf{e}_a)$ where z satisfies $\sum_a z_a = \frac{K-1}{K}$. Thus

$$\phi^K(x) - \phi^K(y) = F_a^K x - F_b^K x = F_a^K\left(z + \frac{1}{K}\mathbf{e}_a\right) - F_b^K\left(z + \frac{1}{K}\mathbf{e}_b\right)$$

$$\bar{\omega}_x^K P_{xy}^K = \frac{1}{\kappa^K} \binom{K}{Kx_1, Kx_2, \ldots, Kx_m} e^{\frac{1}{\epsilon}\phi^K(x)} x_a$$

$$\times \frac{e^{\frac{1}{\epsilon}F_b^K\left(x + \frac{1}{K}(\mathbf{e}_b - \mathbf{e}_a)\right)}}{\sum_{b' \in A} e^{\frac{1}{\epsilon}F_{b'}^K(x + \frac{1}{K}(\mathbf{e}_{b'} - \mathbf{e}_a))}}$$

$$= \frac{1}{\kappa^K} \binom{K-1}{Kz_1, nz_2, \ldots, nz_m} e^{\frac{1}{\epsilon}[\phi^K(y) + \bar{F}_b^K(z) - \bar{F}_a^K(z)]}$$

$$\times \frac{e^{\frac{1}{\epsilon}\left[F_a^K\left(x + \frac{1}{K}(\mathbf{e}_a - \mathbf{e}_b)\right) - \bar{F}_b^K(z) + \bar{F}_a^K(z)\right]}}{\sum_{b' \in A} e^{\frac{1}{\epsilon}F_{b'}^K(x + \frac{1}{K}(\mathbf{e}_{b'} - \mathbf{e}_a))}}$$

$$= \frac{1}{\kappa^K} \binom{K}{Ky_1, ny_2, \ldots, ny_m} y_b e^{\frac{1}{\epsilon}[\phi^K(y) + \bar{F}_b^K(z) - \bar{F}_a^K(z)]}$$

$$\times \frac{e^{\frac{1}{\epsilon}\left[F_a^K\left(x + \frac{1}{K}(\mathbf{e}_a - \mathbf{e}_b)\right) - \bar{F}_b^K(z) + \bar{F}_a^K(z)\right]}}{\sum_{b' \in A} e^{\frac{1}{\epsilon}F_{b'}^K(x + \frac{1}{K}(\mathbf{e}_{b'} - \mathbf{e}_a))}}$$

$$= \frac{1}{\kappa^K} \binom{K}{Ky_1, ny_2, \ldots, ny_m} y_b e^{\frac{1}{\epsilon}[\phi^K(y)]}$$

$$\times \frac{e^{\frac{1}{\epsilon}\left[F_a^K\left(x + \frac{1}{K}(\mathbf{e}_a - \mathbf{e}_b)\right)\right]}}{\sum_{b' \in A} e^{\frac{1}{\epsilon}F_{b'}^K\left(x + \frac{1}{K}(\mathbf{e}_{b'} - \mathbf{e}_a)\right)}}$$

$$= \bar{\omega}_y^K P_{yx}^K$$

Using Boltzmann-Gibbs learning combined with imitation dynamics with at least one player on each action, Sandholm (2011) shows that the stationary distribution is of order:

$$\frac{e^{\phi^K(\underline{x})/\epsilon}}{\sum_{\underline{x}\in S^K} e^{\phi^K(\underline{x})/\epsilon}}.$$

Proposition 152

The convergence time to a η-equilibrium under the Boltzmann-Gibbs learning is of order $K \log\log(K) + \log(\frac{1}{\eta})$ if the rationality level is sufficiently large.

Proof. The proof uses logarithmic Sobolev inequalities for Markov chains on finite spaces, which shows rates of convergence of Markov chains on finite state spaces to their stationary distributions. We combine this result with the fact that the stationary distribution is an ϵ-best response. Following Diaconis and Saloff-Coste (1996), the total variation of:

$$\| H_t^K - \bar{\omega}^K \|_{TV} = \sum_{A \subseteq S^K} |H_t^K(A) - \bar{\omega}^K(A)| = \frac{1}{2} \sum_{\underline{x}\in S^K} |H_t^K\underline{x} - \bar{\omega}_{\underline{x}}^K|,$$

$$\| H_t^K - \bar{\omega}^K \|_{TV} \leq e^{-c}$$

for $t \geq \frac{1}{\lambda}(\log(\log(\frac{1}{\min_x \bar{\omega}_x^K})) + c)$ where $1 - \lambda$ is the largest eigenvalue of the operator $\frac{P^K + {}^t(P^K)}{2}$ which is greater than $\frac{c_1}{K}e^{-\frac{1}{\epsilon}}$. The stationary distribution $\bar{\omega}^K$ generates an $\frac{\eta}{2}$-equilibrium if ϵ is sufficiently small. It suffices to take:

$$\epsilon \leq \frac{\eta}{2m\log m(M+1)}, \quad M = \sup_{\underline{x}} \phi^K \underline{x}.$$

This completes the proof.

Example 153

A K-player congestion game is a game in which each player's strategy consists of a set of resources, and the cost of the set of strategy depends only on the number of players using each resource, i.e., the cost takes form $c_e(\xi_e(\underline{a}))$, where $\xi_e(\underline{a})$ is the number of players using resource e, and c_e is a non-negative increasing function. A standard example is a network congestion game on a directed graph (e.g., road or transportation network (Beckmann et al., 1956)), in which each player must select a path from some source to some destination, and each edge has an associated delay function that increases with the number of players using it. It is well known that this game admits the potential function $\phi^K(\xi) = -\sum_e \int_0^{\xi_e(\underline{a})} c_e(y)\, dy$. Applying the Boltzmann-Gibbs learning algorithms, one can use the result corresponding to Proposition 152. The convergence time to be η-close to the potential maximizer is in the order of $K \log\log K + \log(\frac{1}{\eta})$ for any initial condition for a sufficiently large level of rationality.

5.5 GAMES WITH COST OF LEARNING

In this section we present an algorithm called "Cost-to-Learn". The purpose of this section is to show that it can be very costly to learn quickly and learning can take some time. When a player changes its action, there is an associated cost. We can think of the cost to acquire a new technology, or to produce a specific product. We illustrate this in a two-player case.

Consider the two-player game with cost functions for player one $\tilde{u}_1(a_1, a_2) = \underline{f}_1(a_1) + \underline{\beta}_1 \underline{c}(a_1, a_2)$, $\tilde{u}_2(a_1, a_2) = \underline{f}_2(a_2) + \underline{\beta}_2 \underline{c}(a_1, a_2)$. The coupling term defines the more or less competitive part of the game. The coefficients $\{\underline{\beta}_j\}_j$ indicate how much a player interacts with the other player. In addition, we assume that there is a cost of learning, and a cost to move from one action to another. Player 1 has a cost of learning of $\underline{\alpha}_1 \underline{h}_1(a_1, a_1')$. Similarly, player 2 has to pay $\underline{\alpha}_2 \underline{h}_2(a_2, a_2')$ to change his action from a_2 to a_2'. The motivations behind this *"cost of learning"* approach is that, in many situations, changing, improving performance, or guaranteeing a quality of service has cost. If only player 1 moves, his objective function becomes $\tilde{u}_1(a_1', a_2) + \underline{\alpha}_1 \underline{h}_1(a_1, a_1')$. Similarly if player 2 is the only mover he/she gets $\tilde{u}_2(a_1, a_2') + \underline{\alpha}_2 \underline{h}_2(a_2, a_2')$. Assume that $\underline{h}_j(x, x) = 0$, $j \in \{1, 2\}$ (no cost to learn if the action remains the same).

We now construct a sequence of actions with a cost to move.

- Start with $(a_{1,0}, a_{2,0})$.
- Denote by $(a_{1,t}, a_{2,t})$ an action profile at stage t, in which there are two stages of a move: player 1 moves from $a_{1,t}$ to $a_{1,t+1}$ by solving the following problem:

$$a_{1,t+1} \in \arg\min_{a_1' \in \mathcal{A}_1} \left\{ \underline{f}_1(a_1') + \underline{\beta}_{1,t} \underline{c}(a_1', a_{2,t}) + \underline{\alpha}_1(t) \underline{h}_1(a_{1,t}, a_1') \right\},$$

and then given $a_{1,t+1}$, Player 2 moves from $a_{2,t}$ to $a_{2,t+1}$ by solving the following problem:

$$a_{2,t+1} \in \arg\min_{a_2' \in \mathcal{A}_2} \left\{ \underline{f}_2(a_2') + \underline{\beta}_{2,t} \underline{c}(a_{1,t+1}, a_2') + \underline{\alpha}_{2,t} \underline{h}_2(a_{2,t}, a_2') \right\}$$

The learning algorithm *Cost-to-Learn* is then given by:

$$\begin{cases} a_{1,t+1} \in \arg\min_{a_1' \in \mathcal{A}_1} \left\{ \underline{f}_1(a_1') + \underline{\beta}_{1,t} \underline{c}(a_1', x_2(t)) + \underline{\alpha}_{1,t} \underline{h}_1(a_{1,t}, a_1') \right\} \\ a_{2,t+1} \in \arg\min_{a_2' \in \mathcal{A}_2} \left\{ \underline{f}_2(a_2') + \underline{\beta}_{2,t} \underline{c}(a_{1,t+1}, a_2') + \underline{\alpha}_{2,t} \underline{h}_2(a_{2,t}, a_2') \right\} \end{cases}$$

Definition 154

A mapping $f: A \longrightarrow \mathbb{R} \cup \{+\infty\}$ is lower semi-continuous at a_0 if $\liminf_{x \longrightarrow a_0} f(\underline{a}) \geq f(a_0)$. f is lower-semi-continuous if it is lower semi-continuous at every point of its domain.

This is equivalent to saying that $\{\underline{a} \in A \mid f(\underline{a}) > \alpha\}$ is an open set for every $\alpha \in \mathbb{R}$. Assume that the mapping $(f_i)_i$ is proper, bounded below and lower semi-continuous.

Lemma 7

If the algorithm *Cost-to-Learn* converges and if $(\underline{\alpha}_{1,t}, \underline{\alpha}_{2,t}) \longrightarrow (\underline{\alpha}_1, \underline{\alpha}_2)$, $(\underline{\beta}_{1,t}, \underline{\beta}_{2,t}) \longrightarrow (\underline{\beta}_1, \underline{\beta}_2)$ when t goes to infinity then the limit (a_1^*, a_2^*) of any sub-sequence $(a_{1,\phi(t)}, a_{2,\phi(t)})$ generated by the algorithm leads to an equilibrium of the game with cost of learning.

Proof. By definition of $a_{1,t+1}$, one has $\underline{f}_1(a_1') + \underline{\beta}_{1,t}\underline{c}(a_1', a_{2,t}) + \underline{\alpha}_{1,t}\underline{h}_1(a_{1,t}, a_1') \geq \underline{f}_1(a_{1,t+1}) + \underline{\beta}_{1,t}\underline{c}(a_{1,t+1}, a_{2,t}) + \underline{\alpha}_{1,t}\underline{h}_1(a_{1,t}, a_{1,t+1})$, $\forall a_1'$. The first term of the inequality goes to $\underline{f}_1(a_1') + \underline{\beta}_1\underline{c}(a_1', a_2^*) + \underline{\alpha}_1\underline{h}_1(a_1^*, a_1')$ when t goes to infinity. By taking lim inf, the second term gives:

$$\geq \liminf_t \left[\underline{f}_1(a_{1,t+1}) + \underline{\beta}_{1,t}\underline{c}(a_{1,t+1}, a_{2,t}) + \underline{\alpha}_{1,t}\underline{h}_1(a_{1,t}, a_{1,t+1}) \right]$$

$$\geq \liminf_t \underline{f}_1(a_{1,t+1}) + \liminf_t \left[\underline{\beta}_{1,t}\underline{c}(a_{1,t+1}, a_{2,t}) \right]$$

$$+ \left[\underline{\alpha}_{1,t}\underline{h}_1(a_{1,t}, a_{1,t+1}) \right]$$

$$\geq \underline{f}_1(a_1^*) + \left[\underline{\beta}_1\underline{c}(a_1^*, a_2^*) \right] + \left[\underline{\alpha}_1\underline{h}_1(a_1^*, a_1^*) \right]$$

$$\geq \underline{f}_1(a_1^*) + \underline{\beta}_1\underline{c}(a_1^*, a_2^*)$$

This means that $\forall a_1'$, $\underline{f}_1(a_1') + \underline{\beta}_1\underline{c}(a_1', a_2^*) + \underline{\alpha}_1\underline{h}_1(a_1^*, a_1') \geq \underline{f}_1(a_1^*) + \underline{\beta}_1\underline{c}(a_1^*, a_2^*)$. Similarly, $\forall a_2'$, $\underline{f}_2(a_2') + \underline{\beta}_2\underline{c}(a_1^*, a_2') + \underline{\alpha}_2\underline{h}_2(a_2^*, a_2') \geq \underline{f}_2(a_2^*) + \underline{\beta}_2\underline{c}(a_1^*, a_2^*)$. This implies that (a_1^*, a_2^*) is an equilibrium point.

Theorem 155: Convergence

Assume that $\underline{h}_j$, $j \in \{1,2\}$ are positive coefficients and all the coefficients $\underline{\alpha}_{j,t}$ and $\underline{\beta}_j(t)$ are positive, and $\underline{\beta}_{1,t} = \underline{\beta}_{2,t} = \underline{\beta}(t)$ is non-increasing, and $\sum_j \inf_{A_j} \underline{f}_j > -\infty$; then, for any sequence of $(a_{1,t}, a_{2,t})_t$ generated by the *Cost-to-Learn* algorithm, one has:

- $\left\{ \xi_t := \underline{f}_1(a_{1,t}) + \underline{f}_2(a_{2,t}) + \underline{\beta}(t)\underline{c}(a_{1,t}, a_{2,t}) \right\}_t$ is non-increasing and has a limit when t goes to infinity.
- The partial sum of the series defined by the sequence of cost of learning, i.e.:

$$\sum_{t=0}^{T} \left[\underline{\alpha}_{1,t}\underline{h}_1(a_{1,t}, a_{1,t+1}) + \underline{\alpha}_{2,t}\underline{h}_2(a_{2,t}, a_{2,t+1}) \right]$$

is convergent.

Proof. We write the two inequalities defined by the algorithm:

$$\underline{f}_1(a_{1,t+1}) + \underline{\beta}(t)\underline{c}(a_{1,t+1},a_{2,t}) + \underline{\alpha}_{1,t}\underline{h}_1(a_{1,t},a_{1,t+1})$$
$$\leq \underline{f}_1(a_{1,t}) + \underline{\beta}(t)\underline{c}(a_{1,t},a_{2,t}), \tag{5.18}$$
$$\underline{f}_2(a_{2,t+1}) + \underline{\beta}(t)\underline{c}(a_{1,t+1},a_{2,t+1}) + \underline{\alpha}_{2,t}\underline{h}_2(a_{2,t},a_{2,t+1})$$
$$\leq \underline{f}_2(a_{2,t}) + \underline{\beta}(t)\underline{c}(a_{1,t+1},a_{2,t}). \tag{5.19}$$

By adding the two inequalities, one has,

$$\underline{f}_1(a_{1,t+1}) + \underline{f}_2(a_{2,t+1}) + \underline{\beta}(t+1)\underline{c}(a_{1,t+1},a_{2,t+1})$$
$$\quad + \underline{\alpha}_{1,t}\underline{h}_1(a_{1,t},a_{1,t+1}) + \underline{\alpha}_{2,t}\underline{h}_2(a_{2,t},a_{2,t+1})$$
$$\leq \underline{f}_1(a_{1,t+1}) + \underline{f}_2(a_{2,t+1}) + \underline{\beta}(t)\underline{c}(a_{1,t+1},a_{2,t+1})$$
$$\quad + \underline{\alpha}_{1,t}\underline{h}_1(a_{1,t},a_{1,t+1}) + \underline{\alpha}_{2,t}\underline{h}_2(a_{2,t},a_{2,t+1})$$
$$\leq \underline{f}_1(a_{1,t}) + \underline{f}_2(a_{2,t}) + \underline{\beta}(t)\underline{c}(a_{1,t},a_{2,t})$$

This implies that the sequence $\left\{ \xi_t := \underline{f}_1(a_{1,t}) + \underline{f}_2(a_{2,t}) + \underline{\beta}(t)\underline{c}(a_{1,t},a_{2,t}) \right\}_t$ is non-increasing and has a limit. We now prove that the partial sum of the series defined by the sequence of cost of learning is convergent.

$$\xi_{T+1} - \xi_0 = \sum_{t=0}^{T} \left[\underline{\alpha}_{1,t}\underline{h}_1(a_{1,t},a_{1,t+1}) + \underline{\alpha}_{2,t}\underline{h}_2(a_{2,t},a_{2,t+1}) \right]$$
$$= \underline{f}_1(a_1(0)) + \underline{f}_2(a_2(0)) + \underline{\beta}(0)c(a_1(0),a_2(0))$$
$$\quad - \underline{f}_1(a_{1,t+1}) - \underline{f}_2(a_{2,t+1}) - \underline{\beta}(T+1)\underline{c}(a_{1,t+1},a_{2,t+1})$$
$$\leq \underline{f}_1(a_1(0)) + \underline{f}_2(a_2(0)) + \underline{\beta}(0)c(a_1(0),a_2(0))$$
$$\quad - \left[\sum_{j=1}^{2} \inf_{A_j} \underline{f}_j \right]$$

This completes the proof.

Note that the quantity $\underline{f}_1 + \underline{f}_2 + \underline{\beta}\underline{c}$ defines a potential function of the game without cost of learning.

5.6 LEARNING BARGAINING SOLUTIONS

In the wireless literature learning algorithms are known to often converge to inefficient solutions. However, these algorithms can also be used to learn efficient solutions. The goal of this section is to show how learning can be used to obtain a cooperative solution – the Nash bargaining solution (see Chapter 2 for a definition).

The bargaining solution is a natural framework that allows one to define and design fair assignments between players which will play the role of bargainers.

It is characterized by a set of axioms that are appealing in defining fairness, or by a maximization of log-concave function on the set of feasible utilities. In wireless networks, Nash bargaining is interesting since it can be seen as a natural extension of the proportional fairness criterion, which is probably the most popular fairness notion in networks. Users are faced with the problem of negotiating for a fair point in the set of feasible utilities. If no agreement can be reached by the users, the game outcome is the status quo and is given by the disagreement utilities. In contrast with most of the non-cooperative solutions (Nash equilibrium, Stackelberg solution, etc.), the Nash bargaining solution is known to be Pareto optimal. A fictitious play-based learning algorithm for dynamic Nash bargaining is presented below.

5.6.1 Fictitious Play-Based Learning for Bargaining Solutions

Consider a dynamic bargaining game with K players interacting infinitely many times. Time is discrete and the time space is $\mathbb{N}$. At each stage t

- each player j selects an action $a_j \in \mathcal{A}_j$;
- if the action profile $(a_j, \mathbf{a}_{-j})$ meets the constraints of set $\mathcal{C}$ then player j receives $\tilde{u}_j(a_j)$ where $\tilde{u}_j$ is a strictly increasing function, $\tilde{u}_j(0) = 0$;
- if the constraints are not satisfied, player j receives $\tilde{u}_j(0)$.

and the system goes to the next stage $t + 1$. The long-term utility of this dynamic game is defined as a discounted utility:

$$F_{j,\delta} = \sum_{t=1}^{\infty} \lambda(1 - \lambda)^{t-1} \tilde{u}_j(a_{j,t}) \, \mathbb{1}_{\{(a_{j,t}, \mathbf{a}_{-j,t}) \in \mathcal{C}\}} \qquad (5.20)$$

where λ is the discount factor.

5.6.2 Learning in Two-Player Dynamic Bargaining

We introduce a learning rule to play this dynamic bargaining game using Nash bargaining formulation. We consider strategic Nash bargaining described as follows:

- Two players $1, 2$. At $t = 0$, players bargain for the expected discounted utility.
- At time $t > 0$, each player j reacts to the frequency of use of action of the other player. From time $t = 2$, both players use a simple learning rule, and make their decisions according to fictitious play.
- The discounted utility is:

$$F_{\lambda,1} = \sum_{t=1}^{\infty} \lambda(1 - \lambda)^{t-1} \tilde{u}_1(a_{1,t}) \, \mathbb{1}_{(a_{1,t}, a_{2,t}) \in \mathcal{C}},$$

$$F_{\lambda,2} = \sum_{t=1}^{\infty} \lambda(1 - \lambda)^{t-1} \tilde{u}_2(a_{2,t}) \, \mathbb{1}_{(a_{1,t}, a_{2,t}) \in \mathcal{C}}$$

- At time $t > 0$, each player j reacts to the frequency of use of action of the other player.

- Denote $\bar{f}_{j,t}$ as the relative mean use of actions by player j up to $t-1$.
- According to fictitious play, players choose, $a_{1,t}$ (resp. $a_{2,t}$):

$$a_{1,t} \in \arg\max_{a_1} \sum_{a_2, \bar{f}_{2,t}(a_2)>0} \bar{f}_{2,t}(a_2)\tilde{u}_1(a_1)\mathbb{1}_{\mathcal{C}}(a_1,a_2)$$

- That is, in each period, each player chooses his best response to the observed empirical frequency of his opponent's choices.

We analyze the fictitious play outcome for the constraint set:

$$\mathcal{C} = \{a = (a_1,a_2) \mid 0 \le a_1 \le \bar{a}_1,\ 0 \le a_2 \le \bar{a}_2,\ 0 \le a_1 + a_2 \le v(\{1,2\}),$$

with $\bar{a}_1 + \bar{a}_2 > v(\{1,2\}), \bar{a}_j < v(\{1,2\})$.

Theorem 156

- For any $(a_{1,1},a_{2,1}) \in \mathcal{C}$ the following holds: (i) $a_{1,2} = v(\{1,2\}) - a_{2,1}$, (ii) $a_{2,2} = v(\{1,2\}) - a_{1,1}$
- For any $t \ge 3$, $a_{1,t}$ must be either $a_{1,1}$ or $(v(\{1,2\}) - a_{2,1})$, and $a_{2,t} \in \{a_{2,1}, v(\{1,2\}) - a_{1,1}\}$
- $\bar{f}_{1,t}(a_{1,1}) + \bar{f}_{1,t}(v(\{1,2\}) - a_{2,1}) = 1$, $\bar{f}_{2,t}(a_{2,1}) + \bar{f}_{2,t}(v(\{1,2\}) - a_{1,1}) = 1$
- $(a_1,a_2) \in \arg\max_{a_1'} [\tilde{u}_1(a_1')]^{w_1} (\tilde{u}_2(v(\{1,2\}) - a_1'))^{w_2}$, $a_2 = \phi(a_1)$, ϕ defines the Pareto frontier and satisfies the implicit equation:

$$[\tilde{u}_1(a_1)]^{w_1} [\tilde{u}_2(v(\{1,2\}) - a_1)]^{w_2} = [\tilde{u}_1(v(\{1,2\}) - \phi(a_1))]^{w_1} [\tilde{u}_2(\phi(a_1))]^{w_2}$$

Example 157: Rate Control in AWGN Channel

We consider a system consisting of one receiver and its uplink additive white Gaussian noise (AWGN) multiple access channel with two senders. The signal at the receiver is given by $Y = \xi + \sum_{j=1}^m h_j X_j$ where X_j is the transmitted signal of user j, and ξ is zero-mean Gaussian noise with variance N_0. Each user has an individual power constraint $\mathbb{E}(X_j^2) \le p_j$. The optimal power allocation scheme for Shannon capacity is to transmit at the maximum power available, i.e., p_j, for each user. Hence, we consider the case in which maximum power is used. The decisions of the users then consist of choosing their communication rates, and the receiver's role is to decode, if possible. The capacity region is a set of all vectors $\mathbf{a} \in \mathbb{R}_+^2$ such that senders $j \in \{1,2\}$ can reliably communicate at rate a_j, $j \in \{1,2\}$. The capacity region $\mathcal{C}$ for this channel is the set:

$$\mathcal{C} = \left\{ \mathbf{a} \in \mathbb{R}_+^2 \ \middle| \ \sum_{j \in J} a_j \le \log\left(1 + \frac{\sum_{j \in J} p_j w_j}{N_0}\right), \ \forall \emptyset \subsetneq J \subseteq \{1,2\} \right\} \qquad (5.21)$$

where $w_j = |h_j|^2$. Let $\bar{a}_1 = \log\left(1 + \frac{p_1 w_1}{N_0}\right)$, $\bar{a}_2 = \log\left(1 + \frac{p_2 w_2}{N_0}\right)$, and $v(\{1,2\}) = \log\left(1 + \sum_{j \in \{1,2\}} \frac{p_j w_j}{N_0}\right)$. Under the constraint $\mathcal{C}$, if sender j wants to communicate at a higher rate, one of the other senders has to lower his rate; otherwise, the capacity constraint is violated. Let:

$$r_j = \log\left(1 + \frac{p_j w_j}{N_0 + \sum_{j' \in \{1,2\}, j' \ne j} p_{j'} w_{j'}}\right) \qquad (5.22)$$

denote the transmission rate of a sender when the signals of the other sender are treated as noise.

The strategic bargaining game is given by $\left(\mathcal{K} = \{1,2\}, \mathcal{C}, (\mathcal{A}_j)_{j \in \mathcal{K}}, (\tilde{u}_j)_{j \in \mathcal{K}} \right)$, where the set of senders $\mathcal{K}$ is also the set of players; $\mathcal{A}_j, j \in \mathcal{K}$, is the set of actions; $\mathcal{C}$ is the bargaining constraint set; and $\tilde{u}_j, j \in \mathcal{K}$, are the utility functions. We define $\tilde{u}_j : \prod_{j \in \mathcal{K}} \mathcal{A}_j \longrightarrow \mathbb{R}_+$ as follows:

$$\tilde{u}_j(a_j, a_{-j}) = \mathbb{1}_{\mathcal{C}}(\alpha)\tilde{g}_j(a_j) \qquad (5.23)$$

$$= \begin{cases} \tilde{g}_j(a_j) & \text{if } (a_j, a_{-j}) \in \mathcal{C} \\ 0 & \text{otherwise} \end{cases}, \qquad (5.24)$$

where $\tilde{g}_j : \mathbb{R}_+ \longrightarrow \mathbb{R}_+$ is a positive and strictly increasing function.

Using the fictitious play learning algorithm and theorem 156, one can deduce the following results:

- For any $(a_{1,1}, a_{2,1}) \in \mathcal{C}$ the following holds: The "optimal" rates are the first iteration at the maximal face of rate region $\mathcal{C}$, i.e., (i) $a_{1,2} = v(\{1,2\}) - a_{2,1} \geq r_1$, (ii) $a_{2,2} = v(\{1,2\}) - a_{1,1} \geq r_1$.
- Starting the third step of the fictitious play algorithm, the rates chosen by sender one alternating between $a_{1,t}$ must be either $a_{1,1}$ or $(v(\{1,2\}) - a_{2,1})$, and similarly for sender 2: $a_{2,t} \in \{a_{2,1}, v(\{1,2\}) - a_{1,1}\}$.
- The frequencies of use of the rate satisfy $\bar{f}_{1,t}(a_{1,1}) + \bar{f}_{1,t}(v(\{1,2\}) - a_{2,1}) = 1, \bar{f}_{2,t}(a_{2,1}) + \bar{f}_{2,t}(v(\{1,2\}) - a_{1,1}) = 1$. This is equivalent to the time sharing solution between the two boundary points $(a_{1,1}, v(\{1,2\}) - a_{1,1})$ and $(v(\{1,2\}) - a_{2,1}, a_{2,1})$. Note that this solution is known to be a maxmin fair solution and is Pareto optimal. The corner points $(r_1, \bar{a}_2)$ and $(\bar{a}_1, r_2)$ are particular Pareto optimal solutions.

5.7 SUMMARY AND CONCLUDING REMARKS

While game theory can help to understand and model the behavior of players in interactive situations, a fundamental question about the standard methods of game-theoretic prediction remains on how to conform in an equilibrium. Why should we expect players' behavior to conform with some notions of equilibrium play? The traditional approach to this problem is to ask whether players can arrive at equilibrium under epistemic conditions or through a purely adaptive and learning process with observation capabilities. In this chapter, we have presented various learning schemes that are partially distributed; that is, a player knows some information about himself but also about the other players. Here is a short list of key points to keep in mind concerning the learning schemes that were presented.

- The learning algorithms presented in Chapters 5 and 6 have been classified into **partially and fully** distributed algorithms. Although this classification is not much used in the literature of learning, it is relevant in wireless networks, where distinguishing between individual channel state information and global channel state information is generally important. If each player implements his updating rule after explicitly receiving signals or some data about the other players, the learning algorithm is said to be partially distributed, and fully distributed otherwise. Note that the frontier is not always clear. For instance, if a player is fed back with his own SINR and knows his individual channel state information, he has some information about the others.

- All learning algorithms presented in this chapter are partially distributed. Going further with classifying learning algorithms, one may distinguish those where the **performance metric** is known and considered by the player, from those where the performance metric is not known or even considered by the player. In this chapter, the utility function was always known to every player and all players were **rational** since they aimed to maximize it. In Chapter 6, where reinforcement algorithms are considered, players do not evaluate their performance but rather respond to utility stimuli according to a fixed rule (like an automaton).

- As with many learning algorithms in games, the learning algorithms presented do not in general converge[3]. **Convergence** is typically guaranteed in some special classes of games. As far as wireless networks are concerned, the need for convergence is not always obvious. It may be that the average performance of a convergent algorithm which operates over a short time window is worse than a non-convergent algorithm. Convergence is therefore useful when the players have the time to converge, and when predicting the steady state is important (e.g., because the corresponding performance has to be predicted). Convergence rate/speed is therefore another aspect to be considered.

- Some learning schemes can be used in scenarios where the **action or strategy spaces** are continuous (compact typically) while others are only defined for discrete action sets. Most learning schemes are designed for discrete action sets, which allows one to address quite a large number of practical wireless problems (channel selection problems, modulation-coding schemes, power control with a finite number of power levels, etc.). On the other hand, the best response dynamics (and therefore fictitious play-based algorithms) can be applied to both discrete and continuous action sets; in this respect, the famous iterative water-filling algorithm assumes compact action spaces. In Chapter 9, another type of learning technique (based on the replicator dynamics) working for compact actions spaces will be presented.

- The preceding items indicate that among the features to be considered for a learning algorithm we find: information assumptions, observation assumptions, behavior assumptions, convergence properties, and types of action spaces. Let us summarize what we have learned about the learning algorithms that have been considered, concerning these issues.

  - In its standard version, the best response dynamics assume that every player can observe at each stage the actions/strategies played by the others. Players are assumed to know their utility function and to be rational, and

---

[3]Note that what is meant by convergence has to be specified. Indeed, in addition to the fact that different types of convergence may be considered (almost sure convergence, convergence in distribution, etc.), different types of quantities may be considered (actions, probability distributions, utilities, etc.)

action/strategy spaces can be discrete or continuous. The ABRD converges to a pure Nash equilibrium in weakly acyclic games and in potential games. Although no results have been provided here, the BRD generally converges very rapidly, in comparison with reinforcement algorithms (more will be said about this in Chapter 6).

- Fictitious, play-based algorithms are built from the BRD idea, but empirical frequencies are considered instead. Therefore, it is not surprising that FP assumes that: every player can observe the actions of the others, every player knows his (expected) utility, and every player is rational. Standard and Brown FP algorithm converge to a Nash equilibrium in potential games but never converge in games with cycles (such as the Rock-Paper-Scissors game). The presented modified version of FP exploits the knowledge of the joint empirical frequency of the action profiles, leading to a stronger result: coarse correlated equilibria can be reached after convergence.

- A partially distributed version of the logit or Boltzmann-Gibbs learning algorithm (also called smoothed FP, e.g., see Young (2004)) has been presented. Players have discrete actions spaces and need to know their own utility function and the strategies played by the others. However, they do want to maximize their utility functions, which is why they are called players with bounded rationality. Rather, the BG distribution amounts to maximizing a weighted sum of the utility and the entropy, the objective being to explore the action space. The latter aspect may be important to avoid cycles and therefore non-convergence. When convergent, the Boltzmann-Gibbs learning algorithm converges to a logit equilibrium, which is an ϵ-Nash equilibrium.

- Learning algorithms can also be used to learn efficient game outcomes. We have shown how FP can be exploited to learn a Nash bargaining solution.

The other leading approach used to justify equilibrium play is to ask whether players can learn equilibrium strategies during the course of a dynamical interaction process under limited observation capabilities: typically, each player is only assumed to observe a numerical value of his own utility function. This leads to fully distributed learning algorithms, which is the purpose of the next chapter.

Fully Distributed Learning Algorithms

6.1 INTRODUCTION

In this chapter, the focus is on fully distributed learning algorithms for stochastic games with incomplete information and independent state transitions. As mentioned in the preceding chapter, among the features to be considered for a learning algorithm, we find information assumptions, observation assumptions, behavior assumptions, convergence properties, and types of action spaces. One of the reasons for considering fully distributed learning algorithms, in comparison with partially distributed learning schemes, is to assume that the players use less information about the other players and less memory in terms of game history. The learning schemes considered in this chapter are: trial-and-error learning algorithms, reinforcement learning algorithms, regret matching-based learning, (fully distributed) Boltzmann-Gibbs learning algorithms, evolutionary dynamics based learning in heterogeneous population games, and learning algorithms for satisfaction solutions. There are many reasons for selecting these learning algorithms; here are a few of them.

- Trial-and-error learning algorithms correspond to a quite simple scheme which may be easily imagined by a wireless designer who does not know about the literature on learning.
- Most of them show that there is no need to assume that the players are rational (in the sense of a given performance metric) and rationality is common knowledge. For instance, when implementing the described reinforcement learning algorithms, a player updates his learning rule as an automaton without maximizing a given performance metric. Of course, depending on how the automaton is programmed, this may amount to maximizing a certain long-term utility function but in any case, the latter is not known to the player. One important message when using such learning schemes is that Nash equilibria can be the final result of multiple interactions between learning automata.
- Learning algorithms can converge to various types of steady states; the Nash equilibrium is just one of them. Regret-matching learning algorithms, for which a performance criterion is explicitly considered (namely, the regret), may converge to correlated equilibria. Another type of steady state considered in this chapter is satisfaction solutions or satisfaction equilibria (Perlaza et al., 2010). This solution concept is especially relevant in wireless communications where a certain quality

of service (QoS) target has to be reached, rather than of maximizing a certain utility function.

- Most of these algorithms rely on observing the realization or instantaneous numerical value of the individual utility function, which is less restrictive than knowing the function itself. Additionally, noisy observations can be easily accommodated.
- As evolutionary dynamics based learning in heterogeneous population games, the case where players implement different learning algorithms can be easily analyzed if the number of players is large.

Before describing the aforementioned algorithms, we will first show that the problem of fully distributed learning can be formulated as a particular stochastic game. In the previous chapter, the strategic form associated with the considered learning schemes was rather obvious, but when automata are considered, formulating the problem as a game is not always obvious. This is why we have separated fully distributed learning and stochastic games. Considering the evolution of the literature on the problem of playing/learning strategies in repeated interactive situations indicates that fully distributed learning schemes are typically based on mild information and behavior assumptions, but generally lead to inefficient outcomes and take some time to converge. On the other hand, models which assume that players have much more information about the structure of the interactive situation may lead to efficient solutions: the repeated game model with perfect monitoring (see Chapter 3) is a special case of dynamic games for which our observation is well illustrated. Since there are some efforts to improve the efficiency of the steady states of learning schemes by adding more priors and at the same time, there are some efforts to find theorems (e.g., Folk theorems) for repeated/stochastic games by relaxing some information/observation assumptions, the question is: Given the fact that eventually both approaches are about repeating interactive situations, and also using automatons as players, is it possible to unify both approaches? This question also explains why we want to formulate fully distributed learning problems as games.

The chapter is organized as follows. In Section 6.2 we present the general game model for which fully distributed learning is developed. Then we focus on learning by experimentation and trial and error in Section 6.3. After that in Section 6.4 we present reinforcement learning in robust games. Section 6.5 develops regret minimization based learning for correlated equilibrium distribution, Section 6.6 presents a specific class of reinforcement learning based on Boltzmann-Gibbs distribution developed for games with vector of utilities, and Section 6.7 presents the case of heterogeneous games in large populations. Learning algorithms for satisfaction solutions are provided in Section 6.8. The chapter uses some stochastic approximation techniques as provided in the Appendices.

6.2 THE GENERAL GAME-THEORETIC SETTING

As mentioned in the previous section, our current goal is to show that the learning problems addressed in this chapter can also be formulated as a game (like partially distributed learning schemes which were presented in Chapter 5). More precisely, they can be described by a particular model of stochastic games (as seen in Chapter 3). We consider a system with a maximum of K potential players. Each player j has a finite number of actions (pure strategies), denoted by A_j. The utility of player j is a random variable $U_{j,t} \in \mathbb{R}$ with realized value denoted by $u_{j,t} \in \mathbb{R}$ at time t. The random variable is indexed by $\underline{w}$ in some set $\mathcal{W} \subseteq \mathbb{R}^d$, $d \in \mathbb{N}, d \geq 1$, which is referred to as the state of the game. The state at time t is $\underline{w}_t \in \mathcal{W}$. We assume that the players do not know the state of the game, and that they do not know the expression of their utility functions. At each time t, each player chooses an action and receives a numerically noisy value of his utility (the perceived utility) at that time. The perceived utility can be interpreted as a measured metric in the random environment. We assume that the transition probabilities between the states are independent and unknown to the players. The game can be viewed and described as a stochastic game with incomplete information, which is given by:

$$\mathcal{G} = \left(\mathcal{K}, (A_j)_{j \in \mathcal{K}}, \mathcal{W}, (U_j(\underline{w}, .))_{j \in \mathcal{K}, \underline{w} \in \mathcal{W}} \right) \tag{6.1}$$

where:

- $\mathcal{K} = \{1, 2, \ldots, K\}$ denotes the set of potential players. This set is unknown to the players.
- for $j \in \mathcal{K}$, A_j is the set of all the possible actions of player j. In each state, the same action spaces are available. Player j does not know the action spaces of the other players.
- none of the players has any information about the current state. Indeed, we assume that the state space $\mathcal{W}$ and the probability transition on the states are both *unknown* to the players. The state is represented by independent random variable (the transitions between the states are independent of the chosen actions). We assume that the current state w_t is unknown by the players.
- Utility functions are unknown: $U_j : \mathcal{W} \times \prod_{j' \in \mathcal{K}} \mathcal{X}_{j'} \longrightarrow \mathbb{R}$, where $\mathcal{X}_j := \Delta(A_j)$ is the set of probability distributions over A_j i.e.,

$$\mathcal{X}_j := \left\{ \underline{x}_j : x_j(a_j) \in [0, 1], \sum_{a_j \in A_j} x_j(a_j) = 1 \right\} \tag{6.2}$$

- We denote the stage game by $\mathcal{G}(\underline{w}_t) = \left(\mathcal{K}, (A_j)_{j \in \mathcal{K}}, (U_j(\underline{w}_t, .))_{j \in \mathcal{K}} \right)$.

6.2.1 Description of the Dynamic Game

The dynamic game is described as follows.

- At stage $t = 1$, each player j chooses an action $a_{j,1} \in A_j$ and perceives a numerical value which corresponds to a realization of the random variables depending on the actions of the other players, the state of nature, etc. He initializes his estimate of his utility vector to $\hat{\underline{u}}_{j,1}$.
- At stage t, each player $j \in \mathcal{K}$ chooses an action based on his own experience. Then, player j receives a feedback $u_{j,t}$. Based on $u_{j,t}$, player j updates his estimation

vector $\hat{u}_{j,t}$ and builds a strategy $\underline{x}_{j,t+1}$ for the next stage. The strategy $\underline{x}_{j,t+1}$ is a function only of $\underline{x}_{j,t}$, $\hat{u}_{j,t}$ and the $u_{j,t}$. Note that the exact value of the state of nature $\underline{w}_t$ at time t and the past strategies $\underline{x}_{-j,t-1} := (\underline{x}_{k,t-1})_{k\neq j}$ of the other players and their past utilities $\mathbf{u}_{-j,t-1} := (u_{k,t-1})_{k\neq j}$ are unknown to player j at time t. The game moves to $t+1$.

Remark This model is clearly a stochastic game (Shapley, 1953b) with incomplete information and independent state transitions. It includes some interesting well-studied learning problems. A basic example is the class of matrix games with i.i.d. random entries in the form $\tilde{U}(\underline{a}) = \tilde{D}(\underline{a}) + \tilde{S}$ (a deterministic part and a stochastic part with $\mathbb{E}(\tilde{S}) = 0$.)

6.2.1.1 *Private Histories*

As seen in Chapter 3, the (private) history of player j at stage t comprises all observations made up to stage t. As each player is assumed to knows his own actions only and observe the realizations of his utility, his history is merely given by:

$$h_{j,t} = (a_{j,1}, u_{j,1}, a_{j,2}, u_{j,2}, \ldots, a_{j,t-1}, u_{j,t-1}), \tag{6.3}$$

which belongs to the set of private histories of player j at stage t:

$$\mathcal{H}_{j,t} := (\mathcal{A}_j \times \mathbb{R})^{t-1}. \tag{6.4}$$

6.2.1.2 *Behavioral Strategy*

A behavioral strategy for player j is a sequence of mappings $(\tilde{\tau}_{j,t})_{t\geq 0}$ with:

$$\tilde{\tau}_{j,t} : \begin{vmatrix} \mathcal{H}_{i,t} & \longrightarrow & \Delta(\mathcal{A}_j) \\ \underline{h}_{j,t} & \longmapsto & x_j(t). \end{vmatrix} \tag{6.5}$$

The set of behavioral strategies of player j is denoted by Σ_j. The set of complete histories of the dynamic game after t stages is $\mathcal{H}_t = (\mathcal{W} \times \prod_j \mathcal{A}_j \times \mathbb{R}^K)^{t-1}$; it describes the set of active players, the states, the chosen actions and the received utilities for all the players at all past stages before t. A strategy profile $\tilde{\tau} = (\tilde{\tau}_j)_{j\in\mathcal{K}} \in \prod_j \Sigma_j$ and an initial state $\underline{w} \in \mathcal{W}$ induce a probability distribution $P_{\underline{w},\tilde{\tau}}$ on the set of plays $\mathcal{H}_\infty = (\mathcal{W} \times \prod_j \mathcal{A}_j \times \mathbb{R}^K)^\infty$. Given an initial state $\underline{w}$ and a strategy profile $\tilde{\tau}$, the utility of player j is the superior limit of the Cesaro-mean utility $\mathbb{E}_{\underline{w},\tilde{\tau}} v_{j,T} = \mathbb{E}\left(\frac{1}{T}\sum_{t=1}^{T} u_{j,t}\right)$. We assume that $\mathbb{E}_{\underline{w},\tilde{\tau}} v_{j,T}$ has a limit and can be expressed in terms of the stationary distribution as $u_j(\underline{x}_j, \underline{x}_{-j}) = \mathbb{E}_{\underline{w},\underline{x}} \tilde{U}_j(\underline{w}, \underline{a})$. The main idea to understand here is that we associate a static game with the dynamic game. A concrete consequence of this is that, under appropriate assumptions, the limiting player's behaviors may be given by the Nash equilibria of the associated static game. We call this game the expected robust game, and define it as follows.

Definition 158: Expected Robust Game

We define the expected robust game as:

$$\bar{\Gamma} = \left(\mathcal{K}, (\mathcal{X}_j)_{j \in \mathcal{K}}, \mathbb{E}_{\underline{w},\underline{x}} \tilde{U}_j(\underline{w}, .)\right). \tag{6.6}$$

A state-independent equilibrium of the dynamic game is defined as follows.

Definition 159

A vector $\underline{x}$ is a state-independent (Nash) equilibrium if it is independent of the states, satisfies $\underline{x} \in \prod_{j \in \mathcal{K}} \mathcal{X}_j$, and an equilibrium of the expected game $\bar{\Gamma}$, i.e., for all $j \in \mathcal{K}$:

$$\mathbb{E}_{\underline{w},\underline{x}} \tilde{U}_j(\underline{w}, a) \geq \mathbb{E}_{\underline{w},\underline{x}'_j,\underline{x}_{-j}} \tilde{U}_j(\underline{w}, a), \ \forall \underline{x}'_j \in \mathcal{X}_j \tag{6.7}$$

Theorem 160

Assume that $\mathcal{W}$ is a compact set. Then, there is at least one state-independent equilibrium.

The above result ensures the existence of a state-independent equilibrium under compactness of the state space. We therefore focus on the learning aspects of such an equilibrium. In many situations, the updating scheme that will be used has the following form:

$$\text{Newestimate} \longleftarrow \text{Oldestimate} + \text{Stepsize (Target} - \text{Oldestimate)}$$

where the target play the role of the current strategy. The expression [Target − Oldestimate] is an error in the estimation. It is reduced by taking a step size toward the target. The target is presumed to indicate a desirable direction in which to move.

A simple pseudo-code for the simultaneous-act games is as follows:

```
for i = 1 to max_of_iterations
        for j = 1 to number_of_players
            action_profile.append(select_action(player-j))
        endfor
        for j = 1 to number_of_players
            playerj.utility = assign_utility(action_profile)
        endfor
        for j = 1 to number_of_players
            playerj.update_strategies
        endfor
    end
```

We would like to mention once again that, usually, game-theoretic modeling assumes rational players, and that rationality is common knowledge. In the fully distributed learning schemes described in this chapter, each player is assumed to select a learning pattern, but does not need to know whether the other players are rational or not.

6.3 TRIAL-AND-ERROR LEARNING: LEARNING BY EXPERIMENTING

We now present a completely uncoupled (fully distributed) learning rule such that, when used by all players in a finite game, stage-by-stage play comes close to pure Nash equilibrium play a high proportion of the time, provided that the game is generic, and has such a pure equilibrium. The interactive learning by experimentation is in discrete time, $t \in \mathbb{N}, t \geq 1$. At each time t, a player j has his own state $s_{j,t}$ which contains his own decision $a_{j,t}$ and his own perceived utility $u_{j,t}$ i.e., $s_{j,t} = (a_{j,t}, u_{j,t})$. At time $t+1$, each player does some experiment, with some probability $\epsilon_j \in (0,1)$. The player keeps the action $a_{j,t}$ at time $t+1$ when he does not experiment, otherwise he/she plays $a_j \in \mathcal{A}_j$ drawn uniformly at random. If the received utility after experimentation $u_{j,t+1}$ is strictly greater than the old utility $u_{j,t}$, then player j adopts the new state $s_{j,t+1} = (a_{j,t+1}, u_{j,t+1})$, otherwise the player j comes back to his old state $s_{j,t}$.

Theorem 161: Marden et al. (forthcoming)

Given a finite K-player weakly acyclic game $\mathcal{G}$ (see Chapter 5), if all players use simple experimentation with experimentation rate $\epsilon > 0$, then for all sufficiently small ϵ a Nash equilibrium is played at least $1 - \epsilon$ of the time.

We present a convergence result for the *Trial and Error Learning* proved by Young (2009) for the class $\mathcal{G}_\mathcal{A}$. This result generalizes Theorem 161.

Theorem 162: Young (2009)

Let $\mathcal{G}_\mathcal{A}$ be the set of all K-player games on finite action space $\mathcal{A}$ that possess at least one pure Nash equilibrium. If all players use interactive trial and error learning, and the experimentation rate $\epsilon > 0$ is sufficiently small, then for almost all games in $\mathcal{G}_\mathcal{A}$ a (Nash) equilibrium is played at least $1 - \epsilon$ proportion of the time.

We apply this result to discrete power allocation problems without information exchange on total interference and without knowing the utility functions.

Example 163: Learning in Discrete Power Allocation Games

Consider a finite set of transmitters sending information to a finite set of receivers using orthogonal frequencies. Each transmitter (player) is able to transmit to all the channels. The channel gain between a transmitter j and channel ℓ is denoted by $w_{jj'\ell}$. We assume channel gains change more slowly than the time scaling of the transmission. Each transmitter j selects a vector of powers in a finite set $\mathcal{P}_j$. An action for transmitter j is therefore an element of $a_j \in \mathcal{P}_j$. The SINR of transmitter j to channel ℓ is then given by $\text{SINR}_{j,l}(\underline{a}) = \frac{a_{jl}w_{jjl}}{N_{0,l} + \sum_{j' \neq j} a_{j'l}w_{j'jl}}$ where $N_{0,l}$ is the noise at channel l, $w_{j'jl}$ is the channel gain at link l. The utility of player j is his total sum-rate, i.e., $\tilde{U}_j(\underline{a}) = \sum_l b_l \log_2\left(1 + \text{SINR}_{j,l}(\underline{a})\right)$. This strategic power allocation game has been well-studied, and is known to have a pure Nash equilibrium, and it is in the class of weakly acyclic games. We can therefore apply learning by experimentation, and Theorem 162 says that a pure Nash equilibrium will be visited at least $1 - \epsilon$ fraction of the times for any $\epsilon > 0$. Note that the theorem does not indicate how many iterations are needed to be arbitrarily close to a Nash equilibrium utility. Intuitively, the number of iterations depends on the precision ϵ and the structure of the observed utilities.

Example 164: Learning in Medium Access Control Games

Consider the following multiple access game between two mobiles (players). Each mobile can transmit with one of the power levels $p_0 < p_1 < p_2$. Mobile 1 chooses a row, mobile 2 chooses a column. The action p_0 corresponds to zero power which means "does not transmit". The actions p_1, p_2 with $p_1 < p_2$ are strictly positive power levels. Assume that there is a power consumption cost represented by a mapping $c(.)$. If $c \equiv 0$, by applying learning by trial and error, one finds that an equilibrium configuration with utilities $(0,0), (1,0)$ or $(0,1)$ will be visited at least $1 - \epsilon$ fraction of the time. Assume now that $c(.)$ is non-decreasing: $c(p_0) = 0 < c(p_1) < c(p_2) < 1$. It is easy to see that the pure Nash equilibrium disappears (see Table 6.1). The game has a unique mixed Nash equilibrium. The issue of the learning by trial and error algorithm starting from the collision status (p_2, p_2) is a non-equilibrium. We will examine in the next section learning schemes for mixed equilibria.

Table 6.1 Multiple Access Game between Two Mobiles			
	p_2	p_1	p_0
p_2	$(-c(p_2), -c(p_2))$	$(1 - c(p_2), -c(p_1))$	$(1 - c(p_2), 0)$
p_1	$(-c(p_1), 1 - c(p_2))$	$(-c(p_1), -c(p_1))$	$(1 - c(p_1), 0)$
p_0	$(0, 1 - c(p_2))$	$(0, 1 - c(p_1))$	$(0, 0)$

6.4 REINFORCEMENT LEARNING ALGORITHMS

Reinforcement learning was originally studied in the context of single-player environments. A player receives a numerical utility signal, which he seeks to maximize. The environment provides this signal as a feedback for the sequence of actions that has been taken by the player. Learners relate the utility signal to actions previously taken in order to learn a strategy that maximizes their expected future reward. In the multi-player setting, we define a dynamic environment by a dynamic game and each reinforcement learner updates his strategy and/or estimations. Therefore, in a wireless setting, the environment of a player changes not only because of exogenous factors such as the time-varying nature of the channels, but also because the other players react to the action played by each other; indeed, the act of learning changes the thing to be learned. The main challenge is thus to know whether feeding back to the players only the realizations of their utility is enough to drive such a seemingly complex interactive situation to a steady state, or at least to a predictable evolution of the state.

6.4.1 Bush & Mosteller Reinforcement Learning Algorithms

Reinforcement learners use their experience to choose or avoid certain actions based on their consequences. Actions that led to satisfactory outcomes (i.e., outcomes that met or exceeded aspirations) will tend to be repeated in the future, whereas choices that led to unsatisfactory experiences will be avoided. We will now consider the dynamics of a (Bush and Mosteller, 1955) stochastic model of reinforcement learning. In this model, players decide what action to take stochastically: each player's strategy is defined by the probability of undertaking each of the two actions available to them. After every player has selected an action according to its probability, every player receives the corresponding utility and revises her strategy. The revision of strategies takes place following a reinforcement learning approach: players increase their probability of undertaking a certain action if it led to utilities above their aspiration level, and decrease this probability otherwise. When learning, players in the Bush & Mosteller model use only information concerning their own past choices and perceived utilities, and ignore all the information regarding the utilities and choices of the other players, i.e., it is a *fully distributed learning algorithm*.

We will now describe the algorithm formally. Consider a finite game in strategic form $G = (\mathcal{K}, (\mathcal{A}_j), (\tilde{U}_j)_j)$. Denote by $\mathcal{A} = \prod_{j \in \mathcal{K}} \mathcal{A}_j$ the set of strategy profiles, and by $x_j(a_j)$ player j's probability of undertaking action $a_j \in \mathcal{A}_j$. At each time t, each player j chooses an action $a_{j,t}$ and computes her stimulus $\bar{s}_{j,t}$ for the action just chosen $a_{j,t}$ according to the formula:

$$\bar{s}_{j,t} = \frac{u_{j,t} - M_j}{\sup_a |U_j(\underline{a}) - M_j|} \tag{6.8}$$

where $u_{j,t}$ denotes the perceived utility at time t of player j, and M_j is an aspiration level of player j. Hence the stimulus is always a number in the interval $[-1, 1]$. Note that player j is assumed to know the denominator $\sup_a |U_j(\underline{a}) - M_j|$. Secondly, having calculated their stimulus $\bar{s}_{j,t}$ after the outcome a_t, each player j updates her probability $x_j(s_j)$, of undertaking the selected action a_j as follows:

$$x_{j,t+1}(a_j) = \begin{vmatrix} x_{j,t}(a_j) + \lambda_j \bar{s}_{j,t}(1 - x_{j,t}(a_j)) & \text{if } \bar{s}_{j,t} \geq 0 \\ x_{j,t}(a_j) + \lambda_j \bar{s}_{j,t} x_{j,t}(a_j) & \text{if } \bar{s}_{j,t} < 0 \end{vmatrix} \tag{6.9}$$

where λ_j is player j's learning rate ($0 < \lambda_j < 1$). Thus, the higher the stimulus magnitude or the learning rate, the larger the change in probability. The updated probability for the action not selected is derived from the constraint that probabilities must add up to one.

Below we summarize the main assumptions made in the Bush & Mosteller (BM) model, in terms of the nature of the utilities, the information players require, and the computational capabilities that they have.

- Utilities: Utilities and aspiration thresholds are not interpreted as von Neumann-Morgenstern utilities, but as a set of variables measured on an interval scale that is used to calculate stimuli.
- Information: Each player j is assumed to know his action set $\mathcal{A}_j$, and the maximum absolute difference between any utility she might receive and her aspiration threshold. Players do not use any information regarding the other players.
- Computational capabilities and memory: Players are assumed to know their own mixed strategy $\underline{x}$ at any given time. Each player is assumed to be able to conduct the arithmetic operations.

The main issue now is to know the asymptotic behavior of Equation (6.9). For this, some notions of dynamical are needed. The purpose of what follows is to review and illustrate these ideas and to show how convergence can be studied. Let X_t be the state of a system in stage t. X_t is a random variable and q is a particular value of that variable; the sequence of random variables $\{X_t\}_{t \geq 1}$ constitutes a discrete-time Markov process. We say that the Markov process X_t is ergodic if the sequence of stochastic kernels defined by the t-step transition probabilities of the Markov process converges uniformly to a unique limiting kernel that is independent of the initial state. The expected motion (also called drift) of the system in state x is given by a function vector $K(\underline{x})$ whose components are, for each player, the expected change in the probabilities of undertaking each of the possible actions, i.e., $K(\underline{x}) = \mathbb{E}(X_{t+1} - X_t \mid X_t = \underline{x})$. This leads to the deterministic dynamical system described by the ordinary differential equation $\underline{\dot{x}} = K(\underline{x})$. We define a *self-reinforcing rest point* (S2RP) as an absorbing state of the system where both players receive a positive stimulus. An S2RP corresponds to a pair of pure strategies ($x_j(a_j)$ is either 0 or 1) such that its certain associated outcome gives a strictly positive stimulus to both players.

We now present theoretical results on the asymptotic behavior of the BM learning algorithm. Note that with low learning rates the system may take an extraordinarily long time to reach its long-run behavior.

. . . .

Proposition 165: Izquierdo et al. (2007)

In any two-by-two game, assuming that players' aspirations are different from their respective utilities ($\tilde{U}_j(\underline{a}) \neq M_j$ for all j and $\underline{a}$) and below their respective maxmin levels: the process X_t converges to an S2RP with probability 1 (the set formed by all S2RPs is asymptotically reached with probability 1). If the initial state is completely mixed, then every S2RP can be asymptotically reached with positive probability.

....

Proposition 166: Izquierdo et al. (2007)

In any two-by-two game, assuming that players' aspirations are different from their respective utilities and above their respective maxmin:

a. If there is any S2RP then the process X_t converges to an S2RP with probability 1 (the set formed by all S2RPs is asymptotically reached with probability 1). If the initial state is completely mixed, then every S2RP can be asymptotically reached with positive probability.

b. If there is no S2RP then the BM learning process X_t is ergodic with no absorbing state.

c. S2RPs whose associated outcome is non-Nash equilibrium are unstable for the ODE.

d. All S2RPs whose associated outcome is a strict Nash equilibrium where at least one unilateral deviation leads to a satisfactory outcome for the non-deviating player are asymptotically stable.

Example 167

Consider a discrete non-cooperative uplink power control game between mobile users (players). The utility function of a player is the number of successfully transmitted bits per joule consumed at the transmitter. The utility of a mobile user has one of the following forms.

- $\tilde{U}_j(\underline{a}) = \frac{LR}{M} \frac{F_j(\text{SINR}_j(\underline{a}))}{c_j(a_j)}$ where a_j is in a finite set of powers $\mathcal{A}_j$; L is the information bits in frame (packets) of $M > L$ bits at a rate R b/s using p_j watts of power. $c_j(p_j)$ is the energy consumption, $\text{SINR}_j(\underline{a}) = \frac{W}{R} \frac{a_j|h_j|^2}{\sum_{i \neq j} a_i|h_i|^2 + N_0}$, W is the available spread-spectrum bandwidth (in Hz), N_0 is the additive white Gaussian noise (AWGN) power at the receiver (in watts), and $\{|h_j|^2\}$ is the set of path gains from the mobile to the base station. If the function c is the identity function, and the functions $\{F_j\}_j$ are identical to F, then one obtains the well-known *energy efficiency function*.

- $\tilde{U}_j(\underline{a}) = G_j(\text{SINR}_j(\underline{a})) - c_j(a_j)$. This utility is sometimes referred to as the *pricing-based utility function*. The case where the mapping $G_j(\underline{a}) = W \log(1 + \text{SINR}_j(\underline{a}))$ is known as the Shannon capacity.

We know that every finite strategic-form game has at least one mixed strategy Nash equilibrium. That is, in both cases of utility functions U, the resulting games have at least one equilibrium in mixed strategies. We now apply the Bush & Mosteller learning algorithm in these power control games. The algorithm works as follows:

BM1 Set the initial probability vector $\underline{x}_{j,0}$, for each j.
BM2 At every stage t, each mobile chooses a power according to his action probability vector $\underline{x}_{j,t}$. Thus, the j-th mobile chooses the power a_j at instant t, based on the probability distribution $x_{j,t}(a_j)$. Each mobile obtains a utility. The utility to mobile j is $u_{j,t}$. Each mobile updates his action probability according to the Bush & Mosteller learning rule.
BM3 If the stopping criterion are met, stop; else, go to step BM2.

A particular case of the Bush & Mosteller model has been applied to discrete power control in Xing and Chandramouli (2008). Their model is obtained for $M_j = \min \tilde{U}_j$, $\lambda_j = b$ and positive utilities.

6.4.2 Cross Learning Algorithm

Another simple learning technique is the cross learning algorithm (Borgers and Sarin, 1993). When action j' is selected and a normalized utility $u_{j',t} \in [0,1]$ is received at time t, then strategy $\underline{x}$ is updated according to the following equation:

$$x_{j',t+1} = (1 - u_{j',t})x_{j',t} + u_{j',t}, \ x_{i',t+1} = (1 - u_{i',t})x_{i',t}, \text{ for } j' \neq i' \tag{6.10}$$

The information needed to update this algorithm is the numerical value (which can be noisy) of a player's own-utility and own-action. So this learning algorithm is fully distributed. Borgers and Sarin (1993) have shown that the strategy changes induced by this learner converge under infinitesimal time steps to the replicator dynamics. Since the replicator may not converge, the cross learning algorithm does not converge in general.

6.4.3 Arthur's Model of Reinforcement Learning

Stochastic approximation examines the behavior of a learning process by investigating an ordinary differential equation derived from the expected motion of the learning process. One classic result is that if the ODE has a global attractor, the learning process converges with probability one to that point. However, most of the ODE approximations of reinforcement learning do not have any global attractors. These are evolutionary replicator dynamics, adaptive dynamics, best response dynamics, projection, mean field game dynamics, etc. The models of reinforcement learning we study here are based on those studied by Arthur (1993) and Borgers and Sarin (1993). Consider a finite game $\mathcal{G} = (\mathcal{K}, (\mathcal{A}_j)_{j \in \mathcal{K}}, (\tilde{U}_j)_{j \in \mathcal{K}})$ in strategic form: $\mathcal{K}$ is the set of players, $\mathcal{A}_j$ is the set of strategies of player j, and $\tilde{U}_j : \prod_{j'} \mathcal{A}_{j'} \rightarrow \mathbb{R}$ is a random variable with expectation $\tilde{u}_j(a) = \mathbb{E}(\tilde{U}_j(a))$ which represents player j's utility function. We assume that the random variable $\tilde{U}_j$ is almost surely bounded and strictly positive. Note that it is usually assumed that utilities are deterministic functions of strategy profiles, but we relax this assumption in this section. We assume that the players are repeatedly playing the same game $\mathcal{G}$. At each time t, under *reinforcement learning*, each player j is assumed to have a tendency $\alpha_{j,t}(a_j)$ to each action $a_j \in \mathcal{A}_j$. Let $x_{j,t}(a_j)$ be the probability placed by player j on action a_j at time t. In the models of reinforcement learning, we consider that these probabilities are determined by the choice rule $p_{j,t}(a_j) = g_j(s_{j1}, \ldots, s_{j,m_j}, \underline{x}_{jt})$ where $m_j = |\mathcal{A}_j|$.

Here we examine the case where the mapping g_j can be written as:

$$\frac{\alpha_{j,t}(a_j)^\gamma}{\sum_{s'_j}(\alpha_{j,t}(s'_j)^\gamma)}, \ \gamma \geq 1. \tag{6.11}$$

To complete the reinforcement learning model, we need to define how to update the tendencies $\underline{x}_t$.. In this simple model this is expressed as, if player j takes action a_j in period t, then his tendency for a_j is increased by an increment equal to his realized utility. All other tendencies are unchanged. Let $u_{j,t}$ denote the random utility obtained by player j in period t. Thus, we can write the updating rule:

$$x_{j,t+1}(a_j) = x_{j,t}(a_j) + u_{j,t} \, \mathbb{1}_{\{a_{j,t+1}=a_j\}} \tag{6.12}$$

This iterative formulation is normalized by dividing by $\frac{C(n+1)}{nC+u_{j,t}}$ for a factor $C > 0$. The reinforcement learning becomes:

$$x_{j,t+1}(a_j) = \frac{C(n+1)}{nC+u_{j,t}}\left[x_{j,t}(a_j) + u_{jt}\,\mathbb{1}_{\{a_{j,t}=a_j\}}\right] \qquad (6.13)$$

The normalized Arthur model can be written:

$$x_{j,t+1}(a_j) = x_{j,t}(a_j) + \lambda_t u_{j,t}\left(\mathbb{1}_{\{a_{j,t}=a_j\}} - x_{j,t}(a_j)\right), \quad \forall j,\ \forall a_j \in \mathcal{A}_j, \forall t,\ x_{j,t} \in \Delta(\mathcal{A}_j) \qquad (6.14)$$

In order to preserve the simplex of deterministic trajectory, it suffices to have the condition $\lambda_t u_{j,t} \in [0,1]$. We compute the drift (the expected changes in one-step):

$$
\begin{aligned}
F_{j,a_j}(x) &= \lim_{\lambda_t \to 0} \frac{1}{\lambda_t}\mathbb{E}\left(x_{j,t+1}(a_j) - x_{j,t}(a_j)|x_t = x\right) \\
&= \lim_{\lambda_t \to 0} \frac{1}{\lambda_t}\sum_{s_j' \in \mathcal{A}_j}\mathbb{E}\left(x_{j,t+1}(a_j) - x_{j,t}(a_j) \mid x_t = x, a_{j,t} = s_j'\right) x_{j,t}(s_j') \\
&= \sum_{s_j' \neq a_j} x_{j,t}(s_j')\mathbb{E}\left(-u_{j,t}| x_t = x, a_{j,t} = s_j'\right) x_{j,t}(a_j) \\
&\quad + x_{j,t}(a_j)\mathbb{E}\left(u_{j,t}| x_t = x, a_{j,t} = a_j\right)(1 - x_{j,t}(a_j)) \\
&= x_{j,t}(a_j)\left[\mathbb{E}\left(u_{j,t}| x_t = x, a_{j,t} = a_j\right) - \sum_{s_j' \in \mathcal{A}_j} x_{j,t}(s_j')\mathbb{E}\left(-u_{j,t}| x_t = x, a_{j,t} = s_j'\right)\right] \\
&= x_{j,t}(a_j)\left[u_j(e_{a_j}, x_{-j,t}) - \sum_{s_j' \in \mathcal{A}_j} x_{j,t}(s_j')u_j(e_{s_j'}, x_{-j,t})\right] \qquad (6.15)
\end{aligned}
$$

The autonomous ODE $\dot{x}_{j,t}(a_j) = F_{j,a_j}(x_t)$ is a (multi-type) replicator equation.

We now use standard techniques from stochastic approximation theory (Benaim et al., 2006; Borkar, 1997; Kushner and Clark, 1978; Ljung, 1977). The theory of stochastic approximation largely works by predicting the behavior of stochastic processes by using an ordinary differential equation or inclusion that is formed by taking the expected motion of the stochastic process. For example, consider a stochastic process of the form:

$$\underline{x}_{t+1} = \underline{x}_t + \lambda_t^1 f(\underline{x}_t) + \lambda_t^2(\underline{x}_t) + O((\lambda_t^1)^2) \qquad (6.16)$$

where λ_t^1 is the step-size of the process and λ_t^2 is the random component of the process with $\mathbb{E}(\lambda_t^2 \mid \underline{x}_t) = 0$.

Remark A particular case of (6.16) is the model of Arthur (1993) and Borgers and Sarin (1993). Sastry et al. (1994b) used this stochastic learning procedure, and proved the weak convergence (convergence in law) to an ordinary differential equation.

However, the non-Nash rest points have not been classified in their work. Their reinforcement learning algorithm is a stochastic approximation of the replicator dynamics which is known may not lead to equilibria. See also (Hofbauer and Sigmund, 1998; Sandholm, 2011; Tembine, 2009a; Weibull, 1997) for limit cycle trajectories of the replicator dynamics. Notice that, for finite games, the faces of the simplex are invariant under replicator dynamics but the extreme points need not be equilibria. In the receiver selection game described in example 146, the vector $(1,0)$ which corresponds to the pure strategy $R1$ is a rest point of the replicator dynamics but it is not an equilibrium. Reinforcement learning (using stochastic approximation) has been applied to various problems such as load balancing (Bournez and Cohen, 2009), network security (Zhu et al., 2010b), and cognitive radio (Perlaza et al., 2010a).

The mean or averaged ordinary differential equations (ODEs) derived from the stochastic process above would be a variant of the replicator dynamics:

$$\dot{x}_j(a_j) = x_j(a_j) \left[u_j(e_{a_j}, x_{-j}) - \sum_{s'_j \in A_j} x_j(s'_j) u_j(e_{s'_j}, x_{-j}) \right] =: f_{j,a_j}(x) \qquad (6.17)$$

which can be written as $\dot{x}_j = R(x_j)u_j(., x_{-j})$ where $R(x_j)$ is a symmetric matrix given by $a_j \neq a'_j$, $R(x_j)_{a_j a'_j} = -x_j(a_j)x_j(a'_j), R(x_j)_{a_j a'_j}(x_j) = x_j(a'_j)(1 - x_j(a'_j))$.

We also associate a variant known as the adjusted replicator dynamics which has the form:

$$\dot{x}_{j,a_j} = \frac{f_{j,a_j}(x)}{\sum_{s'_j} x_j(s'_j) u_j(e_{s'_j}, x_{-j})} = K_{j,a_j}(x). \qquad (6.18)$$

But despite having ODEs with the same local stability properties, the two processes potentially have different asymptotic behavior. This crucially shows the limits to predicting the behavior of a stochastic process using only ordinary differential equations. The following result provides a non-convergence to linearly unstable equilibria with probability zero. See also Benaïm and Faure (2010), Sastry et al. (1994b), and Thathachar and Sastry (1991).

··

Theorem 168: Hopkins and Posch (2005)

Let x^* be a rest point for the replicator dynamics, or equivalently the adjusted replicator dynamics, and $\gamma = 1$. If

- x^* is not a Nash equilibrium,
- x^* is a Nash equilibrium that is linearly unstable under the adjusted replicator dynamics, or
- there are two strategies, and x^* is a Nash equilibrium that is linearly unstable under the replicator dynamics

then, for the reinforcement learning process defined by updating strategy (6.11) and updating tendency (6.12), from any initial condition with all tendencies positive, that is $x_{j,0}(a_j) > 0$ for each player j and all strategies a_j, $P(\lim_t x_t = x^*) = 0$.

6.5 REGRET MATCHING-BASED LEARNING: LEARNING CORRELATED EQUILIBRIA

In this section we present a procedure for converging to the set of correlated equilibria in terms of an empirical distribution of play, based on Foster and Vohra (1997), Hart and Mas-Colell (2000) no-regret dynamics. Compared to the previous, fully distributed algorithms, this algorithm will be based on frequencies of play and may lead to a correlated equilibrium as a steady state. To make the section sufficiently self-contained we review correlated equilibria and some known related results.

6.5.1 Review of Correlated Equilibrium

In this section we present a procedure for converging to the set of correlated equilibria in terms of empirical distribution of play. The section is based on Foster and Vohra (1997), Hart and Mas-Colell (2000) no-regret dynamics. Compared to the previous fully distributed algorithms, this algorithm will be based on frequencies of play. The notion of correlated equilibrium (CE) was introduced by Aumann (1974) and can be described as follows: Assume that, before the game is played, each player receives a private signal which does not affect the utility functions (the signals are utility-irrelevant). The player may then choose his action in the game depending on this signal. A correlated equilibrium of the original game is a Nash equilibrium of the game with the signals. Considering all possible signal structures generates all correlated equilibria. If the signals are stochastically independent across the players, it is a Nash equilibrium in mixed or pure strategies of the original game. But the signals could well be correlated, in which case new equilibria may be obtained. Equivalently, a correlated equilibrium is a probability distribution on action profiles, which can be interpreted as the distribution of play instructions given to the players by some "device" or "referee". Each player is privately given recommendations for his own play only; the joint distribution is known to all of them. Also, for every possible recommendation that a player receives, the player realizes that the instruction provides a best response to the random estimated play of the other players, assuming they all follow their recommendations. In Foster and Vohra (1997), the authors obtained a procedure converging to the set of correlated equilibria. Hart and Mas-Colell (2000) gives a simple procedure called *no-regret dynamics* that generates trajectories of play that almost surely converge to the set of correlated equilibria.

The setting is the following. Consider a finite game $\mathcal{G}$ in strategic form.

Definition 169

A probability distribution μ on $\prod_j \mathcal{A}_j$ is a correlated equilibrium if, for every $j \in \mathcal{K}$, every $a_j \in \mathcal{A}_j$, and every $a_j \in \mathcal{A}_j$ we have:

$$\sum_{\underline{a}_{-j}} \mu(a) \left[\tilde{U}_j(a_j, \underline{a}_{-j}) - \tilde{U}_j(a_j, \underline{a}_{-j}) \right] \geq 0 \qquad (6.19)$$

Note that every Nash equilibrium is a correlated equilibrium. Indeed, Nash equilibria correspond to the special case where μ is a product measure, that is, the play of the different players is independent. Also, the set of correlated equilibria of finite games is non-empty, closed, and convex.

Example 170

By reconsidering the MAC game described in Chapter 5, one can observe that the set of correlated equilibria of the MAC game include distributions that are not in the convex hull of the Nash equilibrium distributions.

6.5.2 Regret Matching-Based Learning

Suppose that the game $\mathcal{G}$ is played many times $t \in \mathbb{N}$. Fix an initial distribution x_0 chosen arbitrarily at the first stage. At time $t+1$, given a history of play $\underline{h}_t \in (\prod_j \mathcal{A}_j)^t$, each player j chooses $a_{j,t+1}$ according to a probability distribution $x_{j,t+1} \in \mathcal{X}_j$. Given two different strategies a_j^* and a_j, define the difference in average up to t as:

$$D_j(t+1, a_j^*, a_j) = \frac{1}{t} \sum_{t'=1,\ a_{j,t'}=a_j^*}^{t} \left[\tilde{U}_j(a_j, \underline{a}_{-j,t}^*) - \tilde{U}_j(\underline{a}_t^*) \right]. \tag{6.20}$$

Define the average regret at period t for not having played, every time that a_j^* was played in the past, the different strategy a_j as:

$$R_j(t+1, a_j^*, a_j) = \max(0, D_j(t+1, a_j^*, a_j)). \tag{6.21}$$

Fix $\alpha > 0$ to be a large enough number. Let a_j^* be the last strategy (of reference) chosen by player j, i.e., $a_j^* = a_{j,t}$. Then the probability distribution $x_{j,t+1}$ used by j at time $t+1$ is defined by the following:

$$x_{j,t+1}(a_j) := \begin{vmatrix} \frac{1}{\alpha} R_j(t+1, a_j^*, a_j) & \forall\, a_j \neq a_j^* \\ 1 - \sum_{\substack{a_j \in \mathcal{A}_j, \\ a_j \neq a_j^*}} x_{j,t+1}(a_j) \end{vmatrix} \tag{6.22}$$

Note that for a large enough α, this defines a probability distribution. The probabilities of switching to different strategies are proportional to their regrets relative to the current strategy. In particular, if the regrets are small, then the probability of switching from current play is also small.

For every stage, let $f_t(\cdot)$ be the empirical distribution of the actions profiles played up to time t;

$$f_t(\underline{a}) = \frac{1}{t} \sum_{t'=1}^{t} \mathbb{1}_{\{\underline{a}_{t'}=\underline{a}\}} \tag{6.23}$$

> **Theorem 171: Hart and Mas-Colell (2000)**
>
> If every player plays according to the no-regret dynamics, then the empirical distributions of play $f_t(\cdot)$ converge as t goes to infinity to the set of correlated equilibrium distributions of the game $\mathcal{G}$.

Note that this theorem does not state that the sequence of frequencies $f_t(\cdot)$ converges to a point.

Example 172: Learning CE in Cognitive Radio Networks

We consider a cognitive radio network with free spectrum bands. In such a context, the secondary (unlicensed) users have the choice between accessing the licensed spectrum bands owned by primary users (PUs) which are charged as a function of the enjoyed signal-to-interference-plus-noise ratio (SINR) or received power at primary users (PUs), and switching to the unlicensed spectrum bands which are free of charge but become more crowded when more secondary users (SUs) operate in them. Consequently, the SUs should balance between accessing the free spectrum bands which probably have more interference, and paying for communication gains by staying with the licensed bands. In this example, we illustrate the case of one free spectrum band available to all SUs and $K - 1$ licensed bands processed by primary users. We denote the set of bands by $\{1, 2, \ldots, K - 1, K\}$. A secondary user SU has an additional choice of switching to band n and the corresponding utility is $\tilde{U}_j(\text{SINR}_j)$. A secondary user j faces the choice of accessing the licensed band with a minimum effective price, and the free band n. The primary channel selection game has at least one equilibrium (in pure or mixed strategies).

We then address the problem of how to reach an equilibrium. We first notice that when using a myopic best response update in the PU selection game, one may have a cycling situation (the algorithm is not guaranteed to converge to a NE). In fact, during the course of unused primary channel selection, the secondary users may notice that the utility of accessing a licensed spectrum is higher than staying in the free spectrum, and thus switch to the licensed spectrum accordingly. Since the SUs do this simultaneously, the free spectrum becomes under-loaded and the SUs will switch back to the free spectrum in the next iteration. This phenomenon, in which a player keeps switching between two strategies, is known as a cycling effect, as mentioned above. To eliminate this, we use no-regret learning to converge to a correlated equilibrium (with probability 1).

6.6 BOLTZMANN-GIBBS LEARNING ALGORITHMS

6.6.1 Boltzmann-Gibbs Distribution

The Boltzmann-Gibbs-based learning scheme (called also softmax) is particularly interesting because it gives a good approximation of the maxima and equilibria after a reasonably small number of iterations. If the utility estimation is good enough, then approximated equilibria are obtained. Let $\hat{\mathbf{u}}_{j,t} \in \mathbb{R}^{|\mathcal{A}_j|}$ be the estimated utility of player j at time t. The Boltzmann-Gibbs distribution:

$$\tilde{\beta}_{j,\epsilon}(\hat{\mathbf{u}}_{j,t})(a_j) = \frac{e^{\frac{1}{\epsilon_j}\hat{u}_{j,t}(a_j)}}{\sum_{a'_j \in \mathcal{A}_j} e^{\frac{1}{\epsilon_j}\hat{u}_{j,t}(a'_j)}}, \; a_j \in \mathcal{A}_j, j \in \mathcal{K} \tag{6.24}$$

is known to be close to equilibrium when $\frac{1}{\epsilon_j}$ is large (see Chapter 5), which can be interpreted as the rationality level of player j. The case where the ϵ_j are identically equal to ϵ, has been introduced in Willard (1978) and in that context, the factor $\frac{1}{\epsilon}$ is interpreted as a temperature parameter. As explained in Chapter 5, the Boltzmann-Gibbs distribution is a Nash equilibrium of the perturbed game with utilities: $u_j(\underline{x}_j, \underline{x}_{-j}) + \epsilon_j H(\underline{x}_j)$, where $H(\underline{x}_j) = -\sum_{a_j \in A_j} x_j(a_j) \log(x_j(a_j))$ is the entropy function. The corresponding Nash equilibria of this perturbed game are called logit or Boltzmann-Gibbs equilibria.

6.6.2 The Setting

Although the expression of the Boltzmann-Gibbs distribution is the same as for the partially distributed version presented in Chapter 5, some differences need to be emphasized. We assume that each player does not know his utility function, but instead has an estimation of the average utility of the alternative actions. He makes a decision based on this rough information by using a randomized decision rule to revise his strategy. The effect on the utilities of the chosen alternative are then observed, and used to update the strategy for that particular action. Each player only experiments with the utilities of the selected action on that stage, and uses this information to adjust his strategy. This scheme is repeated several times, generating a discrete-time stochastic learning process. Basically we have three parameters: the current information state, the rule for revising strategies from the state to randomized actions, and a rule to update the new state. Although players only observe the utilities of the specific action chosen at the given stage, the observed values depend on some specific dependency parameters determined by the other player's choices revealing implicit information about the system. *The natural question is whether such a learning algorithm, based on a minimal piece of information, is sufficient to induce coordination and to make the system stabilize to an equilibrium.*

The answer to this question is positive for dynamic congestion games, routing games with parallel links, and access point selection games under the Boltzmann-Gibbs learning scheme.

The congestion games case study. We consider a dynamic congestion game with a finite set of players and a finite set of common action space. Denote by $u_{\underline{a}}^w(k, t)$ the average w-weighted utility for the action a when k players have chosen this action at time t. The more an action is used the less the weighted utility is bigger i.e.,

$$\forall t, \ u_a^w(k, t) \geq u_a^w(k+1, t), \ k \leq K,$$

where K is the total number of players. An estimation utility of player j is a vector $\hat{\mathbf{u}}_j = (\hat{u}_{j,a}^{i'})_{i',\underline{a}}$, where $\hat{\mathbf{u}}_j^{i'}$ represents player j's estimate of the average utility of the action a for the objective i'. Player j can then compute the weighted average estimated utility $\hat{\mathbf{u}}_j^w = (\hat{u}_{j,a}^w)_a$. Player j can then update his strategy by using the

Boltzmann-Gibbs scheme: use the action a with probability $x_{j,a}(\hat{u}) := \dfrac{e^{\frac{1}{\epsilon_j} \hat{u}_{j,a}^w}}{\sum_{a'} e^{\frac{1}{\epsilon_j} \hat{u}_{j,a'}^w}}$, where ϵ_j is a positive parameter. Let $v_t \geq 0$ be a step size satisfying $\sum_{t \geq 0} v_t = +\infty$, $\sum_{t \geq 0} v_t^2 < \infty$. We now describe the learning algorithm:

Algorithm 1: Stochastic Learning Algorithm based on Boltzmann-Gibbs dynamics

> **forall the** *Players* **do**
> > Initial Boltzmann distribution $\underline{x}_{j,0}$;
> > Initialize to some estimations $\hat{u}_{j,a}^{i'}(k_0, 0)$;
>
> **end**
> **for** *t=1* **to** *max* **do** do;
> **foreach** *Player j* **do**
> > Observe his current utility;
> > Update via Boltzmann-Gibbs dynamics $\underline{x}_{j,t+1}$;
> > Compute the distribution over $a_{j,t+1}$ and $k_{a_{j,t+1}}$ from $\underline{x}_{t+1}$;
> > Update his estimation via $\hat{u}_{j,a,t+1} = \hat{u}_{j,a,t} + v_{t+1}\left(W_{j,a,t+1} - \hat{u}_{j,a,t}\right)$;
> > Estimate the random utilities $\hat{u}_{j,a_{j,t+1}}(k_{a_{j,t+1}}, t+1)$;
>
> **end**

The learning process may then be described as follows. At stage $t+1$ the past estimation $\hat{u}_{j,a}^w(k_{a,t}, t)$ determines the transition probabilities $\underline{x}_{j,t}(a) = x_{j,a}(\hat{u}_{j,a}^w(k_{a,t}, t))$ which are used by player j to experiment by using a random action $a_{j,t+1}$. The action profile of all the players determines a total random number $k_{a,t+1}$ of players j such that $a_{j,t+1} = a$. The weighted utility of a is then $\hat{u}_{j,t+1}^w(k_{a_{j,t+1}, t+1}) = u_{j,a}^w(k, t+1)$ if $a_{j,t+1} = a, k_{a,t+1} = k$. Finally, each player j observes only the utility of the chosen alternative $a_{j,t+1}$ and updates his/her estimations by averaging:

$$\hat{u}_{j,a,t+1}^{i'} = (1 - v_{t+1})\hat{u}_{j,a,t}^{i'} + v_{t+1}u_{j,a,t+1}^{i'} \text{ if } a_{j,t+1} = a.$$

Otherwise the estimation is unchanged: $\hat{u}_{j,a,t+1}^{i'} = \hat{u}_{j,a,t}^{i'}$. Define:

$$W_{j,a,t+1} = \begin{cases} u_{j,a,t+1} & \text{if } a_{j,t+1} = a \\ \hat{u}_{j,a,t} & \text{otherwise} \end{cases}$$

The learning algorithm can be rewritten as:

$$\hat{u}_{t+1} - \hat{u}_t = v_{t+1}\left(W_{t+1} - \hat{u}_t\right)$$

This process has the form of a stochastic learning algorithm with the distribution of the random vector W_t being determined by the individual updating rules which depend upon the estimations. Assuming that the utility functions are bounded, the sequences generated by the learning algorithm are also bounded. Hence the asymptotic behavior of the learning algorithm can be studied by analyzing the continuous adaptive dynamics of the drift $\mathbb{E}(W_{t+1} \mid \hat{u}_t)$. It is easy to see that the expectation of W given $\hat{u}$ is:

$$\mathbb{E}(W_{j,a} \mid \hat{u}) = \underline{x}_{j,a}(\hat{u})\bar{u}_{j,a} + (1 - \underline{x}_{j,a}(\hat{u}))\hat{u}_{j,a}.$$

Hence, the expected change in one stage can be written as:

$$\mathbb{E}\left(W_{j,a} - \hat{u}_{j,a}\right) = \underline{x}_{j,a}(\hat{u})\bar{u}_{j,a} + (1 - \underline{x}_{j,a}(\hat{u}))\hat{u}_{j,a} - \hat{u}_{j,a}$$

$$= \underline{x}_{j,a}(\hat{u})[\bar{u}_{j,a} - \hat{u}_{j,a}].$$

We deduce the following lemma:

Lemma 8

The stochastic learning algorithm generates the continuous time dynamics given by:

$$\frac{d}{dt}\hat{u}_{j,a,t} = \underline{x}_{j,a,t}(\hat{u})\left(\bar{u}_{j,a}(t) - \hat{u}_{j,a,t}\right) \qquad (6.25)$$

where $\bar{u}_{j,a}(t) = \mathbb{E}\left(u_{j,a}\left(\sum_{j'} Ber_{a,t}^{j'}, t\right) \mid Ber_{a,t}^{j} = 1\right)$ represents the average utility observed by player j when he/she chooses action a and the other players choose it with probabilities $x_{j',a}$ and $Ber_{a,t}^{j}$ denotes a Bernoulli random variable with the parameter $\mathbb{P}(Ber_{a,t}^{j} = 1) = x_{j,a}$.

Proposition 173

The Boltzmann-Gibbs-based stochastic learning algorithm almost surely converges to Nash equilibria.

Proof. The proof follows the same lines as in Benaïm and Faure (2010), Benaim et al. (2006) or Theorem 8 in Cominetti et al. (2007). Using Proposition 4.1 and 4.2 of Benaïm and Faure (2010) with the limit set Theorem 5.7 in Benaïm and Faure (2010), it follows that the symmetric Nash equilibrium is a global attractor, and the function $y \longmapsto \| y - x \|_{\infty}$ is a Lyapunov function for the dynamics in (6.25).

6.6.3 Transition from Micro to Macro in Congestion Games

Define the scaled utility functions as $u_a^{K,i'}\left(\frac{k_a}{K}, t\right)$ and the mean profile $X^K = \frac{1}{K}\sum_{j=1}^{K} \mathbb{1}_{\{a_{j,t}=a\}}$. Assuming that the second moment of the number of players that use the same action is finite[1], the mean process converges weakly to a mean field limit, hence the solution of the system of ordinary differential equations given by the drift $\frac{1}{\Delta_K}f^K(x(t))$ where $f^K(x(t)) = \mathbb{E}\left(X^K(t + \Delta_K) - X^K(t)|X^K(t) = x(t)\right)$ is the expected change in the system in one-stage with duration Δ_K. For example, the Boltzmann-Gibbs dynamics (also called logit dynamics or smooth best response dynamics) is given by:

$$\dot{x}_{ja}(t) = \sum_{a'} x_{j,a'}(t)\rho_{a'a}^{j}(x(t),t) - x_{ja}(t) \qquad (6.26)$$

---

[1]Notice that the number of players that can interact can be very large.

where $\rho_{a'a}^j(x(t)) = \dfrac{e^{\frac{u_{j,a}(x(t),t)}{\epsilon_j}}}{\sum_{\bar{a}} e^{\frac{u_{j,\bar{a}}(x(t),t)}{\epsilon_j}}}$. Players from class j can learn $x(t)$ via the ordi-

nary differential equation (ODE) and can use an action a with probability $x_{j,a}$. Note that our study can be extended to the case where one has both atomic and non-atomic players by considering the weighted population profile: $\tilde{X}^K(t)_{(j,a)} = \frac{1}{\sum_j \gamma_j^K} \sum_j \gamma_j^K \delta_{a_{jt}=a}$ where γ_j^K is the weight ("power") of j in the population of size n.

6.7 EVOLUTIONARY DYNAMICS-BASED LEARNING IN HETEROGENEOUS NETWORKS

We now study how to combine various learning schemes based on mean field game dynamics (see (Sandholm, 2011; Tembine, 2009a,b; Tembine et al., 2009b)). Different learning and adaptive algorithms have been studied in the literature (Bournez and Cohen, 2009; Long et al., 2007; Sandholm, 2011; Srivastava et al., 2005). In all these references the players have to follow the same rule of learning; i.e., they have to learn in the same way. We now ask the following question: what happens if the players have different learning schemes? Our aim here is to propose a class of learning schemes in which players use less information about the other players and less memory about the history of the game, and need not use the same learning scheme. Below we list some mean field limits that we borrow from evolutionary game dynamics (Zhu et al., 2009a,b). They are of the form:

$$\dot{x}_a^\theta(t) = \sum_{\bar{a}} x_{\bar{a}}^\theta(t)\rho_{\bar{a},a}^\theta(x(t)) - x_a^\theta(t) \sum_{\bar{a}} \rho_{a,\bar{a}}^\theta(x(t)) \tag{6.27}$$

where θ denotes the class of players and ρ is the migration rate between strategies.

- *Excess utility dynamics:* Brown–von Neumann–Nash dynamics is one well-known form of excess utility dynamics. It is obtained for:

$$\rho_{a',a}^{1,\theta}(x(t)) = \max(0, u_a^\theta(x(t)) - \sum_{\bar{a}} x_{\bar{a}}^\theta(t)u_{\bar{a}}^\theta(x(t)). \tag{6.28}$$

The set of the rest points of the BNN dynamics is exactly the set of (Nash) equilibria.
- *Imitation of neighbors: replicator dynamics* is obtained for:

$$\rho_{a',a}^{2,\theta}(x(t)) = x_a^\theta(t)\max(0, u_a^\theta(x(t)) - u_{a'}^\theta(x(t))). \tag{6.29}$$

The set of the rest points of replicator dynamics contains the set of equilibrium states, but it can be much bigger since it is known that replicator dynamics may not lead to equilibrium. Typically the corners are rest points and the faces of the simplex are invariant, but may not be an equilibrium.

- *Boltzman-Gibbs dynamics*; also called *smooth best response* dynamics or logit dynamics is obtained for:

$$\rho_{a',a}^{4,\theta}(x(t)) = \frac{e^{\frac{u_a^\theta(x(t))}{\epsilon}}}{\sum_{\bar{a}} e^{\frac{u_{\bar{a}}^\theta(x(t))}{\epsilon}}}, \ \epsilon > 0. \tag{6.30}$$

Using the logit map, it can be shown that the time average of the replicator dynamics is a perturbed solution of the best reply dynamics. The rest points of the smooth dynamics are approximated equilibria. More details on logit learning can be found in Fudenberg and Levine (1998).

- *Pairwise comparison dynamics:* As an example the generalized Smith dynamics is obtained for $\rho_{a',a}^{5,\theta} = \max(0, u_a^\theta(x(t)) - u_{a'}^\theta(x(t)))^\gamma$, $\gamma \geq 1$. The set of rest points of generalized Smith dynamics is exactly the set of equilibria. Sandholm (2011) proved that under this class of dynamics, global convergence holds in stable games and in potential games, etc. Extension to evolutionary game dynamics with *migration between locations of players* and application to hybrid power control in wireless communications can be found in Tembine (2009a), Tembine et al. (2008b).

- *Best response dynamics* is obtained for ρ^6 equal to the best reply to $x(t)$. The set of rest points of best response dynamics is exactly the set of equilibria. More details on best response dynamics can be found in Gilboa and Matsui (1991).

- *Ray-projection dynamics* is a myopic adaptive dynamic in which a subpopulation grows when its expected utility is less than the ray-projection utility of all the other classes. It is obtained from $\rho_{a',a}^{3,\theta}(x(t)) = \Lambda_a^\theta(x(t))$ where Λ is a continuous map.

  Notice that replicator dynamics, best response dynamics, and logit dynamics can be obtained as a special case of the ray-projection dynamics.

Consider a population in which the players can adopt different learning schemes in $\{\rho^1, \rho^2, \rho^3, \ldots\}$ (finite, say with size κ). Then, based on the the composition of the population and the use of each learning scheme, we build a hybrid game dynamic. The in-incoming and outgoing flow, as well as the inter-neighborhood flow are expressed in terms of the weighted combination of different learning schemes, which are picked from the set $\{\rho^1, \rho^2, \rho^3, \ldots\}$.

Define the property *weighted equilibrium stationarity* (WES) as follows:

Every rest point of the hybrid game dynamics generated by the weighted utility is a weighted equilibrium, and every constrained weighted equilibrium is a rest point of the dynamics.

Note that this property is not satisfied by the well-known replicator dynamics, as it is known that replicator dynamics may not lead to Nash equilibria. We have the following result:

Proposition 174

Let $\tilde{\lambda}^j$ be the proportion of the players who adopt the learning scheme ρ^j. If all the learning schemes contained in the support of $\tilde{\lambda} = (\tilde{\lambda}^1, \ldots, \tilde{\lambda}^\kappa) \in \mathbb{R}_+^\kappa$ satisfy the property (WES), then the hybrid mean field limit game dynamics generated by these learning schemes satisfies also the weighted equilibrium stationarity property.

Proof. If x is an equilibrium then x is a rest point of all the dynamics in the support of $\tilde{\lambda}$. We will now prove that any rest point of the combined dynamics in the support of $\tilde{\lambda}$ is an equilibrium. Suppose that this is not the case. Then there exists at least one j such that x is not a rest point of the dynamics generated by the learning scheme ρ^j which satisfies (WES). This means that x is not an equilibrium. We conclude any rest point of combined dynamics is an equilibrium. This completes the proof.

Remark (How to eliminate the rest points that are not equilibria?) Consider the family of learning schemes generated by $\rho_{a',a}^{\gamma,\theta} = \max(0, u_a^\theta(x(t)) - u_{a'}^\theta(x(t)))^\gamma, \gamma \geq 1$. It is easy to see that this family satisfies the property (WES). We deduce that if the population is constituted of 99% of players using a learning scheme via ρ^γ, and 1% of the population use a replicator-based learning scheme, then the resulting combined dynamics will satisfy the property (WES). We conclude that every rest point of the replicator dynamics which is a non-Nash equilibrium will be eliminated using this new combined dynamics. This says that players can learn in a bad way, but if the fraction of good learners is non-zero then the remaining points of the resulting combined dynamics will be equilibria.

6.8 LEARNING SATISFACTION EQUILIBRIUM

In this section we provide a fast and fully distributed learning algorithm for a particular constrained game. The algorithm does not require knowledge of the past actions used by other players, but is completely based on the perceived utilities. We show that the algorithm converges to a satisfaction equilibrium or solution, and explicitly gives the convergence time for a specific class of utility function.

Consider ℓ resources. The set of actions is denoted by $\mathcal{A} = \{1, 2, \ldots, l\}$. Each resource j is associated with a utility $f_j^K : \mathbb{R}^l \longrightarrow \mathbb{R}$. We consider a finite number of players, as well as the asymptotic case (the mean field limit where the number of players tends to infinity). The fraction of players assigned to resource j is denoted by x_j^K. The vector $\underline{x}^K = (x_j^K)_j$ satisfies $\forall j, x_j^K \geq 0, \sum_j x_j^K = 1$. A system state at time t is a vector $\underline{x}^K(t) = (x_j^K(t))_j$. The induced utility by the players on resource j is $f_j^K(\underline{x}^K)$. For the asymptotic case we will examine the case where $f_j(\underline{x}) = f_j(x_j)$, i.e., the utility at resource j depends only on the load x_j. We assume that there is a minimum utility requirement of each player, so each player is satisfied if his utility is at least U. If we refer to the throughput, this gives a minimum throughput requirement to each player: $f_j(\underline{x}) \geq U, \forall j$. This gives a constraint satisfaction problem. A system state x that satisfies $\forall j \in \mathcal{A}, f_j(\underline{x}) \geq U$ is a *satisfaction solution*. Since the throughput decreases with the load, there is a threshold x_j^* such that if $f_j(x) < U$ then $x_j > x_j^*$ (the QoS constraint is violated). The maximum load of channel j such that the constraint is satisfied is therefore $M_j = \sup\{x_j, f_j(x_j) \geq U\}$ which is typically $f_j^{-1}(U)$ for bijective utilities.

We now present the fully distributed learning algorithm for the satisfaction solution.

Algorithm 2: Distributed Learning Algorithm

> **forall the** *players* **do**
> | Initialize to some assignment;
> **end**
> **for** *t=1* **to** *max* **do** do;
> **foreach** *player i* **do**
> | Denote by $j_t = j$ the resource where player i is connected to ;
> | Observe the utility $f_j(x)$;
> | If $f_j(x) < U$ then with probability $(U - f_j)/Af_j$ choose randomly j', and set
> | $j_{t+1} = j'$;
> **end**

Assume that the constraint set $\{x \mid \forall j, \ f_j(x) \geq U, \ x_j \geq 0, \ \sum_j x_j = 1\}$ is non-empty. We will show that the learning algorithm converges to a system state where every player has at least the utility U in the long run. To that end, we introduce a Lyapunov function, which is zero only at a satisfaction solution, and show that the Lyapunov function decreases towards zero. Consider the mapping $L(x) = \sum_{j \in A} \max(0, x_j - M_j)$. The following holds:

- $L(x) \geq 0$
- $[L(x) = 0, \ x \in \Delta(A)] \iff x$ is a satisfaction solution.
- Denote by SSS the set of satisfaction solutions. If x_t is the current system (different to SSS) and x_{t+1} denotes the next state generated by the learning algorithm, then $L(x_{t+1}) - L(x_t) < 0$.

We now determine explicitly the convergence of the learning algorithm for any initial condition. We first define an approximate η-satisfaction solution: Let $B_{\eta'}(x) = \{j, f_j(x) < (1 - \eta')U\}$,

The system x is the η-satisfaction solution for η'-sufficient small, $\sum_{j \in B_{\eta'}(x)} x_j \leq \eta$.

....

Proposition 175

For any initial system state x_0, the time to reach an approximate η-satisfaction solution is in the order of $\frac{|A|}{\eta^2}$.

....

Proof. Consider a state x which is not an ϵ-satisfaction solution, i.e., $\sum_{j \in B_\epsilon(x)} x_j \geq \epsilon$.

It is sufficient to establish the result for the channel j such that $f_j(\underline{x}) < (1 - \frac{\epsilon}{M+1})U$ to the Lyapunov function. The fraction of players leaving these actions is:

$$w_* = \sum_{j \in B_\epsilon(\underline{x})} x_j \frac{U - f_j(\underline{x})}{A f_j(\underline{x})} \geq \epsilon \frac{\epsilon U}{A(1 - \frac{\epsilon}{M+1})U}.$$

$V(\underline{x}) \geq \epsilon$ implies that there exists j such that $M_j - x_j \geq \frac{\epsilon}{m}$ and $\frac{w_*}{\ell}$ of the players migrate to j. The Lyapunov function is reduced at least by $D = \frac{\epsilon^2}{\ell A(1 - \frac{\epsilon}{M+1})}$. Thus, $V(\underline{x}_T) = \sum_{t=1}^{T}(V(\underline{x}_t) - V(\underline{x}_{t-1})) + V(\underline{x}_0) \leq -TD + V(x_0)$. Hence, $V(\underline{x}_T) \leq \epsilon$ when:

$$T \geq (V(\underline{x}_0) - \epsilon)\frac{1}{D} = (V(\underline{x}_0) - \epsilon)\frac{\ell A(1 - \frac{\epsilon}{M+1})U}{\epsilon^2 U}$$

which is in the order of $V(\underline{x}_0)\frac{\ell A}{\epsilon^2}$.

Example 176: QoS – Based Dynamic Spectrum Access

Consider a dynamic spectrum access under a quality of service requirement. Denote by ℓ the total number of channels in the system. x_j denotes the load at channel j. Each player's satisfaction level is U (which can represent a throughput). Assume a large number of players spatially distributed in the system. Assume that the throughput at channel j, f_j depends only on the load x_j. Assuming that the throughput U is feasible, we apply the learning algorithm and an η-satisfaction solution is obtained after, at most, $M\frac{\ell}{\eta^2}$ iterations.

6.9 SUMMARIZING REMARKS AND OPEN ISSUES

6.9.1 Summarizing Chapters 5 and 6

Tables 6.2 and 6.3 summarize some important results from Chapters 5 and 6. All the acronyms used in it are defined or redefined in Table 6.3, and some comments are reviewed when relevant.

To conclude this comparison analysis, Figure 6.1 is provided to illustrate, in a given wireless scenario (a power allocation problem in 2-band 2-user Gaussian interference channels where individual transmission rates are optimized), the behavior of some learning algorithms in terms of their convergence rate (Rose et al., 2011). In Fig. 6.1, we have evaluated the number of stages required to achieve convergence as a function of the signal to noise ratio (SNR). The term convergence refers to the fact that from one stage to another the probability distribution profile does not change significantly: the Euclidean distance between consecutive profiles is smaller than a given threshold. For this figure, we used the term smoothed fictitious play for BGL2 and JUSTE-RL is a version of BGL where both strategies and utilities are reinforced over time (Perlaza et al., 2010).

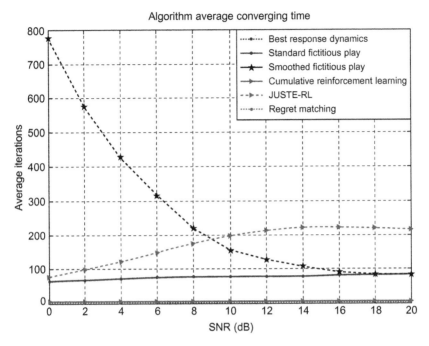

FIGURE 6.1

Average number of stages required to observe convergence to an equilibrium in a channel selection game versus the signal to noise ratio (SNR). Convergence is said to be effective when the Euclidean distance between two consecutive (probability distribution) vectors at stage t and $t+1$ is lower than $\epsilon = 10^{-4}$ (Rose et al., 2011).

6.9.2 Discussion and Open Issues

We have presented various learning algorithms for wireless games with different information levels: both partially distributed and fully distributed learning algorithms. Convergence, non-convergence, stability and applicability to wireless engineering have been discussed. We have also studied the convergence time of some specific learning algorithms. However, some interesting questions remain open:

- For what classes of wireless games can one design *distributed fast algorithms* that get close to equilibrium? If convergence occurs, what are the theoretical bounds to the rate at which convergence to equilibrium can occur?

    We have given examples of finite games in which the iterative best response algorithm converges in a finite number of steps (using improvement path techniques), but in general the convergence time of learning algorithms (when they converge) can be arbitrarily large.

    What are the speed of convergence of the learning schemes (when they converge)? The learning algorithms may not converge, but in some special cases

Table 6.2 Summary

Learning Scheme	CA/DA	FD/PD	Suff. Cond. on $\mathcal{G}$ for Conv.	Convergence	Steady State
ABRD	Both	PD	Weakly acyclic	Finite time	BP/NE/LC
			Dominance solvable	Finite time	BP/NE/LC
			Potential	Finite time	BP/NE/LC
			Supermodular	Finite time	BP/NE/LC
BFPL	Both	PD	Potential	In frequencies	NE
			Zero-sum	unbounded time	
			generic $2 \times m$	unbounded time	
			Dominance solvable	unbounded time	
MFPL	Both	PD	Potential	In frequencies	CCE
			Zero-sum	unbounded time	
			Dominance solvable		
BGL1	DA	PD	Potential	unbounded time	LC, LE
			Dominance solvable		
			Generic 2×2		
BGL2	DA	FD	Congestion	Utility dyn. $O\left(\frac{1}{\eta}\right)$	LC, LE, NE
			Generic 2×2	unbounded time	
			Dominance solvable		
NBSL	Both	PD	Potential	unbounded time	PO
TEL	DA	FD	Generic game with NE	In frequencies	high proportion of NE
ARL	DA	FD	Generic 2×2	Replicator dyn.	boundary/NE/LC
			Potential	unbounded time	
			Cooperative dynamics		
			Dominance solvable		
BRL	DA	FD	Potential	Replicator dyn.	boundary/NE/LC
			Cooperative dynamics	unbounded time	
			Generic 2×2		
			Dominance solvable		
RML	DA	FD		In frequencies	CE
				unbounded time	
SEL	DA	FD	Potential	Finite time	SE
			Supermodular	Finite time	SE

where convergence does occur, we have observed different speed of convergence to a rest point when different learning rates are used by the players. We have observed that increasing the learning rates can accelerate the convergence to rest points. However, for some learning algorithms such as logit response learning,

Table 6.3 List of Acronyms

Acronym	Meaning
ABRD	Asynchronous best response dynmics
BFPL	Brown's fictitious play learning
MFPL	Modified fictitious play learning
BGLn	Boltzmann-Gibbs learning version n
NBSL	Nash bargaining solution learning
TEL	Trial-and-error learning
ARL	Arthur's model based reinforcement learning
BRL	Borger and Sarin's model based reinforcement learning
RML	Regret matching-based learning
SEL	Satisfaction equilibrium learning
CA	Continuous action (set/space)
DA	Discrete action (set/space)
FD	Fully distributed (learning)
PD	Partially distributed (learning)
BP	Points which are on the boundary of the unit simplex
CCE	Coarse correlated equilibrium
CE	Correlated equilibrium
LC	Limit cycle
LE	Logit equilibrium
NE	Nash equilibrium
PO	Pareto optimal
SE	Satisfaction equilibrium

the time it takes to converge (close) to the stationary distribution is exponential with respect to the number of players.

- What type of learning and adaptive behavior do players use in large interactive systems? We have presented a learning framework in which if players use different learning schemes, they learn in different ways. This new approach has the advantages of capturing many observed scenarios. We gave qualitative properties to the combined dynamics obtained when the players learn in different ways. In particular, in the presence of good learners, bad learners, slow learners or quick learners, our approach can be useful if the fraction of players that are good learners is not too small. However, the analysis under these mixtures of different learners seems more complicated.
- Are equilibrium approaches the right concepts for predicting long-term behavior in wireless games? Since most equilibrium concepts are known to be inefficient in

known scenarios (power allocation, access control, spectrum management, routing, etc.), and most proposed learning algorithms fail to converge, the outcome will not be played. It is hence reasonable to ask if the proposed solution concepts are well adapted. This aspect corresponds to an active area and many researchers are currently focusing on set-valued solution concepts (instead of point-solution concepts), robust games, game situations, adding both endogenous and exogenous context of the classical formulation, non-equilibrium approach, etc.

Another interesting direction is the design of fully distributed learning algorithms leading to *global optima* in simple classes of wireless games.

Case Studies

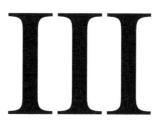

Energy-Efficient Power Control Games

7.1 INTRODUCTION

Many wireless communications systems are optimized in terms of quality of service (QoS), which can include for example, performance criteria such as transmission rate, reliability, latency, or security. It turns out that applications where a trade-off between QoS and energy consumption has to be found have become more and more important, especially over the past decade. Wireless sensor networks (e.g., see Akyildiz et al. (2002)) and ad-hoc networks (e.g., see ElBatt and Ephremides (2004)) are two good examples that illustrate the importance of finding such a trade-off. As in other areas, like economics, finance, or physics, the way of measuring this trade-off, as considered in this chapter, corresponds to the benefit to cost ratio. In wireless communications, the transmission rate and transmit power are two common measures of the benefit and cost of a transmission, respectively. Therefore, the ratio of transmission rate (say in bit/s) to transmit power (in J/s) appears to be a quite natural energy efficiency measure of a communication system. In Goodman and Mandayam (2000) and Shah et al. (1998) the authors define the energy-efficiency as the net number of information bits that are transmitted without error per unit time (goodput) to the transmit power level. This chapter considers this pragmatic performance measure for analyzing the problem of distributed power control in wireless systems. The wireless system under consideration is a distributed multiple access channel. By definition (Cover and Thomas, 2006), the multiple access channel consists of a network of several transmitters and one receiver. The network is said to be distributed (or decentralized) in terms of decision: the receiver does not dictate to the users their transmit power level. Indeed, based on given knowledge (his own uplink channel typically), each user can freely choose his power control policy, in order to selfishly maximize his individual energy-efficiency. As noticed by Goodman and Mandayam (2000) and Shah et al. (1998), a relevant approach to this problem is to formulate it as a strategic-form non-cooperative game for which the set of players is the set of transmitters, the sets of actions are the intervals corresponding to the transmit power ranges, and the players' utility functions are their energy-efficiencies.

This case study, "Energy-efficient power control games", comprises six main parts (without counting the part dedicated to the system model). In Sections 7.3 and 7.4, a static or one-shot game model is exploited to analyze the distributed power

Game Theory and Learning for Wireless Networks

183

control problem where the transmit power level is updated once per block. A block duration corresponds to a time window over which the channel is assumed to be constant. More precisely, a block is defined as a sequence of M consecutive symbols which comprises a training sequence; that is a certain number of consecutive symbols are used to estimate the channel, the signal-to-interference plus noise ratio (SINR), etc. Note that the game is assumed to be played once per block and in an independent manner from block to block. Therefore, transmitters are assumed to maximize their instantaneous energy-efficiency for each block. In Section 7.3, it is shown that the unique outcome of the game is a pure Nash equilibrium (NE). To play their equilibrium action, we will see that under some conditions, only individual channel state information (CSI) is needed at each transmitter. In Section 7.4, a pricing mechanism is introduced (Saraydar et al., 2002) in order to improve the transmitters' utilities at equilibrium. However, it will be seen that implementing this pricing mechanism requires global channel state information at all the transmitters.

In Section 7.5, some hierarchy is introduced into the system. In this context, terminals can be leaders or followers in the sense of Stackelberg (1934). Hierarchy is naturally present in contexts where there are primary (licensed) users and secondary (unlicensed) users who can sense their environment, as they are equipped with cognitive radios (Akyildiz et al., 2006; Fette, 2006; Haykin, 2005); in this case a two-level Stackelberg game is sufficient to describe the system. Hierarchy is also natural when the users have access to the medium in an asynchronous manner. In the latter case, the Stackelberg approach can be interpreted as a multi-stage (dynamic) game where leaders play first and followers, who observe the actions played by the leaders, play next. In fact, what is presented in Section 7.5 is a more complex hierarchy structure since a multi-level Stackelberg game is considered. Such a structure can be a given feature of the network, or it can be introduced in a distributed network to improve the network equilibrium efficiency. It can therefore be seen as an alternative approach to pricing. However, in contrast to pricing, only individual CSI is needed in order for the transmitters to play their equilibrium actions. In Sections 7.6 and 7.7, a dynamic formulation of the problem is exploited. Again, this can be seen as a way of improving system performance, or/and taking into account some features of the network. Two scenarios are investigated, depending on whether the transmitters can update their power level several times per block or not. The two scenarios will be referred to as fast and slow power control. In Sections 7.6.1 and 7.6.2 it is shown that, under some conditions, both scenarios can be modeled by a repeated game (with discounting) where the static game studied in Section 7.3 is repeated and played over a block or time window that is smaller than the channel coherence time. In Section 7.7, only the general case of slow power control (in which the channel can vary from stage to stage) is considered. The general model for studying this problem is the one of stochastic games (e.g., see Dutta (1995) for stochastic games with perfect monitoring). In terms of CSI, Section 7.7 requires global CSI at the transmitters whereas only individual CSI is needed to implement the proposed (non-stochastic) repeated equilibrium strategy.

7.2 GENERAL SYSTEM MODEL

We consider a distributed multiple access channel with a finite number of users, which is denoted by K. The network is said to be distributed in the sense that the receiver (e.g., a base station) does not dictate to the transmitters (e.g., mobile stations) their power control policy. Rather, all the transmitters choose their policy by themselves in order to selfishly maximize their energy-efficiency; in particular they can ignore some specified centralized policies. We assume that the users transmit their data over quasi-static or block fading channels at the same time and on the same frequency band; recall that a block is defined as a sequence of M consecutive symbols which comprises a training sequence. The receiver knows on each block all the uplink channel gains (coherent communication assumption) whereas each transmitter has only access to the knowledge of his own channel. The latter assumption is realistic at least in two common scenarios: (a) the uplink-downlink channel reciprocity is valid and the receiver sends training signals to the transmitters; (b) the uplink channel varies slowly and the receiver implements a reliable feedback mechanism to inform the transmitters with their channel state. The signal model used corresponds to the information-theoretic channel model used for studying multiple access channels (Cover, 1975; Wyner, 1974); e.g., see Belmega et al. (2009b) for more comments on the multiple access technique involved. What matters is that this model is simple to present, captures the different aspects of the problem (the SINR structure in particular), and also can be readily applied to specific systems such as code division multiple access (CDMA) systems (Goodman and Mandayam, 2000; Lasaulce et al., 2009b) or multi-carrier CDMA systems (Meshkati et al., 2006). The equivalent baseband signal received by the receiver can be written as:

$$y(n) = \sum_{i=1}^{K} h_i(n)x_i(n) + z(n) \qquad (7.1)$$

where $i \in \mathcal{K}$, $\mathcal{K} = \{1, \dots, K\}$, $x_i(n)$ represents the symbol transmitted by transmitter i at time n, $\mathbb{E}|x_i|^2 = p_i$, z is assumed to be distributed according to a zero-mean Gaussian random variable of variance σ^2, and each channel gain h_i varies over time, but is assumed to be constant over each block. For each transmitter i, the channel gain modulus is assumed to lie in a compact set $|h_i|^2 \in \left[\eta_i^{\min}, \eta_i^{\max}\right]$ in all sections except in Section 7.7. The assumption of limited support for $|h_i|$ is both practical and important to guarantee the existence of the equilibrium derived in Section 7.6. For example, the lower bound on the channel gain $\eta_i^{\min}$ may translate in wireless communications to the fact that the receiver has a finite sensitivity: below a certain threshold in terms of signal-to-interference plus noise ratio, communication is cut off and the transmitter is therefore no longer active. The upper bound $\eta_i^{\max}$ may model the existence of a minimum distance between the transmitter and receiver. Transmit power levels p_i will be assumed to lie in a compact set $\mathcal{P}_i = [0, P_i^{\max}]$ in Sections 7.3–7.6 and in a discrete set $\mathcal{P}_i = \left\{P_i^1, \dots, P_i^{m_i}\right\}$ (with $P_i^{m_i} = P_i^{\max}$ and $m_i \in \mathbb{N}^*$) in Sections 7.7 and 7.8.

In Section 7.7 the discrete set assumption is used to apply the results of Fudenberg and Yamamoto (2009) and Hörner et al. (2009) without introducing additional technicalities. Finally, the receiver is always assumed to implement single-user decoding; i.e., each flow of symbols is decoded independently of the others, which are considered as noise. For more advanced receivers such as those implementing successive interference cancellation, the reader is referred to Bonneau et al. (2008) and Lasaulce et al. (2009b).

7.3 THE ONE-SHOT POWER CONTROL GAME

Here we study the scenario, introduced by Goodman and Mandayam (2000), where each transmitter updates his power level once per block in order to maximize his instantaneous utility. In order to define the corresponding static power control game some notation is in order. We denote by R_i the information transmission rate (in bit/s) for user i, and by f an efficiency function representing the block success rate, which is assumed to be sigmoidal and identical for all the users; recall that a sigmoidal or S-shape function is convex up to a point and then concave from this point. For a given block, the SINR at receiver $i \in \{1, \ldots, K\}$ is denoted by SINR_i and written as:

$$\mathrm{SINR}_i = \frac{\eta_i p_i}{\sum_{j \neq i} \eta_j p_j + \sigma^2} \tag{7.2}$$

where $\eta_i = |h_i|^2$ and $p_i \in [0, P_i^{\max}]$. With these notations, the static power control (PC) game, called $\mathcal{G}$, is defined in its normal or strategic form as follows.

Definition 177: Static Power Control Game

The static power control game is a triplet $\mathcal{G} = (\mathcal{K}, \{\mathcal{P}_i\}_{i \in \mathcal{K}}, \{u_i\}_{i \in \mathcal{K}})$ where $\mathcal{K} = \{1, \ldots, K\}$ is the set of players, $\mathcal{P}_1, \ldots, \mathcal{P}_K$ are the sets of actions, and $u_1, \ldots, u_K$ are the utility functions of the different players, which are defined by:

$$u_i(p_1, \ldots, p_K) = \frac{R_i f(\mathrm{SINR}_i)}{p_i} \quad [\text{bit/J}]. \tag{7.3}$$

We suppose that the game is played with complete information and rational players, and that rationality is common knowledge. In other words, every transmitter knows $\mathcal{G}$ (this assumption calls for certain comments, which will be given in Sec. 7.3.3), every transmitter chooses his action from his best response set, and all the transmitters know the others do so and so on. At least three questions arise naturally from these assumptions. Is there a solution to the conflict of interest associated with the distributed power control problem? Is this solution unique? Is it possible to determine it? Answering these questions corresponds to solving the existence, uniqueness, and determination problems of a pure Nash equilibrium respectively. To this end, we exploit the theorems provided in Chapter 2.

7.3.1 **Existence of a Pure Nash Equilibrium**

Not surprisingly, the existence issue is related to the choice of the efficiency function f. It turns out that, under reasonable assumptions for f, the Debreu–Fan–Glicksberg existence theorem can be applied. First, for all $i \in \mathcal{K}$, each action space $[0, P_i^{\max}]$ is compact and convex. Second, if for all $i \in \mathcal{K}$, $\lim_{p_i \to 0} u_i(p_i, \underline{p}_{-i}) = 0$ (where $\underline{p}_{-i}$ is the vector of actions played by the players other than i), the utility function u_i is continuous in the action (power) profile $\underline{p} = (p_1, \ldots, p_K)$ on $\mathcal{P} = \bigotimes_{i=1}^{K} \mathcal{P}_i = \mathcal{P}_1 \times \cdots \times \mathcal{P}_K$

where the notation $\bigotimes$ refers to the Cartesian product of sets. This sufficient condition on f is reasonable, since it assumes $f(0) = 0$ and is met for two important efficiency functions: the one used in Goodman and Mandayam (2000) and Shah et al. (1998) which approximates the block success rate by $f(x) = (1 - e^{-x})^M$; and also the one proposed by Belmega and Lasaulce (2009) where $1 - f(x)$ corresponds to the outage probability in information theory (Ozarow et al., 1994); that is, $f(x) = e^{-\frac{c}{x}}$ where $c = 2^R - 1$, R being the target rate in the outage probability (see Belmega and Lasaulce (2009) for more details). As a third step towards applying the Debreu–Fan–Glicksberg theorem, the quasi-concavity of $u_i(\underline{p})$ w.r.t. p_i needs to be verified. In fact, this follows directly from the sigmoidness assumption for f. In Rodriguez (2003) it is proved that, if $f(x)$ is sigmoidal, then $\frac{f(x)}{x}$ is quasi-concave. The graph of the function $u(x) = \frac{(1-e^{-x})^M}{x}$ is shown in Figure 7.1 for $M = 100$.

7.3.2 **Determination and Uniqueness of the Pure Nash Equilibrium**

As mentioned in Chapter 2, it seems natural to know whether there is a unique equilibrium and then determine the corresponding solution(s). However, this is not always relevant. For the static game under study (for which its existence has been proved), it is easier to determine the possible equilibrium solutions first and then prove uniqueness. In our setup, Nash equilibria are the solutions of the following system of equations:

$$\begin{cases} \frac{\partial u_1}{\partial p_1}(\underline{p}) = 0 \\ \quad \vdots \\ \frac{\partial u_K}{\partial p_K}(\underline{p}) = 0. \end{cases} \qquad (7.4)$$

By calculating and setting these derivatives to zero, one easily finds that the optimal action for player i is to tune his transmit power level such that his SINR equals β^*, which is the unique (non-zero) solution of the equation $xf'(x) - f(x) = 0$ (uniqueness follows from the sigmoidness of f). Using the expressions of the SINR of the different users, we find that the corresponding equilibrium is given by:

$$\forall i \in \{1, \ldots, K\}, \ p_i^{\text{NE}} = \frac{\sigma^2}{|h_i|^2} \beta^* \mu_K \qquad (7.5)$$

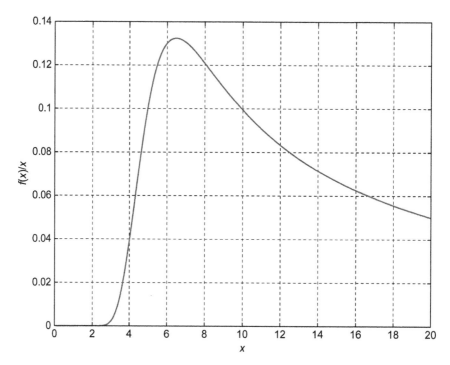

FIGURE 7.1

Graph of the function $u(x) = \frac{(1-e^{-x})^M}{x}$. The function u is quasi-concave and has a single maximum.

where $\mu_K = \frac{1}{1-(K-1)\beta^*}$ is a penalty term due to multiple access interference. Note that we have implicitly assumed that this solution is in the interval $[0, P_i^{\max}]$. Otherwise p_i^{NE} has to be replaced with $\min\{p_i^{\mathrm{NE}}, P_i^{\max}\}$. We refer to this scenario as the saturated regime: the network is said to operate in the saturated regime if at least one terminal transmits at its maximal power. We will come back to this scenario in Sec. 7.3.3. At this point, we have determined one pure Nash equilibrium, which is $\underline{p}^{\mathrm{NE}} = (p_1^{\mathrm{NE}}, \dots, p_K^{\mathrm{NE}})$. Is it the only pure equilibrium of $\mathcal{G}$? The following proposition gives the answer.

Proposition 178: Equilibrium Uniqueness

The game $\mathcal{G}$ has a unique pure NE.

Proof. To prove the uniqueness of the Nash equilibrium, we apply Theorem 74 in Chapter 2, which has been derived by Yates (1995) and more recently re-used by Saraydar et al. (2002). We know from Theorem 74 that if the vector-valued best-response (BR) function $\underline{BR}(\underline{p}) = \left(BR_1(\underline{p}_{-1}), \ldots, BR_K(\underline{p}_{-K}) \right)$ is monotonic and scalable, then there is a unique Nash equilibrium. Knowing that each user tunes his transmit power such that his SINR equals to β^*, it is easy to express the best responses of the transmitters. Here we have that the best response function:

$$\forall i \in \{1, \ldots, K\}, \; BR_i(\underline{p}_{-i}) = \frac{\beta^*}{|h_i|^2} \left(\sigma^2 + \sum_{j \neq i} |h_j|^2 p_j \right). \tag{7.6}$$

Again, this holds under the non-saturated regime assumption, where no transmitter transmits at $P_i^{\max}$. Clearly we have that:

1. $\underline{p}' \leq \underline{p} \Rightarrow \underline{BR}(\underline{p}') \leq \underline{BR}(\underline{p})$ since $\frac{\partial BR_i(\underline{p})}{\partial p_j} = \beta^* \left| \frac{h_j}{h_i} \right|^2 \geq 0$ (monotonicity);
2. $\forall \alpha > 1, \; \underline{BR}(\alpha \underline{p}) < \alpha \underline{BR}(\underline{p})$ (scalability);

where the assumed order $\underline{p}' \leq \underline{p}$ means that $\forall i \in \{1, \ldots, K\}, p_i' \leq p_i$.

7.3.3 Comments on the Information Assumptions

It would be legitimate for the reader to ask what the transmitters need to know. Indeed, in Sec. 7.3, we have mentioned that all transmitters need to know $\mathcal{G}$ (and therefore have global CSI) whereas the equilibrium solution (7.5) only requires individual CSI at the transmitters. Notice that the complete information assumption is a general and an a priori assumption. In the specific game under study, less information is needed, provided the transmitters are rational and rationality is common knowledge. Specifically, in the non-saturated regime mentioned in the preceding section, individual CSI is sufficient. Let us explain why this is the case. We want to show that, in the non-saturated regime, only $|h_i|$ needs to be known by transmitter i. In the existing literature (Goodman and Mandayam, 2000; Lasaulce et al., 2009b; Meshkati et al., 2006, etc.) this game feature is observed once the equilibrium solution has been determined. It turns out that it is due to the intrinsic structure of the one-shot game and not only of the specific solutions analyzed in the existing literature. This can therefore be confirmed before deriving the equilibrium solution. A simple way of proving this statement is to consider a game where utilities are normalized:

$$\frac{u_i(p_1, \ldots, p_K)}{|h_i|^2} = \frac{R_i f \left(\frac{p_i |h_i|^2}{\sum_{j \neq i} p_j |h_j|^2 + \sigma^2} \right)}{p_i |h_i|^2}. \tag{7.7}$$

By making the change of variables $a_i = p_i|h_i|^2$ the normalized utility function becomes:

$$\widehat{u}_i(a_1,\ldots,a_K) = \frac{R_i f\left(\dfrac{a_i}{\displaystyle\sum_{j\neq i} a_j + \sigma^2}\right)}{a_i} \tag{7.8}$$

with $a_i \in \mathcal{A}_i = \left[0, \frac{\eta_i^{\max}}{\eta_i^{\min}} \frac{\sigma^2}{1-(K-1)\beta^*}\right]$. It is seen from the normalized utilities that channel gains only play a role when players de-normalize their actions by computing $p_i = \frac{a_i}{|h_i|^2}$, which shows that only individual CSI is needed to play the game. On the other hand, if one of the transmitters transmits at full power, this attractive property is partially lost. For example, in the 2-player case, if player 2 transmits at his maximum power $p_2 = P_2^{\max}$, player 1 transmits at $p_1^{NE} = \beta^* \frac{\sigma^2}{|h_1|^2} + \beta^* \left|\frac{h_2}{h_1}\right|^2 P_2^{\max}$. Therefore transmitter 1 needs to have global CSI ($|h_1|, |h_2|$). The interest in designing a system such that a non-saturated equilibrium is obtained is therefore obvious.

7.4 LINEAR PRICING-BASED POWER CONTROL

As mentioned in Goodman and Mandayam (2000), the Nash equilibrium of the one-shot power control game (7.5) is generally inefficient in the sense of Pareto that is, it is possible to find power profiles which Pareto dominate $\underline{p}^{NE}$. This is why Saraydar et al. (2002) proposed a pricing mechanism to obtain improvements for the transmitters' utilities with respect to the case with no pricing. Although pricing has not been studied in the preceding chapters it can be understood very easily. For example, consider a routing problem in which users (say drivers) have to choose between two routes A and B. If the number of drivers on road A is found to be too high (by a central authority), in the sense that it leads to an inefficient situation (too many traffic jams or accidents), the central authority can decide to charge a toll for driving on route A. This toll system implements a pricing mechanism. A wireless counterpart of routing problems where Shannon transmission rate type utilities are considered, will be studied in Chapter 9. For the one-shot power control game a simple pricing mechanism is presented in Saraydar et al. (2002), in which utility function is modified as:

$$\widetilde{u}_i(\underline{p}) = u_i(\underline{p}) - \alpha p_i, \tag{7.9}$$

which corresponds to implementing a linear pricing technique.

For instance, the parameter $\alpha \geq 0$ (called an externality in game theory) can be tuned so as to reach a desired gain in terms of the sum of utilities, and/or follow a certain fairness criterion between the transmitters. The existence and uniqueness of a pure Nash equilibrium in the modified game must be demonstrated. In general, the quasi-concavity property is not preserved by addition of a linear function, which is what happens here for $\widetilde{u}_i$, $i \in \mathcal{K}$. Therefore, the Debreu–Fan–Glicksberg existence theorem cannot be used anymore. But, as noticed in Saraydar et al. (2002), by making an additional change, the resulting static game becomes super-modular, which allows one to apply an existence theorem for super-modular games (Theorem 59 in Chapter 2). Indeed, the trick proposed in Saraydar et al. (2002) is to assume that each transmitter operates in the interval $\widetilde{\mathcal{P}} = [P_i^c, P_i^{\max}]$, where P_i^c is the power level which transmitter i requires to operate at an SINR equal to β_0, β_0 being the unique inflection point of f. It is simple to check that if $\forall i \in \mathcal{K}, \forall \underline{p}_{-i} \geq \underline{p}'_{-i}$, then the quantity $\widetilde{u}_i(\underline{p}) - \widetilde{u}_i\left(p_i, \underline{p}'_{-i}\right)$ is non-decreasing in p_i on $\widetilde{\mathcal{P}}_i$. The interpretation is that if the other transmitters, $-i$, increase their power (generating more interference), then transmitter i has to increase his. The game:

$$\widetilde{\mathcal{G}}_\alpha = (\mathcal{K}, \{\widetilde{\mathcal{P}}_i\}_{i \in \mathcal{K}}, \{\widetilde{u}_i\}_{i \in \mathcal{K}}) \tag{7.10}$$

is therefore super-modular, which ensures the existence of at least one pure Nash equilibrium, following Theorem 59 in Chapter 2.

The uniqueness problem is not trivial, and so far only simulations (Saraydar et al., 2002) have been used to analyze this issue. Another disadvantage of pricing-based, energy-efficient power control is that global CSI is required at the transmitters, which is one of the reasons why the authors of Lasaulce et al. (2009b) proposed an alternative way of tackling the problem, by introducing some hierarchy in the system.

7.5 THE HIERARCHICAL POWER CONTROL GAME

The distributed multiple access channel comprises $K + 1$ terminals – i.e. K transmitters and 1 receiver. In Sections 7.3 and 7.4 the receiver was not a player of the game (i.e., it cannot take any decisions), and the problem is symmetric for the transmitters in terms of available information (individual CSI in particular). In this section, we first consider the situation where the receiver is still not a player, but some transmitters are able to observe the actions of the others while others do not (Sec. 7.5.1). Hierarchy is therefore introduced in the sense that some transmitters have more or less information than others. As a second step (Sec. 7.5.2), the receiver will be considered as a player of the game who has to rank the transmitters of the network. In the existing literature, three scenarios have been addressed.

1. Scenario 1 (Lasaulce et al., 2009b): the power control game comprises one leader and $K - 1$ followers. The leading transmitter cannot observe the actions of the other transmitters. In contrast, all the followers can observe the action chosen by the leader.

2. Scenario 2 (He et al., 2010b): this time the number of leaders is arbitrary (say K_1) and the number of followers is also arbitrary $K_2 = K - K_1$. This two-level game generalizes Scenario 1.

3. Scenario 3 (He et al., 2010b): there are K levels of hierarchy, with one transmitter per level. This is the scenario reported in this chapter.

The motivations for such a framework are as follows (Lasaulce et al., 2009b):

- To obtain an equilibrium profile which is more efficient (in the sense of Pareto) than the Nash equilibrium profile of $\mathcal{G}$.
- As this can be reached by implementing a pricing mechanism, the second motivation is to avoid the four main drawbacks of the pricing-based power control policy, namely the goal is that:
  - the additional assumptions made on the action spaces in order to have the super-modularity property in $\widetilde{G}_\alpha$ be removed;
  - the system is predictable analytically; that is the uniqueness of the game outcome is ensured;
  - the optimal power control policy be determined explicitly;
  - the transmitters exploit individual CSI only (at least in the non-saturated regime).

A general motivation for this approach is to design networks where intelligence is split between the receiver (e.g., the base station) and transmitters (e.g., the mobile stations) in order to find a desired trade-off between the global network performance reached at the equilibrium and the amount of signaling needed to make it work. This cost induced by the proposed formulation is not evaluated here and both this evaluation and the mentioned tradeoff are problems left to the interested reader.

7.5.1 Introducing Hierarchy Among the Transmitters

In this section, we propose a Stackelberg formulation of the power control game where there are K different levels. Without loss of generality (but possibly with loss of optimality) we assume that transmitter K can observe the action played by transmitters $K-1, K-2, \ldots, 1$, transmitter $K-1$ can observe the action played by transmitters $K-2, K-3, \ldots, 1$, $\ldots$, and transmitter 1 cannot observe any transmitter. The receiver is not a player of the game here. In this respect, we always assume that single-user decoding is implemented at the receiver. The motivations for using single-user decoding are precisely that the receiver has to remain neutral in the game, and/or to limit the receiver complexity, and/or to minimize the possible signaling cost induced by a more advanced receiver. For example, in Lasaulce et al. (2009b) the authors assume successive interference cancellation, which naturally introduces hierarchy between transmitters via the decoding order used by the receiver. Under the assumptions made, the utility of transmitter 1 only depends on p_1 since transmitter 1 knows that transmitter 2 observes his action and reacts to this observation accordingly. So $p_2 = p_2(p_1)$, the utility of transmitter 2 only depends on (p_1, p_2), $\ldots$, and the utility of transmitter K depends on the whole action profile $(p_1, \ldots, p_K)$. Knowing that the reaction of a player is always a scalar-valued function, the corresponding utility functions can be defined by:

$$
\begin{cases}
u_K^S(p_1, p_2, \ldots, p_{K-1}, p_K) = u_K(p_1, p_2, \ldots, p_{K-1}, p_K) \\
u_{K-1}^S(p_1, p_2, \ldots, p_{K-1}) = u_{K-1}[p_1, p_2, \ldots, p_{K-1}, p_K(p_1, p_2, \ldots, p_{K-1})] \\
\qquad\qquad \vdots \\
\quad u_1^S(p_1) \qquad = u_1[p_1, p_2(p_1), p_3(p_1, p_2(p_1)), \ldots, p_K(p_1, p_2(p_1), \ldots, \\
\qquad\qquad\qquad p_{K-1}(p_{K-2}(\ldots(p_1))))]
\end{cases}
$$

$$(7.11)$$

where the functions $p_2(p_1)$, $p_3(p_2(p_1))$, etc. are respectively the reaction of transmitter 2 to transmitter 1, etc.

A multi-level Stackelberg solution (SS) is defined as a Nash equilibrium in the game:

$$
\mathcal{G}^S = (\mathcal{K}, \{\mathcal{P}_i\}_{i \in \mathcal{K}}, \{u_i^S\}_{i \in \mathcal{K}})
\tag{7.12}
$$

where for all players, the reaction of each player is taken in his best response correspondence.

Definition 179: Multi-Level Stackelberg Solution

A multi-level Stackelberg solution is defined as a pure Nash equilibrium of $\mathcal{G}^S$; that is $\underline{p}^{SS}$ is a Stackelberg solution if it is a solution of the system of equations:

$$
\begin{cases}
p_K(p_1, \ldots, p_{K-1}) \in \arg\max_{p_K} u_K^S(p_1, \ldots, p_K) \\
p_{K-1}(p_1, \ldots, p_{K-2}) \in \arg\max_{p_{K-1}} u_{K-1}^S(p_1, \ldots, p_{K-1}) \\
\qquad\qquad \vdots \\
p_1 \in \arg\max_{p_1} u_1^S(p_1).
\end{cases}
\tag{7.13}
$$

It turns out that none of the theorems provided in Chapter 2 seems to be appropriate to prove the existence and uniqueness of a multi-level Stackelberg solution of the power control game $\mathcal{G}^S$. This shows us a case where available theorems are not exploitable, and therefore an ad hoc proof needs to be found. This is what the authors of Lasaulce et al. (2009b) did by introducing sufficient conditions to ensure the existence and uniqueness of the Stackelberg solution. These conditions will not be reported here. What will be mentioned is that the corresponding conditions are met for the two efficiency functions $f(x) = (1 - e^{-x})^M$ and $f(x) = e^{-\frac{c}{x}}$ that we have already discussed. Under the sufficient conditions of He et al. (2010b), there is a unique multi-level Stackelberg solution $\underline{p}^{SS}$, and this equilibrium can be compared with the Nash equilibrium obtained in the one-shot game of Sec. 7.3, that is, $\underline{p}^{NE}$. Let us express the Stackelberg solution profile and then compare it with the one-shot game Nash equilibrium profile. By calculating the different derivatives $\frac{\partial u_i^S}{\partial p_i}$ and setting them to zero it can be shown that (He et al., 2010b):

$$p_i^{SS} = \frac{\sigma^2}{|h_i|^2} \frac{\beta_i \prod_{j \neq i}(1+\beta_j)}{\prod_{j=1}^{K}(1-\beta_j b_j)} \tag{7.14}$$

where β_i is the unique solution of the equation $x(1-b_i x)f'(x) - f(x) = 0$, $b_K = 0$, and for all $i \in \{1,\ldots,K-1\}$, b_i is given by:

$$b_i = \prod_{j=i+1}^{K}\left(\frac{1+\beta_j}{1-\beta_j b_j}\right) - 1. \tag{7.15}$$

We see that each terminal only needs individual CSI to play his best action in the multi-level power control game. By calling a hierarchy configuration a choice consisting of designating which transmitter is number 1, which transmitter is number 2, etc., the following proposition can then be proved (He et al., 2010b).

. . . .

Proposition 180: Stackelberg Solution vs Nash Equilibrium

For any hierarchy configuration, $\underline{p}^{SS}$ Pareto dominates $\underline{p}^{NE}$; that is, for all $i \in \mathcal{K}$, $u_i(\underline{p}^{SS}) \geq u_i(\underline{p}^{NE})$.

An interesting observation that follows from this proposition is that not only the leader gains in terms of utility by being in a hierarchical network. Any follower (independent of his rank), also prefers the hierarchical game to the one-shot game. Note that this result is not general for multi-level hierarchical games; in economics, for instance, in the case of a Cournot duopoly (Hamilton and Slutzky, 1990), only the leader can benefit from the introduction of a hierarchy. Even more surprisingly, in our context, a user prefers to be a follower to a leader. In cognitive networks with primary and secondary users, only terminals that are equipped with a cognitive radio will have this privilege. In a conventional cellular network, the designation of a leader can be made by the base station. To make it simple, consider the example of one leader and $K-1$ followers (Lasaulce et al., 2009b). By broadcasting p_1, the receiver discloses the transmitter who allows a certain global performance metric to be maximized and the power level he should play. The users, who are designated as followers, and who are selfish, rational, and can not coordinate with each other, are going to play their best response $p_i(p_1)$. Knowing this, the user who is designated as a leader has to transmit at p_1 to maximize his utility. Of course, an important question for the network designer or owner is whether there is an optimal hierarchy configuration. This is addressed in the next section.

7.5.2 **When the Receiver Becomes a Player**

The problem of determining the optimal hierarchy configuration amounts to considering the receiver as a player, who can assign some ranks to the transmitters and optimize his own utility function. A common choice of utility function is the network social welfare:

$$w = \sum_{i=1}^{K} u_i. \tag{7.16}$$

Let us denote by π the one-to-one mapping (permutation) which assigns a given rank $\pi(i)$ to transmitter i:

$$\pi : \begin{vmatrix} \mathcal{K} & \rightarrow & \mathcal{K} \\ i & \mapsto & \pi(i). \end{vmatrix} \tag{7.17}$$

The receiver is therefore the super-leader, who optimizes his utility function $w(\pi)$ where π is in a $K!$-element set. He then declares his choice to the transmitters who play the game described in Sec. 7.5.1. The authors of He et al. (2010b) have shown the following result.

> **Proposition 181: Optimal Hierarchy Configuration**
>
> The best mapping π in the sense of social welfare is to rank the transmitters in reverse order of their rate-channel gain product $R_i|h_i|^2$; in particular, the leader is the transmitter who has the lowest rate-channel gain product.

Proof. Let $w^{(i)}$ be the social welfare when user i is chosen to be the leader (resp. follower) and p_i^L (resp. p_i^F) be his transmit power at the SE. We have that:

$$w^{(i)} - w^{(j)} = R_i \frac{f(\gamma^*)}{p_i^L} + R_j \frac{f(\beta^*)}{p_j^F} - R_i \frac{f(\beta^*)}{p_i^F} - R_j \frac{f(\gamma^*)}{p_j^L} \tag{7.18}$$

$$= \frac{R_j|h_j|^2 - R_i|h_i|^2}{|h_i|^2} \left[\frac{f(\beta^*)}{p_i^F} - \frac{f(\gamma^*)}{p_i^L} \right] \tag{7.19}$$

$$= \frac{R_j|h_j|^2 - R_i|h_i|^2}{|h_i|^2} \left(u_{i,F}^{SE} - u_{i,L}^{SE} \right) \tag{7.20}$$

$$> 0$$

where equality (9.30) follows from $|h_i|^2 p_i^L = |h_j|^2 p_j^L$ and $|h_i|^2 p_i^F = |h_j|^2 p_j^F$. As any user always has a better utility by being chosen as a follower instead of a leader (Lasaulce et al., 2009b), we see that the difference is non-negative if and only if $R_i|h_i|^2 \leq R_j|h_j|^2$, which concludes the proof.

As discussed in Lasaulce et al. (2009b) other choices are possible for the receiver's utility and can lead to different conclusions. The equivalent virtual multiple input multiple output (MIMO) network performance is one of these, but will not be considered here.

7.6 REPEATED POWER CONTROL GAME

So far, we have only considered one-shot power control games: for a given vector of channel gains, each transmitter chooses his power level in order to maximize his instantaneous utility. In this section and the next, we consider a more general situation; all the transmitters can update their power levels several times within a block or a duration less than the channel coherence time. The corresponding power control type is called "distributed fast power control", generalizing the more conventional distributed slow power control for which the power levels are updated only once per block. In both cases, we want to take into account the fact that the players (namely the transmitters) interact several times, either within a block or from block to block. This feature introduces new types of behaviors (cooperation, punishment, etc.) with respect to the one-shot game.

Conventionally, for i.i.d. channels, power control schemes are designed such that the power levels are chosen in an independent manner from block to block. In distributed networks with selfish transmitters, the point of view has to be re-considered even if the channels are i.i.d. due to the fact that long-term interaction may change the behavior of selfish transmitters. The framework considered here is therefore the one of dynamic games. More specifically, we will analyze the case of repeated games (RG). Formally, from a game-theoretic viewpoint we assume that: the games are with complete information; transmitters are rational; rationality is common knowledge; transmitters are only informed with their individual CSI and a public signal. However, the proposed power control scheme can be implemented even if all these assumptions do not hold. As it will be seen, a transmitter only needs to know his own CSI and the power of the received signal for each block. This means that the framework in terms of information assumptions is very similar to that for iterative water-filling algorithms (Scutari et al., 2009; Yu et al., 2004). These comments also indicate that game theory can be used as a way of inventing new power control schemes, even if, in practice, some assumptions such as "rationality is common knowledge" may not hold.

7.6.1 Fast Power Control

Here, we assume that transmitters update their power levels several times within a block or a time window which has a duration of less than the channel coherence time. In the one-shot power control game, each transmitter observes the channel gain associated with the current block, i.e., $|h_i|$, and updates his power level once according to (7.5) in order to maximize his instantaneous utility. In repeated games, players want to maximize their averaged utility. To define the latter quantity, we first need to define a game stage, the game history, and the strategy of a transmitter. Game stages correspond to instants at which transmitters can update their power levels. In the case of fast power control, they coincide with sub-blocks (comprising one or several symbols). We assume that, for a given game stage $t \geq 1$, each transmitter knows and takes into account all the past realizations of a public signal before choosing his power level $p_i(t)$. More specifically we assume that every player knows at each stage of the game the following public signal:

$$\omega(t) = \sigma^2 + \sum_{i=1}^{K} |h_i|^2 p_i(t) \tag{7.21}$$

where $|h_i|$, $i \in \mathcal{K}$, does not vary from stage to stage. By noticing that $\forall i \in \mathcal{K}$, $p_i(t)$ $|h_i|^2 \times \frac{\mathrm{SINR}_i(t)+1}{\mathrm{SINR}_i(t)} = \omega(t)$ we see that each transmitter can construct the public signal from the sole knowledge of his action, individual channel gain, and individual SINR.

Note that this framework does not correspond to that of repeated games with perfect monitoring, for which each player knows and takes into account the actions of all the players that have been played in the past: in our setup only a weighted sum of the actions needs to be known and not the individual actions themselves. The proposed framework is the one of repeated games with a public signal (e.g., see Tomala (1998)) in which the public signal is a deterministic function of the played actions. By denoting $\Omega = \left[\sigma^2, \sigma^2 + \sum_{i=1}^{K} \eta_i^{\max} P_i^{\max} \right]$, the public signal corresponds to an observation function defined by:

$$\omega : \begin{vmatrix} \bigotimes_{i=1}^{K} \mathcal{P}_i & \rightarrow & \Omega \\ \underline{p}(t) & \mapsto & \omega(t). \end{vmatrix} \tag{7.22}$$

The pair of vectors $(\underline{\omega}_t, \underline{p}_{i,t}) = (\omega(1), \ldots, \omega(t-1), p_i(1), \ldots, p_i(t-1))$ is called the history of the game for transmitter i at time t and lies in the set $\Omega^{t-1} \times \mathcal{P}_i^{t-1}$. This is precisely the history $\left(\underline{\omega}_t, \underline{p}_{i,t} \right)$ which is assumed to be known by the transmitters before playing at stage t; this definition of the history assumes that the transmitters have perfect recall (Kuhn, 1950, 1953). With these notations, a pure strategy of a transmitter in the repeated game can be defined properly (for more information about repeated game models see Chapter 3).

Definition 182: Players' Strategies in the RG

A pure strategy for player $i \in \mathcal{K}$ is a sequence of causal functions $(\tau_{i,t})_{t \geq 1}$ with:

$$\tau_{i,t} : \begin{vmatrix} \Omega^{t-1} \times \mathcal{P}_i^{t-1} & \rightarrow & [0, P_i^{\max}] \\ (\underline{\omega}_t, \underline{p}_{i,t}) & \mapsto & p_i(t). \end{vmatrix} \tag{7.23}$$

The strategy of player i, which is a sequence of functions, will be denoted by τ_i (removing the game stage index in (7.23) is a common way to refer to a player's strategy). The vector of strategies $\underline{\tau} = (\tau_1, \ldots, \tau_K)$ will be referred to as a joint strategy. A joint strategy $\underline{\tau}$ induces a unique action plan $(\underline{p}(t))_{t \geq 1}$. To each profile of powers $\underline{p}(t)$ corresponds a certain instantaneous utility $u_i(\underline{p}(t))$ for player i. In our setup, each player does not care about what he gets at a given stage, but does care what he gets over the whole duration of the game. This is why we consider utility functions that result from averaging over the instantaneous utility. More precisely, we consider a utility function which corresponds to the model of discounted repeated games (DRG) studied in Chapter 3.

Definition 183: Players' Utilities in the DRG

Let $\underline{\tau} = (\tau_1, \ldots, \tau_K)$ be a joint strategy. The utility for player $i \in \mathcal{K}$ is defined by:

$$v_i^{\lambda}(\underline{\tau}) = \sum_{t=1}^{+\infty} \lambda (1-\lambda)^{t-1} u_i(\underline{p}(t)) \tag{7.24}$$

where $\underline{p}(t)$ is the power profile of the action plan induced by the joint strategy $\underline{\tau}$ and $0 < \lambda < 1$ is the discount factor and is known to every player (since the game is with complete information).

In the current wireless literature concerning the problem under consideration, the discounted repeated game model is typically used as follows: in Etkin et al. (2007) the discount factor is used as a way of accounting for the delay sensitivity of the network; in Wu et al. (2009), the discount factor is used to give the transmitters the possibility of valuing short-term and long-term gains differently. Interestingly, Sobel (1992) offers another interpretation of this model. This author sees the discounted repeated game as a finite repeated game (see Chapter 3), where the number of stages or game duration is unknown to the players. The number of stages is therefore considered as an integer-valued random variable which is almost finite definitely and whose law is known by the players. In other words, λ can be seen as the stopping probability at each game stage: the probability that the game stops at stage t is thus $\lambda (1-\lambda)^{t-1}$. The function v_i^{λ} would correspond to an expected utility given the law of T. This shows that the discount factor is also useful to study wireless games in which a player enters/leaves the game. We would also like to mention that it can also model heterogeneous discounted repeated game where players have different discount factors $\lambda_1, \ldots, \lambda_K$; in which case λ represents $\min_i \lambda_i$ (as pointed out in Shapley (1953b)).

To conclude on the choice of the repeated game model, note that it is possible to apply the model of a finitely repeated game to the power control problem, which is done in Treust and Lasaulce (2010a). As already mentioned, this choice is more suited to scenarios where the number of times players interact is finite and known (e.g., the number of training symbols in a block).

Definition 184: Equilibrium Strategies in the DRG

A joint strategy $\underline{\tau}$ supports an equilibrium of the discounted repeated game defined by $\left(\mathcal{K}, (\mathcal{P}_i)_{i \in \mathcal{K}}, \Omega, \omega, (v_i^{\lambda})_{i \in \mathcal{K}} \right)$ if:

$$\forall i \in \mathcal{K}, \forall \tau_i', \ v_i^{\lambda}(\underline{\tau}) \geq v_i^{\lambda}(\tau_i', \underline{\tau}_{-i}) \tag{7.25}$$

where $\underline{\tau}_{-i}$ is the vector of strategies played by the players other than i.

Having defined equilibrium strategies, the natural questions to be asked are: whether there exists an equilibrium strategy, whether it is unique, and how can it/they be determined. The purpose of Folk theorems (Chapter 3) is exactly to answer the first two questions. They characterize the set or region of equilibrium utilities. When the repeated game is played with perfect monitoring, the (standard) Folk theorem states that an outcome of the game is supported by an equilibrium strategy, if and only if it is feasible and individually rational. Clearly, Folk theorems answer the question of existence. In general, there is not a unique equilibrium. Rather, there is usually a number of them, and there can even be an infinite number of possible equilibria. When actions are partially observed, the notions of feasibility and individual rationality have to be re-considered (Lehrer, 1989, 1990, 1992; Tomala, 1998). Folk theorems are therefore more difficult to establish when an arbitrary structure is assumed for the signals observed by the players, and in particular for an arbitrary public signal. When available, Folk theorems state the existence of an equilibrium strategy for a given point in the region of achievable equilibrium utilities but a strategy for reaching this point remains to be found. Therefore, in this chapter, our objective is more modest. We only focus on the existence and determination issues. These are solved by providing a given equilibrium strategy, which is given by the following theorem (Treust and Lasaulce, 2010a).

Theorem 185: Equilibrium Strategy in the DRG

Assume that the following condition is met:

$$\lambda \leq \frac{1 - (K-1)\gamma^*}{(K-1)\gamma^*} \frac{f(\gamma^*)}{f(\beta^*)} - \frac{1 - (K-1)\beta^*}{(K-1)\beta^*} \tag{7.26}$$

where γ^* is the unique solution of $x[1 - (K-1)x]f'(x) - f(x) = 0$. Then, for all $i \in \mathcal{K}$, the following action plan is a subgame perfect Nash equilibrium of the discounted repeated game for a given and arbitrary channel realization $\underline{h}$:

$$\forall t \geq 1, \ \tau_{i,t} = \begin{vmatrix} p_i^{\mathrm{AP}} & \text{if } \omega(t) = \frac{2\sigma^2}{1-(K-1)\gamma^*} \\ p_i^{\mathrm{NE}} & \text{otherwise} \end{vmatrix} \tag{7.27}$$

where the agreement point (AP) is defined by:

$$\forall i \in \mathcal{K}, \ p_i^{\mathrm{AP}} = \frac{\sigma^2}{\eta_i} \frac{\gamma^*}{1 - (K-1)\gamma^*}. \tag{7.28}$$

The proof of this theorem is available in Treust and Lasaulce (2010a). Here we will only comment on the agreement point, the equilibrium condition, the deviation detection mechanism, and the punishment procedure.

Agreement point. The agreement point corresponds to a certain point in the feasible utility region of the static game $\mathcal{G}$. Indeed, we consider a subset of points of this region for which the power profiles $(p_1,\ldots,p_K)$ verify $p_i|h_i|^2 = p_j|h_j|^2$ for all $(i,j) \in \mathcal{K}^2$. The considered subset therefore consists of the solutions for the following system of equations:

$$\forall (i,j) \in \mathcal{K}^2, \quad \frac{\partial u_i}{\partial p_i}(p) = 0 \text{ with } p_i|h_i|^2 = p_j|h_j|^2. \tag{7.29}$$

It turns out that, following the lines of the proof of SE uniqueness in Lasaulce et al. (2009b), it is easy to show that a sufficient condition for ensuring both the existence and uniqueness of the solution to this system of equations is that there exists $x_0 \in]0, \frac{1}{K-1}[$ such that $\frac{f''(x)}{f'(x)} - \frac{2(K-1)}{1-(K-1)x}$ is strictly positive on $]0, x_0[$, and strictly negative on $]x_0, \frac{1}{K-1}[$. This condition is satisfied for the two following efficiency functions: $f(x) = (1 - e^{-x})^M$ (Shah et al., 1998) and $f(x) = e^{-\frac{c}{x}}$ (Belmega and Lasaulce, 2009) with $c = 2^R - 1$ (R is the transmission rate). If this condition should be found to be too restrictive, it is always possible to derive a purely numerical condition, which can be translated into a condition on K, M (in the case of Shah et al. (1998)), or R (Belmega and Lasaulce, 2009). Under the aforementioned condition, the unique solution of (7.29) can be confirmed as:

$$\forall i \in \mathcal{K}, \quad \tilde{p}_i = \frac{\sigma^2}{|h_i|^2} \frac{\tilde{\gamma}}{1 - (K-1)\tilde{\gamma}} \tag{7.30}$$

where $\tilde{\gamma}$ is the unique solution of $x[1 - (K-1) \cdot x]f'(x) - f(x) = 0$.

It is important to distinguish here between the equal-SINR condition imposed in (7.29) and the equal-SINR solution of the one-shot game (7.5). In the first case, the SINR has a special structure imposed by the condition (7.29) which is $\text{SINR}_i = \frac{|h_i|^2 p_i}{\sigma^2 + (K-1)|h_i|^2 p_i}$. Therefore, each transmitter is assumed to maximize a single-variable utility function $u_i(\underline{q}_i)$ with $\underline{q}_i = p_i \times \left(\left|\frac{h_i}{h_1}\right|^2, \left|\frac{h_i}{h_2}\right|^2, \ldots, \left|\frac{h_i}{h_K}\right|^2 \right)$. In the second case (one-shot NE), the solution is the solution of a K-unknown K-equation system, which happens to have the equal-SINR property. The proposed operating point (OP), given by (7.30), is thus fair in the sense of the SINR since $\forall i \in \mathcal{K}$, $\text{SINR}_i = \tilde{\gamma}$. To conclude on this agreement point, let us emphasize that, in the one-shot game, this point does not correspond to an equilibrium, but allows one to devise a cooperation plan in the discounted repeated game. Also, unsurprisingly, only individual CSI is needed to play at the corresponding levels.

Equilibrium conditions. Before starting the game, the players agree on a certain cooperation/punishment plan. Each transmitter $i \in \mathcal{K}$ always transmits at p_i^{AP} if no deviation is detected. Otherwise he plays a punishment level, namely p_i^{NE}. The proposed strategy supports an equilibrium if the gain brought by deviating is less than the loss induced by the punishment procedure applied by the other transmitters

(Friedman, 1971). To make sure that this occurs effectively, the game-stopping probability must be sufficiently low. This explains the presence of the upper bound on λ. To conclude, note that, as mentioned in Chapter 1, the frontier between non-cooperative and cooperative games is not always obvious. Here, repetition may induce cooperation but the players remain selfish, and the game cannot be said to be cooperative in the sense that no coalitions are defined, only individual utilities are considered, and problems such as how to share the gains are not encountered.

Deviation detection mechanism. The cooperation plan is only implementable if the transmitters can detect a deviation from the cooperation plan. It turns out that knowledge of the public signal is sufficient for this purpose. Indeed, when the transmitters play at the AP, the public signal equals $\frac{2\sigma^2}{1-(K-1)\gamma^*}$. Therefore, if one transmitter deviates from this point, the public signal $\omega(t)$ is no longer constant. Of course, if more transmitters deviate from the equilibrium in a coordinated manner this detection mechanism can fail; this is not inherent to the proposed cooperation plan but to the Nash equilibrium definition. If this issue should turn out to be crucial it would be necessary to consider other solution concepts, such as strong equilibria (see Chapter 1).

Punishment procedure. The punishment is that the other transmitters play at the one-shot game Nash equilibrium. Note that usually (e.g. see Sobel (1992)) the punishment procedure consists of playing at the minmax levels (i.e., $p_i = P_i^{\max}$). The drawback for punishing at the one-shot game Nash equilibrium point is that the punishment is less severe. The consequence is that the upper bound of the discount factor has to be smaller than if a punishment procedure at the minmax levels is chosen. On the other hand, the advantage for punishing at the Nash equilibrium point is that the proposed equilibrium strategy is subgame perfect (Selten, 1965, 1975). The subgame perfection property ensures that if a joint strategy $\underline{\tau}$ of the RG supports an equilibrium then, after all possible histories $\left(\underline{\omega}_t, \underline{p}_{1,t}, \ldots, \underline{p}_{K,t}\right)$, the joint strategy $\underline{\tau}(\underline{\omega}_t, \underline{p}_{1,t}, \ldots, \underline{p}_{K,t})$ is still an equilibrium strategy, which makes the distributed network performance predictable.

7.6.2 Slow Power Control in the Worst Case Scenario

In the preceding section, we discussed transmitters updating their power levels several times within a time window that was less than the channel coherence time (a block typically), and a new repeated game being played when a new channel realization was drawn. This regime (of fast power control policies) can be defined mathematically by the condition $f_{PC} > \frac{1}{T_{coh}}$, where f_{PC} is the power control frequency and T_{coh} the channel coherence time. In this section we assume the regime of slow power control policies, which corresponds to $f_{PC} \leq \frac{1}{T_{coh}}$. In this regime a game stage can be a block or a frame comprising several blocks. The main technical difference with respect to the scenario of Sec. 7.6.1 is that the channel state $\underline{h}$ varies from stage to stage. The transmitters' utility depends on the value of $\underline{h}$ at stage t, which is denoted by $\underline{h}(t)$. In contrast with fast power control, the same one-shot game is not repeated from stage to stage. In general, the framework of stochastic repeated games is needed to treat this problem, as considered in the next section. In the specific case of the power control game under study, the conventional framework (in which the same game is repeated) can be sufficient under certain conditions. We have seen in Sec. 7.3 that in the non-saturated regime (i.e., $\forall i \in \mathcal{K}, p_i^{NE} \leq P_i^{max}$), the one-shot game $\mathcal{G}$ is equivalent to the static game:

$$\widehat{\mathcal{G}} = (\mathcal{K}, \{\mathcal{A}_i\}_{i \in \mathcal{K}}, \{\widehat{u}_i\}_{i \in \mathcal{K}}) \tag{7.31}$$

where $\widehat{u}_i(a_1, \ldots, a_K) = \dfrac{R_i f\left(\frac{a_i}{\sum_{j \neq i} a_j + \sigma^2}\right)}{a_i}$ and $\mathcal{A}_i = \left[0, \dfrac{\eta_i^{max}}{\eta_i^{min}} \dfrac{\sigma^2}{1-(K-1)\beta^*}\right]$.

Thanks to this equivalence, it is sufficient to study the discounted repeated game associated with $\widehat{\mathcal{G}}$ and deduce the equilibrium strategies merely by de-normalizing the actions. This analysis is therefore transparent to the channel state. However, the channel state plays a role in the equilibrium existence conditions. The solution proposed by Treust and Lasaulce (2010a) is to consider the worst case propagation scenario for the equilibrium conditions. The corresponding upper bound of λ translates the worst scenario in terms of stability: a transmitter deviates from the cooperation plan at a given stage (say $t = t'$) where his channel is at his maximum (i.e., $|h_i(t')|^2 = \eta_i^{max}$) and undergoes the minimal punishment (i.e., his channel stays at its minimum $|h_i(t)|^2 = \eta_i^{min}$ during the punishment procedure $t > t'$). This is translated by Theorem 186 (see Treust and Lasaulce (2010a) for the proof).

Theorem 186: Equilibrium Strategy in the DRG

Assume that the following condition is met:

$$\forall i \in \mathcal{K}, \lambda \leq \frac{\delta(\beta^*, \tilde{\gamma})}{\delta(\beta^*, \tilde{\gamma}) + \frac{\eta_i^{max}}{\eta_i^{min}}[(K-1)f(\beta^*) - \delta(\beta^*, \tilde{\gamma})]} \tag{7.32}$$

where $\delta(\beta^*, \tilde{\gamma}) = \frac{1-(K-1)\tilde{\gamma}}{\tilde{\gamma}} f(\tilde{\gamma}) - \frac{1-(K-1)\beta^*}{\beta^*} f(\beta^*)$. Then, for all $i \in \mathcal{K}$, the following action plan is a subgame perfect Nash equilibrium of the discounted repeated game:

$$\forall t \geq 1, \ \tau_{i,t} = \begin{vmatrix} p_i^{AP} & \text{if } \omega(t) = \frac{2\sigma^2}{1-(K-1)\gamma^*} \\ p_i^{NE} & \text{otherwise.} \end{vmatrix} \tag{7.33}$$

One of the differences between this equilibrium condition and that obtained in the case of fast power control is that the channel dynamics appear. Specifically, the upper bound on the discount factor depends on the ratio $\frac{\eta_i^{max}}{\eta_i^{min}}$, which we define as the channel dynamics for transmitter i. As a remark, notice that (7.26) is the special case of (7.32) in which $\frac{\eta_i^{max}}{\eta_i^{min}} = 1$. The problem of how much this condition is restrictive can be seen from two complementary perspectives. If the channel gain dynamics is given, the question is whether the maximum (resp. minimum) value of the discount factor (resp. game stages) guarantees the existence of an equilibrium. The values of η_i^{min} and η_i^{max} depend on the propagation scenario and considered technology; in systems like WiFi networks these quantities typically correspond to the path loss dynamics, the receiver sensitivity, and the minimum distance between the transmitter and receiver.

On the other hand, if the discount factor is given (e.g., by the traffic statistics), this imposes lower and upper bounds on the channel gains. It can happen that the admissible range for the discount factor can be incompatible with the channel gain dynamics, in which case the model of RG needs to be refined. The link between the game stopping probability or discount factor and channel gain dynamics is illustrated in Figure 7.2. The corresponding setup is: for random CDMA systems with a spreading factor equal to N, the efficiency function is $f(x) = (1 - e^{-x})^M$; Rayleigh fading channels are assumed. The figure represents the quantity $10\log_{10}\left(\frac{\eta^{\max}}{\eta^{\min}}\right)$ as a function of λ for $M = 2$ and different numbers of transmitters and spreading factors: $(K,N) \in \{(2,2),(4,5),(10,12)\}$. It can be seen that the discounted repeated game model seems to be suitable not only in scenarios where $|h_i|$ models the path

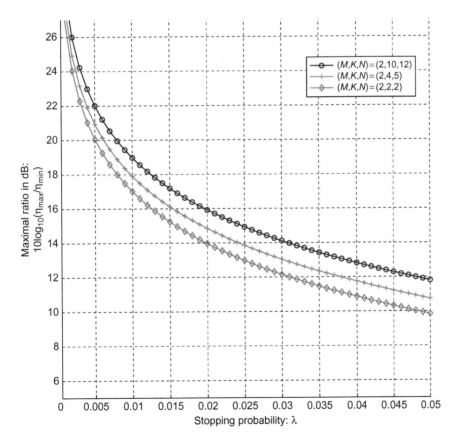

FIGURE 7.2

Admissible channel gain dynamics vs. discount factor for $M = 2$ and $(K,N) \in \{(2,2), (4,5),(10,12)\}$.

loss effects but also the fading effects. Of course if the discount factor is too high, more refined models can be found or designed. If this turns out not to be possible, this can mean that the model of repeated games is inappropriate.

7.7 SLOW POWER CONTROL AND STOCHASTIC GAMES

In the case of the non-saturated regime, we have seen that even though the channel state varies from stage to stage, the static energy-efficient power control game possesses a special property which allows one to model the slow power control problem by a (non-stochastic) repeated game. However, there are at least two reasons for considering a more general model, namely, a stochastic game model (Treust et al., 2010a). First, the normalization "trick" used to exploit repeated games leads to sub-optimal expected utilities. Second, in general, the framework of stochastic games (see Chapter 3) is needed to study problems where the instantaneous utilities depend not only on the played actions, but also on the current channel state. It turns out that finding an equilibrium strategy for the stochastic repeated game associated with the energy-efficient slow power control problem is not easy (Treust et al., 2010a). As a first step, let us explain how Theorem 112 in Chapter 3 can be applied here. This theorem is a Folk theorem for stochastic repeated games with a public signal. Here we provide the conditions under which this theorem can be applied to the problem of energy-efficient power control and comment on them.

1. The action sets must be discrete: $p_i(t) \in \mathcal{P}_i$, with $|\mathcal{P}_i| < +\infty$. Notice that, so far, we have assumed that the power level of transmitter i was in a compact set $[0, P_i^{\max}]$. If the discrete set assumption is inappropriate, Theorem 112 needs to be extended to the continuous case. On the other hand, the discrete set assumption may turn out to be more realistic than the compact set assumption; this is the case in the UMTS (universal mobile telecommunications standard) standard where the power can only be incremented/decremented by steps of 1 dBm.

2. The set of channel states must be discrete: $\eta_i(t) \in \Gamma_i$, with $|\Gamma_i| < +\infty$. Again, there are systems where this assumption is appropriate (e.g., when channel gains are quantized).

3. The signal available at each transmitter must be public. More precisely, each transmitter needs to know the pair of vectors $(\underline{\omega}_t, \underline{\eta}_t)$ where $\underline{\omega}_t = (\omega(1), \ldots, \omega(t-1))$, $\omega(t) = \sigma^2 + \sum_{i=1}^{K} |h_i(t)|^2 p_i(t)$, $\underline{\eta}_t = [\underline{\eta}(1), \ldots, \underline{\eta}(t-1)]$, and $\underline{\eta}(t) = \left(|h_1(t)|^2, \ldots |h_K(t)|^2 \right)$. This means that global CSI is needed. However, the action $p_i(t)$ remains private.

4. The channel transition probability $P_{\underline{\eta}}$ must be irreducible $\forall t \geq 1, P_{\underline{\eta}} \big(\underline{\eta}(t) | \underline{\eta}(t-1) \big) > 0$ and known to the transmitters.

Under these assumptions, it is possible to prove a Folk theorem for the slow power control game (see Treust et al. (2010a) for more details). The technical challenge remains to find strategies for achieving the points in it. In Mériaux et al. (2011b), the authors show that exploiting the agreement point presented in Sec. 7.6.1 and time-sharing leads to distributed power control strategies which perform better than those obtained by modeling the problem by a repeated game. The main ideas are as follows.

1. Each transmitter has only individual CSI (that is $|h_i|^2$ is known) and observes a public signal sent by the receiver. The latter is called the recommendation signal.
2. The public signal recommends that a set of transmitters should transmit over the current block at the levels indicated by the agreement point, and the others are asked to not transmit for this block.
3. In the framework of stochastic games, it can be shown that (Mériaux et al., 2011b) for a sufficiently low discount factor λ, a punishment mechanism can be implemented, and the transmitters have an interest in following the recommendation. This makes time-sharing legitimate in the network since no transmitter benefits from ignoring the recommendations.

Here, we provide some simulations to illustrate these comments (Treust et al., 2010a).

Numerical results. We consider a simple stochastic process with two channel states: $(\eta_1, \eta_2) \in \{(7,1), (1,7)\}$. The transition probability is constant over the channel states: $P_\eta(\cdot) = \left(\frac{1}{2}, \frac{1}{2}\right)$ and its invariant measure is $\mu = \left(\frac{1}{2}, \frac{1}{2}\right)$. Consider the scenario $(K, M, \overline{N}) = (2, 2, 2)$. Figure 7.3 represents the achievable utility region for the two different channel states and the long-term expected utility region. The region at the North-East of the two mixed lines represents the set of (expected) individually rational utilities. Its intersection with the expected achievable utility region describes the set of public perfect equilibrium utilities. Three important points are highlighted in the different scenarios: the expected Nash equilibrium of the one-shot game studied in Goodman and Mandayam (2000), the cooperation point studied in Treust et al. (2010a), and the point at which the expected social welfare is maximized (star). From this figure it can be seen that a significant gain can be obtained by using the repeated games model instead of the one-shot model. Moreover, a significant improvement in terms of expected utilities is a direct consequence of the global CSI compared to the individual CSI.

As a second type of numerical result, the performance gain brought by the stochastic discounted repeated game (SDRG) formulation of the distributed power control problem is assessed. Considering a simple stochastic process where $\eta_i = 2$ and $\eta_j = 1$ for all $j \in \mathcal{K} \setminus \{i\}$ and the i's player is drawn with uniform distribution over the K players. The expected social optimum w_{SDRG} is assessed. Denote by w_{NE} (resp. w_{DRG} and w_{SDRG}) the efficiency of the Nash equilibrium (resp. DRG and SDRG equilibrium) in terms of social welfare. Figure 7.4 represents the quantity $\frac{w_{SDRG} - w_{NE}}{w_{NE}}$ and $\frac{w_{DRG} - w_{NE}}{w_{NE}}$ in percentage terms as a function of the spectral efficiency $\alpha = \frac{K}{N}$, with $N = 128$ and $2 \leq K < \frac{N}{\beta^*} + 1$. The asymptotes $\alpha_{max} = \frac{1}{\beta^*} + \frac{1}{N}$ are indicated as dotted lines for different values $M \in \{10, 100\}$. The gain becomes very significant when the system load is close to $\frac{1}{N} + \frac{1}{\beta^*}$; this is because the power at the one-shot game Nash equilibrium becomes large when the system becomes more and more loaded. As explained in Lasaulce et al. (2009b) for the Stackelberg approach these gains are in fact limited by the maximum transmit power.

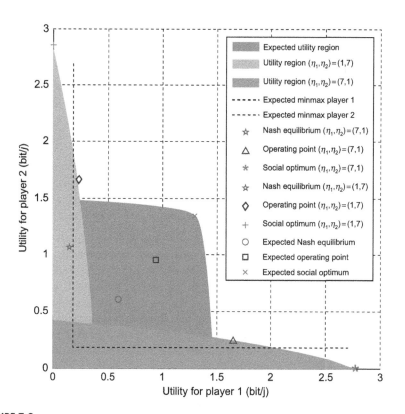

FIGURE 7.3

Expected utility region for $(K,M,N) = (2,2,2)$ by considering different channel configurations.

7.8 RELAXING SOME INFORMATION AND BEHAVIOR ASSUMPTIONS

In Sections 7.3–7.7 the different power control games assumed complete information, rational players, and that rationality was common knowledge. As discussed in Chapters 4 and 5, these assumptions may be questionable in some scenarios. In terms of information assumptions, it is more realistic to assume that a transmitter can only have access to a certain estimate of the individual CSI or global CSI. This framework corresponds to the one of games with incomplete information, presented briefly in Sec. 7.8.1. The authors of Treust and Lasaulce (2010b) have formulated the one-shot power control problem as a Bayesian game; that is the game is played with incomplete information but all transmitters are assumed to be rational and rationality is common knowledge. In Sec. 7.8.2, some results of Treust and Lasaulce (2010b) were reported. The well-known reinforcement learning algorithm, initially introduced by

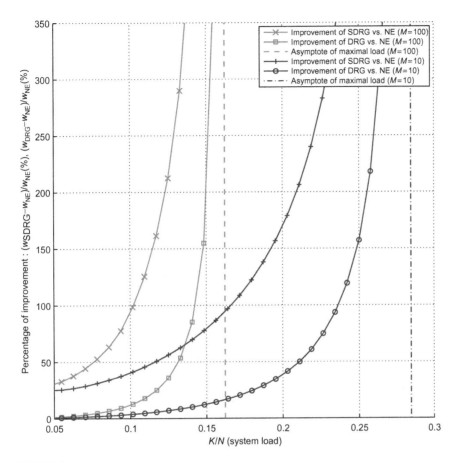

FIGURE 7.4

The gain brought by the repeated-game based cooperation in terms of social welfare (standard and stochastic repeated games) w.r.t. to the purely non-cooperative scenario (Nash equilibrium).

Mosteller and Bush (1955) and re-used more recently in Sastry et al. (1994b) and Xing and Chandramouli (2008), was exploited. The rationality assumption is not necessary in this context. In terms of knowledge, each transmitter only needs to know the value of his utility associated with the last action he played.

7.8.1 Bayesian Power Control Games

To make this chapter sufficiently self-contained, we will review the results presented in Chapter 4. For the sake of simplicity, for all $i \in \mathcal{K}$, each set Γ_i is assumed to be discrete. Additionally, as in Sec. 7.7 the set $\mathcal{P}_i$ is assumed to be discrete. The Bayesian (one-shot) power control game consists of an extension of $\mathcal{G}$ for which a certain observation structure is added:

$$\mathcal{G}^{B} = \left(\mathcal{K}, \{\theta_i\}_{i \in \mathcal{K}}, q, \{\mathcal{P}_i\}_{i \in \mathcal{K}}, \{\Sigma_i\}_{i \in \mathcal{K}} \left\{u_i^{B}\right\}_{i \in \mathcal{K}}\right)$$ (7.34)

where:

- $\theta_i : \begin{vmatrix} \Gamma_1 \times \cdots \times \Gamma_k & \rightarrow & \mathcal{S}_i \\ \underline{\eta} & \mapsto & s_i = \theta_i(\underline{\eta}) \end{vmatrix}$ (7.35)

 is a measurable application from the set of channel states, $\Gamma_i = \{\eta_i^1, \ldots, \eta_i^{m_i}\}$, $m_i \in \mathbb{N}^*$, to the set of observations of transmitter i which is $\mathcal{S}_i$, with $|\mathcal{S}_i| < +\infty$;
- $q(\underline{s}_{-i}|s_i)$ is the probability that transmitters other than i observe $\underline{s}_{-i}$ given that transmitter i observes s_i;
- a strategy for a player i is a function $\sigma_i : \mathcal{S}_i \rightarrow \mathcal{P}_i$ where $\sigma_i \in \Sigma_i$ and Σ_i is the set of possible mappings;
- u_i^{B} is the expected utility for transmitter i which is defined by:

$$u_i^{B}(\sigma_1, \ldots, \sigma_K, s_i) = \mathbb{E}_{\underline{\eta}, \underline{s}_{-i}}\left[u_i\left(\sigma_1(s_1), \ldots, \sigma_K(s_K), \underline{\eta}\right) | s_i\right].$$ (7.36)

For each transmitter i, denote by $\Lambda_i(s_i)$ the ambiguity set associated with the observation s_i. For instance, in the case of perfect global CSI, this set boils down to a singleton: $\Lambda_i(s_i) = \{\theta^{-1}(s_i)\} = \{\underline{\eta}\}$. Using this notation the expected utility can be rewritten as:

$$u_i^{B}(\sigma_1, \ldots, \sigma_K, s_i) = \sum_{\underline{\eta} \in \Lambda_i(s_i)} \frac{P(\underline{\eta})}{\Pr[\underline{\eta} \in \Lambda_i(s_i)]} \sum_{\underline{s}_{-i}} q(\underline{s}_{-i}|s_i) u_i\left(\sigma_1(s_1), \ldots, \sigma_K(s_K), \underline{\eta}\right).$$ (7.37)

where $P(\underline{\eta})$ is the probability that the channel state $\underline{\eta}$ is drawn. Using the fact that channel gains of the different links are generally independent in wireless networks, the expected utility is expressed by:

$$u_i^{B}(\sigma_1, \ldots, \sigma_K, s_i) = \sum_{\underline{\eta} \in \Lambda_i(s_i)} \frac{P(\underline{\eta})}{\Pr[\underline{\eta} \in \Lambda_i(s_i)]} \sum_{\underline{s}_{-i}} \prod_{j \neq i} \Pr[\underline{\eta} \in \Lambda_j(s_j)] \frac{f\left(\frac{\eta_i \sigma_i(s_i)}{\sum_{j \neq i} \eta_j \sigma_j(s_j) + \sigma^2}\right)}{\sigma_i(s_i)}.$$ (7.38)

The next step is to find out whether a Bayesian equilibrium exists, whether it is unique, etc. Therefore the classical machinery described in this chapter has to be put to work, which is left to the reader as a problem to be solved.

7.8.2 Reinforcement Learning for Power Control

In the wireless literature in the broad sense, there are only a few papers exploiting learning theory in games. The authors of Xing and Chandramouli (2008) have applied the learning approach to the problem of energy-efficient power control. They have applied the reinforcement learning algorithm (RLA) initially introduced by

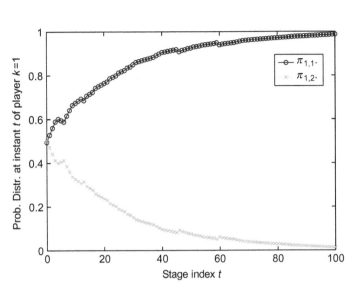

FIGURE 7.5

Convergence of the reinforcement learning algorithm applied to the energy-efficient power control problem with $K = 2$, $|\mathcal{P}_i| = 2$.

Mosteller and Bush (1955) and revisited by Thathachar and Sastry (2003) (see Chapter 5 for more details) in order to propose a distributed (always in the sense of the decision) power control policy for networks with finite numbers of transmitters and power levels. Based on the sole knowledge of the value of his utility at each step, the corresponding algorithm consists, for each transmitter, of updating his strategy – which is his probability distribution over the possible power levels. The authors of Xing and Chandramouli (2008) have conducted the convergence analysis for 2-transmitter 2-action power control games. As explained in Chapter 5, one of the main properties of the RLA is that a game with partial information (namely the value of the instantaneous utility associated with the played action), when played a sufficiently large number of times, can converge to an equilibrium which would be obtained with complete information. Figure 7.5 represents the evolution of the probability distribution (over two power levels) associated with the RLA as a function of the game stage index; $\pi_{i,1}$ is the probability assigned by transmitter i to action 1. Convergence is obtained after about 80 stages. Of course, if the number of possible transmit power levels is higher, convergence becomes slower. Decreasing convergence rates is a challenge for learning algorithm designers.

More will be said about how to apply reinforcement learning in wireless games in Chapter 9.

Power Allocation Games and Learning in MIMO Multiuser Channels*

8.1 INTRODUCTION

This chapter is written in the same spirit as with the preceding one. One of the main objectives is always to illustrate notions such as the Nash equilibrium (NE), to show how NE existence and uniqueness theorems can be applied, and to show how the NE can be determined. The main technical differences with respect to the preceding chapter are:

- other utility functions are considered for the transmitters (which are always the players); namely, transmission rates (Shannon rates in all sections and goodputs in Section 8.7);
- the strategy of a transmitter consists of choosing a power allocation policy (instead of a power control policy in Chapter 7);
- other channel models are assumed, namely multiple input multiple output (MIMO) multiuser channels, and the case of interference (relay) channels;
- much more emphasis is put on learning techniques. In particular, a learning technique is used to learn good strategies when the sets of strategies are discrete and another one is proposed for learning w.r.t. continuous strategy spaces.

The importance of learning techniques has been made obvious in Chapters 5 and 6. Even though they typically only converge for special types of games (dominance solvable games, potential games, supermodular games, etc.) and may require many iterations/stages to do this, they possess the notable feature of assuming transmitters with reduced knowledge; typically, the latter do not know the structure of the game under play. This chapter therefore bridges the gap between the NE concept (Chapters 1 and 2) and learning techniques (Chapters 5 and 6) for well-known wireless problems. In particular, the existence of a unique pure NE may be useful to predict the unique stable outcome of the (seemingly complex) learning processes.

---

*Note: This chapter has been written jointly by Dr Elena-Veronica Belmega and the authors.

8.2 POWER ALLOCATION GAMES IN MIMO MULTIPLE ACCESS CHANNELS

In this section, we consider a multiple access channel (MAC) consisting of several transmitter nodes and a common destination. The receiver node decodes the incoming messages from all the transmitters. We assume that the terminals are equipped with multiple antennas. The received base-band signal can be written as:

$$\underline{Y}(\tau) = \sum_{k=1}^{K} \mathbf{H}_k(\tau)\underline{X}_k(\tau) + \underline{Z}(\tau) \tag{8.1}$$

where $\underline{X}_k(\tau)$ is the $n_{t,k}$-dimensional column vector of symbols transmitted by user k at time τ, $\mathbf{H}_k \in \mathbb{C}^{n_r \times n_{t,k}}$ is the channel matrix of user k and $\underline{Z}(\tau)$ is a n_r-dimensional complex white Gaussian noise distributed as $\mathcal{N}(\underline{0}, \sigma^2 \mathbf{I}_{n_r})$, assuming a number of $n_{t,k}$ antennas at the transmitter $k \in \{1, \ldots, K\}$ and n_r antennas at the common receiver. For the sake of clarity, the time index τ is omitted from our notations.

The transmitters are assumed to use good codes in the sense of their Shannon achievable rates. Also, coherent communications are assumed; i.e., the receiver has perfect channel state information and decodes all its intended messages perfectly. Two different decoding techniques will be studied: single user decoding (SUD), and successive interference cancellation (SIC). The SIC decoding technique involves the existence of a coordination signal representing the decoding order at the receiver. The coordination signal is assumed to be perfectly known at the transmitter side. If it is generated by the receiver, its distribution can be optimized, but this induces a certain cost in terms of downlink signaling. On the other hand, if the coordination signal comes from an external source, e.g., an FM transmitter, the transmitter nodes can acquire the coordination signal for free in terms of downlink signaling. However, this generally involves a certain sub-optimality in terms of uplink rate. In both cases, the coordination signal will be represented by a random variable denoted by $S \in \mathcal{S} \triangleq \{1, \ldots, K!\}$. In a real wireless system, the frequency with which the realizations would be drawn would be roughly proportional to the reciprocal of the channel coherence time (i.e., $1/T_{\text{coh}}$). Note that the proposed coordination mechanism is suboptimal because it does not depend on the realizations of the channel matrices. The corresponding performance loss is in fact very small.

In the following discussion, the static case first is briefly reviewed and then the more challenging case of fast fading links is presented.

8.2.1 Static Links

In this section, the channel gains are assumed to be deterministic constants.

Let us start the analysis with a toy example: the two user single-input single-output (SISO) MAC as given in Belmega et al. (2008b). This example gives useful insights on the general case because it can be solved in closed-form. When SUD is assumed, transmitting at full power strictly dominates all the other strategies. Thus, the unique NE corresponds to the state where all the transmitters saturate their powers. This is a very inefficient operating point because of the high interference level. One way to reduce the interference is to consider SIC decoding. The decoding order is dictated by flipping an unfair coin (Bernoulli random variable of parameter $q \in [0, 1]$). The existence and uniqueness of the NE solution can be proved using Theorems 50 and 70 in Chapter 2. It turns out that, for any value of $q \in [0, 1]$, the performance of the system at the NE achieves the sum-capacity of the SISO MAC channel. In conclusion, the receiver node can choose a certain operating point on the Pareto-optimal frontier by tuning the parameter q. Another way to combat multi-user interference was proposed in Alpcan et al. (2002). The authors introduced a pricing function as a penalty for the interference the users create in the network. In the remainder of this section, the focus is on the general MIMO MAC.

8.2.1.1 *Single User Decoding*
We assume that the receiver applies a SUD scheme. The strategy of user k consists of choosing the input covariance matrix, $\mathbf{Q}_k = \mathbb{E}[\underline{x}_k \underline{x}_k^H]$ in the set:

$$\mathcal{A}_k^{\text{SUD}} = \{\mathbf{Q} \in \mathbb{C}^{n_{t,k} \times n_{t,k}} : \mathbf{Q} \succeq \mathbf{0}, \text{Tr}(\mathbf{Q}) \le \overline{P}_k\}, \tag{8.2}$$

to maximize its achievable rate:

$$\mu_k^{\text{SUD}}(\mathbf{Q}_k, \mathbf{Q}_{-k}) = \log_2 \left| \mathbf{I}_{n_r} + \rho \mathbf{H}_k \mathbf{Q}_k \mathbf{H}_k^H + \rho \sum_{\ell \ne k}^{K} \mathbf{H}_\ell \mathbf{Q}_\ell \mathbf{H}_\ell^H \right|$$

$$- \log_2 \left| \mathbf{I}_{n_r} + \rho \sum_{\ell \ne k} \mathbf{H}_\ell \mathbf{Q}_\ell \mathbf{H}_\ell^H \right|, \tag{8.3}$$

where $\rho = \frac{1}{\sigma^2}$. The existence of the NE follows from Theorem 50. As opposed to the SISO case, the NE is not in general unique. Sufficient conditions to ensure the uniqueness of the NE are: i) $\text{rank}(\mathbf{H}^H \mathbf{H}) = \sum_{k=1}^{K} n_{t,k}$ where $\mathbf{H} = [\mathbf{H}_1, \ldots, \mathbf{H}_K]$; ii) $\sum_{k=1}^{K} n_{t,k} \le n_r + K$. The proof follows from Belmega et al. (2010b) by considering that the static links case corresponds to a single realization of the fast fading links case.

The users converge to one of the Nash equilibrium points by applying an iterative algorithm based on the best-response correspondences:

$$\text{BR}_k(\mathbf{Q}_{-k}) = \arg \max_{\mathbf{Q}_k \in \mathcal{Q}_k} \mu_k^{\text{SUD}}(\mathbf{Q}_k, \mathbf{Q}_{-k}). \tag{8.4}$$

The best-response correspondence of user k is identical to the well-known single-user water-filling solution given in Yu et al. (2004). Using the following algorithm, the users converge to one of the NE points in a distributed manner:

- Initialization: for all $\ell \in \mathcal{K}, \mathbf{Q}_\ell^{[1]} \leftarrow \frac{\overline{P}_\ell}{n_{t,\ell}} \mathbf{I}_{n_{t,\ell}}$ and $k \leftarrow 1$
- At iteration step $t > 1$, only user k updates its covariance matrix:
  $\mathbf{Q}_k^{[t]} \leftarrow \text{BR}_k \left(\mathbf{Q}_{-k}^{[t-1]} \right), \mathbf{Q}_{-k}^{[t]} \leftarrow \mathbf{Q}_{-k}^{[t-1]}$. If $k = K$ then $k \leftarrow 1$, else $k \leftarrow k + 1$.
- Repeat the previous step ($t \leftarrow t + 1$) until convergence is reached.

The players sequentially update their covariance matrices in a certain specified order (e.g., in a round-robin fashion). In Yu et al. (2004), this algorithm was proved to converge to a point maximizing the sum-capacity of the MIMO Gaussian MAC channel from any initial point. However, because of the interference terms, the sum-capacity of the Gaussian MIMO MAC is not achieved at the NE point, similarly to the SISO case.

8.2.1.1.1 Parallel MAC

Before moving on to the SIC decoding scheme, another special but interesting case is discussed: the parallel MAC. The channel matrices are square diagonal matrices where $\forall k \in \mathcal{K}$, $n_r = n_{t,k} = B$ and $\mathbf{H}_k = \mathbf{diag}(h_{k,1}, \ldots, h_{k,B})$. This type of channel is typically used to model uplink communications in wireless networks which are composed of several nodes (receivers, access points, base stations, etc.) that operate in orthogonal frequency bands (Belmega et al., 2008a; Perlaza et al., 2009b). The network is composed of K mobile terminals (MT) and B base-stations (BS) operating in orthogonal frequency bands. In this case, the user action sets are reduced to vector sets instead of matrix sets (i.e., the users choose their power allocation policies from among the available frequencies).

What is interesting about this case is that two different power allocation games can be defined. The difference lies in the choice of the users' action sets: i) the users share their available powers among the BS (i.e., Soft-Handover or BS Sharing Game) or; ii) the users are restricted to transmit to a single BS (i.e., Hard-Handover or BS Selection Game).

For the BS Sharing Game, all the results discussed in the MIMO case apply here. As argued in the general MIMO case, the NE is not in general unique. Sufficient conditions for ensuring the uniqueness of the NE simply become: i) rank($\mathbf{H}^H\mathbf{H}$) = $K \times B$ and; ii) $K \times B \leq K + B$. However, these conditions are very restrictive and are met only in some particular cases:

a. If $K = 1$ and $B \geq 1$, the problem being reduced to a simple optimization solved by water-filling;
b. If $K \geq 1$ and $B = 1$ where the dominant strategy for every user is to transmit with full power at the unique BS available;
c. $K = 2$ and $B = 2$ which is somehow more complicated.

Recently, it has been proved (Mertikopoulos et al., 2011) that the NE is unique with probability 1 for any $K \geq 1$ and $B \geq 1$.

None of the previous results can be applied to the BS Selection Game. The action sets are not compact and convex sets; but are discrete sets; therefore, the properties of concave games cannot be used. The discrete power allocation problem is studied from a routing game theoretical perspective. The game can be proved to be an exact potential one, where the potential function is the network achievable sum-rate. This implies the existence of at least one pure-strategy NE directly. Based on elements of non-oriented graph theory, the exact number of equilibrium points is provided.

Also, an iterative algorithm based on the best-response correspondences is shown to converge to one of the NE. The proof is based on random walks and directed-graph theory. The algorithm is fully distributed, since it requires only local channel state information and the overall interference plus noise power.

When comparing the performance in terms of network sum-rate of the two games, a remarkable observation was made. Assuming $K \geq B$, the performance of the BS Selection Game is superior to that of the BS Sharing Game. This characterizes a Braess-type paradox, i.e., increasing the possible choices of players results in a degeneration of the global network performance.

8.2.1.2 *Successive Interference Cancellation*

In what follows, SIC is applied at the receiver. The strategy of user k consists of choosing the best vector of precoding matrices $\mathbf{Q}_k = \left(\mathbf{Q}_k^{(1)}, \mathbf{Q}_k^{(2)}, \ldots, \mathbf{Q}_k^{(K!)} \right)$ where $\mathbf{Q}_k^{(s)} = \mathbb{E}\left[\underline{X}_k^{(s)} \underline{X}_k^{(s),H} \right]$, for $s \in \mathcal{S}$. For the sake of clarity, we introduce a notation which will be used to replace the realization s of the coordination signal with no loss of generality. Denote by $\mathcal{P}_K$ the set of all possible permutations of K elements, such that $\pi \in \mathcal{P}_k$ denotes a certain decoding order for the K users, and $\pi(k)$ denotes the rank of user $k \in \mathcal{K}$ and $\pi^{-1} \in \mathcal{P}_K$ denotes the inverse permutation (i.e., $\pi^{-1}(\pi(k)) = k$), such that $\pi^{-1}(r)$ denotes the index of the user that is decoded with rank $r \in \mathcal{K}$. Denote by $p_\pi \in [0,1]$ the probability that the receiver implements the decoding order $\pi \in \mathcal{P}_K$, which means that $\sum_{\pi \in \mathcal{P}_K} p_\pi = 1$. In the static case, p_π represents the fraction of time over which the receiver applies the decoding order π. The vector of precoding matrices can be denoted by $\mathbf{Q}_k = \left(\mathbf{Q}_k^{(\pi)} \right)_{\pi \in \mathcal{P}_K}$. The utility function of User k is given by:

$$\mu_k^{\text{SIC}}(\mathbf{Q}_k, \mathbf{Q}_{-k}) = \sum_{\pi \in \mathcal{P}_K} p_\pi R_k^{(\pi)}\left(\mathbf{Q}_k^{(\pi)}, \mathbf{Q}_{-k}^{(\pi)} \right) \tag{8.5}$$

where:

$$R_k^{(\pi)}(\mathbf{Q}_k^{(\pi)}, \mathbf{Q}_{-k}^{(\pi)}) = \log_2 \left| \mathbf{I}_{n_r} + \rho \mathbf{H}_k \mathbf{Q}_k^{(\pi)} \mathbf{H}_k^H + \rho \sum_{\ell \in \mathcal{K}_k^{(\pi)}} \mathbf{H}_\ell \mathbf{Q}_\ell^{(\pi)} \mathbf{H}_\ell^H \right|$$

$$- \log_2 \left| \mathbf{I}_{n_r} + \rho \sum_{\ell \in \mathcal{K}_k^{(\pi)}} \mathbf{H}_\ell \mathbf{Q}_\ell^{(\pi)} \mathbf{H}_\ell^H \right| \tag{8.6}$$

where $\mathcal{K}_k^{(\pi)} = \{ \ell \in \mathcal{K} | \pi(\ell) \geq \pi(k) \}$ represents the subset of users that will be decoded after user k in the decoding order π. An important point to mention here is the power constraint under which the utilities are maximized. Indeed, three different power allocation games can be defined as a function of the power constraint:

- Spatial power allocation (SPA) game:

$$
\mathcal{A}_k^{SIC,SPA} = \left\{ \mathbf{Q}_k = \left(\mathbf{Q}_k^{(\pi)} \right)_{\pi \in \mathcal{P}_K} \;\middle|\; \forall \pi \in \mathcal{P}_K, \mathbf{Q}_k^{(\pi)} \succeq 0, \mathrm{Tr}(\mathbf{Q}_k^{(\pi)}) \le n_t \overline{P}_k \right\}.
$$

$$(8.7)$$

Here, the users are restricted to uniformly allocating their powers over time (independently of the decoding order).

- Temporal power allocation (TPA) game:

$$
\mathcal{A}_k^{SIC,TPA} = \left\{ \mathbf{Q}_k = \left(\alpha_k^{(\pi)} P_k \mathbf{I}_{n_t} \right)_{\pi \in \mathcal{P}_K} \;\middle|\; \forall \pi \in \mathcal{P}_K, \alpha_k^{(\pi)} \ge 0, \sum_{\vartheta \in \mathcal{P}_K} p_\vartheta \alpha_k^{(\vartheta)} \le 1 \right\}.
$$

$$(8.8)$$

Here, the users are restricted to uniformly allocating their powers over their transmit antennas.

- Space-time power allocation (STPA) game:

$$
\mathcal{A}_k^{SIC,STPA} = \left\{ \mathbf{Q}_k = \left(\mathbf{Q}_k^{(\pi)} \right)_{\pi \in \mathcal{P}_K} \;\middle|\; \forall \pi \in \mathcal{P}_K, \mathbf{Q}_k^{(\pi)} \succeq 0, \sum_{\vartheta \in \mathcal{P}_K} p_\vartheta \mathrm{Tr}(\mathbf{Q}_k^{(\vartheta)}) \le n_t \overline{P}_k \right\}.
$$

$$(8.9)$$

This is a generalization of the previous cases, where the users are free to allocate their powers both in time and over the transmit antennas.

The analysis of the NE follows the same lines as the fast fading MIMO MAC studied in Belmega et al. (2009b) and Belmega et al. (2010b). The existence of the NE is assured based on the properties of concave games. For the particular cases of the SPA and TPA games, the NE is proved to be unique (Belmega et al., 2009b). This is no longer true for the space-time power allocation game (Belmega et al., 2010b). Sufficient conditions for ensuring the uniqueness of the NE are: i) $\forall, k \in \mathcal{K}$, $\mathrm{rank}(\mathbf{H}_k^H \mathbf{H}_k) = n_{t,k}$; and ii) $n_{t,k}^2 \le n_r^2 + 1$. In order to determine the covariance matrices at the NE point, an iterative water-filling algorithm similar to the SUD case can be applied. Convergence results for a sequential updating rule can be found in Bertsekas (1995).

8.2.2 Fast Fading Links

Similarly to Sec. 8.2.1, we start by summarizing the two-player SISO case in Lai and El Gamal (2008). Assuming SUD scheme at the receiver, the authors of Lai and El Gamal (2008) proved the existence and uniqueness of the NE point. What is interesting is that at this point, only the user with the strongest channel will transmit, while the other remains silent. The interference is completely canceled. Furthermore, as opposed to the static links case, the system sum-rate at the NE achieves the sum-capacity point of the achievable rate region. For SIC decoding, the authors propose a scheme where the decoding order depends on the fading coefficients. The NE is proved to exist and to be unique. However, only the two corner points of the achievable rate region can be achieved. In order to achieve the other Pareto-optimal points, the authors propose a repeated game formulation.

Here the focus is on the general fast fading MIMO case analyzed in Belmega et al. (2010b) and Belmega et al. (2009b). In Belmega et al. (2009b), only the special cases of TPA and SPA games were investigated assuming SIC decoding. In Belmega et al. (2010b), both decoding techniques were considered (i.e., SUD and SIC decoding). When SIC was assumed, the general STPA game was studied.

In order to take into account the effects of antenna correlation, channel matrices are assumed to be structured according to the unitary-independent-unitary model introduced in Tulino and Verdú (2005):

$$\forall k \in \{1,\ldots,K\}, \ \mathbf{H}_k = \mathbf{V}_k \tilde{\mathbf{H}}_k \mathbf{W}_k, \tag{8.10}$$

where $\mathbf{V}_k$ and $\mathbf{W}_k$ are deterministic unitary matrices that allow one to take into consideration the correlation effects at the receiver and transmitters, respectively. The channel matrix $\tilde{\mathbf{H}}_k$ is an $n_r \times n_{t,k}$ random matrix whose entries are zero-mean independent complex Gaussian variables with an arbitrary variance profile, such that $\mathbb{E}|\tilde{\mathbf{H}}_k(\ell,c)|^2 = \frac{\sigma_k^2(\ell,c)}{n_{t,k}}$. The Kronecker propagation model, for which the channel transfer matrices factorizes as $\mathbf{H}_k = \mathbf{R}_k^{1/2} \mathbf{\Theta}_k \mathbf{T}_k^{1/2}$, is a special case of the UIU model. The variance profile is separable, i.e., $\sigma_k^2(\ell,c) = \frac{d_k^{(T)}(\ell) d_k^{(R)}(c)}{n_t}$, $\mathbf{V}_k = \mathbf{U}_{\mathbf{R}_k}$, $\mathbf{W}_k = \mathbf{U}_{\mathbf{T}_k}$ with $\mathbf{\Theta}_k$ being a random matrix with zero-mean i.i.d. entries. The matrices $\mathbf{T}_k$, $\mathbf{R}_k$ represent the transmitter and receiver antenna correlation matrices, such that $d_k^{(T)}(\ell)$, $d_k^{(R)}(c)$ are their associated eigenvalues, and $\mathbf{U}_{\mathbf{T}_k}$, $\mathbf{U}_{\mathbf{R}_k}$ are their associated eigenvector matrices respectively.

It turns out that the existence of a unique NE can be proved for both decoding techniques. The main difficulty in determining the NE state is that there are no closed-form expressions of the ergodic rates and the optimal power allocation policies. Indeed, the optimal eigenvalues of the transmit covariance matrices are not easy to find, but may be accessible using extensive numerical techniques. Our approach consists of approximating these ergodic rates to obtain expressions that are easier to interpret and to optimize. In order to do this, two extra assumptions will be made:

i. $\mathbf{V}_k = \mathbf{V}, \forall k \in \mathcal{K}$, which means that the receive antenna correlation matrices $\mathbf{R}_k$ (Kronecker model) decompose to the same eigenvector basis $\mathbf{V}$;

ii. $n_{t,k} = n_t$, i.e., the transmitters have the same number of antennas. Our approach consists of two steps.

First, the optimal eigenvectors are determined. Then, based on this result, the different transmission rates are approximated by their large-system (i.e., $n_r \to \infty$, $n_t \to \infty$, $\frac{n_r}{n_t} \to \beta$) deterministic equivalents. The results provided in Tulino and Verdú (2005) for the single-user MIMO channels are exploited to obtain the optimal eigenvalues of the covariance matrices. The corresponding approximations can be found to be accurate even for relatively small numbers of antennas (e.g., see Biglieri et al. (2002) and Dumont et al. (2006) for more details).

8.2.2.1 *Single User Decoding*
When the SUD is assumed at the receiver, each user has to choose the best precoding matrix $\mathbf{Q}_k = \mathbb{E}[\underline{X}_k\underline{X}_k^H]$, in the sense of his utility function:

$$
u_k^{\text{SUD}}(\mathbf{Q}_1, \mathbf{Q}_2) = \mathbb{E}\log\left|\mathbf{I}_{n_r} + \rho\mathbf{H}_k\mathbf{Q}_k\mathbf{H}_k^H + \rho\sum_{\ell \neq k}\mathbf{H}_\ell\mathbf{Q}_\ell\mathbf{H}_\ell^H\right|
$$

$$
- \mathbb{E}\log\left|\mathbf{I}_{n_r} + \rho\sum_{\ell \neq k}\mathbf{H}_\ell\mathbf{Q}_\ell\mathbf{H}_\ell^H\right| \tag{8.11}
$$

The strategy set of user k is given in (8.2).

· ·

Theorem 187

The space power allocation game described by: $\mathcal{G}^{\text{SUD}} = \left(\mathcal{K}, \left\{\mathcal{A}_k^{\text{SUD}}\right\}_{k\in\mathcal{K}}, \left\{u_k^{\text{SUD}}\right\}_{k\in\mathcal{K}}\right)$, where the utility functions $u_k^{\text{SUD}}(\mathbf{Q}_k, \mathbf{Q}_{-k})$ are given in (8.11), has a unique pure-strategy Nash equilibrium.

The proof of the existence of a NE is based on the properties of concave games in Rosen (1965). Proving the diagonal strict concavity condition of Rosen (1965) is sufficient to ensure the uniqueness of the NE point. However, extending Theorem 70 is not trivial and the proof is given in Belmega et al. (2010b).

In order to find the optimum covariance matrices, we proceeded in the same way as described in Lasaulce et al. (2007). First, the optimum eigenvectors were determined, and then the optimum eigenvalues were obtained by approximating the utility functions under the large system assumption.

Since $\mathbf{V}_k = \mathbf{V}$, the results in Tulino and Verdú (2004, 2005) for single-user MIMO channels can be exploited, assuming the asymptotic regime in terms of the number of antennas: $n_r \to \infty$, $n_t \to \infty$, $\frac{n_r}{n_t} \to \beta$. It turns out that no loss of optimality occurs by choosing the covariance matrices $\mathbf{Q}_k = \mathbf{W}_k\mathbf{P}_k\mathbf{W}_k^H$, where $\mathbf{W}_k$ is the same unitary matrix as in (8.10) and $\mathbf{P}_k$ is the diagonal matrix containing the eigenvalues of $\mathbf{Q}_k$. Although this result is easy to obtain, it is instrumental in our context for two reasons. First, the search for the optimum precoding matrices boils down to a search of the eigenvalues of these matrices. Second, as the optimum eigenvectors are known, available results in random matrix theory can be applied to find an accurate approximation of these eigenvalues (Tulino and Verdú, 2004, 2005). The corresponding approximated utility for user k is:

$$\tilde{u}_k^{\mathrm{SUD}}(\mathbf{P}_k, \mathbf{P}_{-k}) = \frac{1}{n_r} \sum_{k=1}^{K} \sum_{j=1}^{n_t} \log_2(1 + K\rho P_k(j)\gamma_k(j))$$

$$+ \frac{1}{n_r} \sum_{i=1}^{n_r} \log_2 \left(1 + \frac{1}{Kn_t} \sum_{k=1}^{K} \sum_{j=1}^{n_t} \sigma_k(i,j)\delta_k(j) \right)$$

$$- \frac{1}{n_r} \sum_{k=1}^{K} \sum_{j=1}^{n_t} \gamma_k(j)\delta_k(j) \log_2 e \tag{8.12}$$

$$- \frac{1}{n_r} \sum_{\ell \neq k} \sum_{j=1}^{n_t} \log_2(1 + (K-1)\rho P_\ell(j)\phi_\ell(j))$$

$$- \frac{1}{n_r} \sum_{i=1}^{n_r} \log_2 \left(1 + \frac{1}{(K-1)n_t} \sum_{\ell \neq k} \sum_{j=1}^{n_t} \sigma_\ell(i,j)\psi_\ell(j) \right)$$

$$+ \frac{1}{n_r} \sum_{\ell \neq k} \sum_{j=1}^{n_r} \phi_\ell(j)\psi_\ell(j) \log_2 e$$

where the parameters $\gamma_k(j)$ and $\delta_k(j)$ $\forall j \in \{1,\ldots,n_t\}$, $k \in \{1,2\}$ represent the unique solution to the system:

$$\begin{cases} & \forall j \in \{1,\ldots,n_t\}, k \in \mathcal{K} : \\ \gamma_k(j) &= \dfrac{1}{Kn_t} \displaystyle\sum_{i=1}^{n_r} \dfrac{\sigma_k(i,j)}{1 + \frac{1}{Kn_t} \displaystyle\sum_{\ell=1}^{K} \sum_{m=1}^{n_t} \sigma_\ell(i,m)\delta_\ell(m)} \\ \delta_k(j) &= \dfrac{K\rho P_k(j)}{1 + K\rho P_k(j)\gamma_k(j)} \end{cases} \tag{8.13}$$

and $\phi_\ell(j)$, $\psi_\ell(j)$, $\forall j \in \{1,\ldots,n_t\}$ represent the unique solution to the system:

$$\begin{cases} & \forall j \in \{1,\ldots,n_t\}, \ell \in \mathcal{K} \setminus \{k\} : \\ \phi_\ell(j) &= \dfrac{1}{(K-1)n_t} \displaystyle\sum_{i=1}^{n_r} \dfrac{\sigma_\ell(i,j)}{1 + \frac{1}{(K-1)n_t} \displaystyle\sum_{r \neq k} \sum_{m=1}^{n_t} \sigma_r(i,m)\psi_r(m)} \\ \psi_\ell(j) &= \dfrac{(K-1)\rho P_\ell(j)}{1 + (K-1)\rho P_\ell(j)\phi_\ell(j)}. \end{cases} \tag{8.14}$$

The optimal eigenvalues are given by the water-filling solutions:

$$P_k^{\mathrm{NE}}(j) = \left[\frac{1}{n_r \lambda_k \ln 2} - \frac{1}{K\rho\gamma_k(j)} \right]^{+}, \tag{8.15}$$

where $\lambda_k \geq 0$ is the Lagrangian multiplier tuned in order to meet the power constraint: $\sum_{j=1}^{n_t} \left[\frac{1}{n_r \lambda_k \ln 2} - \frac{1}{K \rho \gamma_k(j)} \right]^+ = n_t \overline{P}_k$. Notice that in order to solve the system of equations given above, we can use the same iterative power allocation algorithm as described in Lasaulce et al. (2007). The transmitters are assumed to have perfect knowledge of all the channels' distributions. As for the efficiency of the NE point, the SUD decoding technique is sub-optimal w.r.t. the centralized case and the sum-capacity is not reached at the NE similarly to the static MIMO MAC channel.

An important remark needs to be made at this point. The analysis of the NE (i.e., existence and uniqueness issues) has been performed in the finite setting (exact game). The determination of the NE is performed using numerical methods (required to implement water-filling type algorithms) and the approximated utilities in the asymptotic regime. This is motivated by the fact that the ergodic rates are very difficult to numerically optimize. Furthermore, it turns out that the large system approximations of ergodic rates have the same properties as their exact counterparts, as shown recently by Dumont et al. (2010). However, the analysis of the NE for the approximated game is an interesting issue which is left as an extension of this work.

Another rising issue could be the characterization of the correlated equilibria of the game. The results in Neyman (1997) can be applied here directly, since $\mathcal{G}^{\text{SUD}}$ is an exact potential game with a strict concave potential function, i.e., the achievable network sum-rate. The author of Neyman (1997) has proved that for strict concave potential games, the CE is unique, and consists of playing with probability 1 the unique pure NE.

8.2.2.2 *Successive Interference Cancellation*

As mentioned in Sec. 8.2.1, three different power allocation games can be defined as a function of the action sets: SPA, TPA, and the joint STPA, for which the action sets are given by (8.7), (8.8), and (8.9), respectively. In the remainder of this section, the focus is on the general STPA game. The results for the special cases follow directly. The main difference from the static case is in the utility function of the users which are given by the ergodic achievable rates:

$$u_k^{\text{SIC}}(\mathbf{Q}_k, \mathbf{Q}_{-k}) = \sum_{\pi \in \mathcal{P}_K} p_\pi R_k^{(\pi)}\left(\mathbf{Q}_k^{(\pi)}, \mathbf{Q}_{-k}^{(\pi)}\right) \tag{8.16}$$

where:

$$R_k^{(\pi)}\left(\mathbf{Q}_k^{(\pi)}, \mathbf{Q}_{-k}^{(\pi)}\right) = \mathbb{E}\log_2 \left| \mathbf{I}_{n_r} + \rho \mathbf{H}_k \mathbf{Q}_k^{(\pi)} \mathbf{H}_k^H + \rho \sum_{\ell \in \mathcal{K}_k^{(\pi)}} \mathbf{H}_\ell \mathbf{Q}_\ell^{(\pi)} \mathbf{H}_\ell^H \right|$$

$$- \mathbb{E}\log_2 \left| \mathbf{I}_{n_r} + \rho \sum_{\ell \in \mathcal{K}_k^{(\pi)}} \mathbf{H}_\ell \mathbf{Q}_\ell^{(\pi)} \mathbf{H}_\ell^H \right| \tag{8.17}$$

with the same notation as in Sec. 8.2.1.2. For the fast fading case as opposed to the static one, uniqueness is guaranteed for the general joint space–time power allocation game. The obtained results are stated in the following theorem.

. . . .

Theorem 188

The joint space–time power allocation game described by: $\mathcal{G}^{\text{SIC}} = \left(\mathcal{K}, \left\{\mathcal{A}_k^{\text{SIC,STPA}}\right\}_{k\in\mathcal{K}}, \left\{u_k^{\text{SIC}}\right\}_{k\in\mathcal{K}}\right)$, where the utility functions $u_k^{\text{SIC}}(\mathbf{Q}_k, \mathbf{Q}_{-k})$ are given in (8.16), has a unique pure-strategy Nash equilibrium.

The proof is similar to the SUD case and exploits the extended results of Rosen (1965). The difficulty here lies in proving a matrix trace inequality. This inequality is instrumental in proving that the diagonal strict concavity condition holds. Its proof is given in Belmega et al. (2009a) for $K = 2$ and in Belmega et al. (2010) for arbitrary $K \geq 2$.

In order to find the optimal covariance matrices we proceed in the same way as described in Sec. 8.2.2.1. The optimal eigenvectors of the covariance matrix $\mathbf{Q}_k^{(\pi)}$ are given by $\mathbf{U}_{Q_k^{(\pi)}} = \mathbf{W}_k$. And, the optimal eigenvalues $\mathbf{P}_k^{(\pi)}$ can be found using the large-system assumptions. The approximated utility for user k is $\tilde{u}_k^{\text{SIC}}(\{\mathbf{P}_k^{(\pi)}\}_{k\in\mathcal{K}, \pi\in\mathcal{P}_K}) = \sum_{\pi\in\mathcal{P}_K} p_\pi \tilde{R}_k^{(\pi)}\left(\mathbf{P}_k^{(\pi)}, \mathbf{P}_{-k}^{(\pi)}\right)$ where:

$$
\tilde{R}_k^{(\pi)}\left(\mathbf{P}_k^{(\pi)}, \mathbf{P}_{-k}^{(\pi)}\right) = \frac{1}{n_r} \sum_{\ell\in\mathcal{K}_k^{(\pi)}\cup\{k\}} \sum_{j=1}^{n_t} \log_2\left(1 + (N_k^{(\pi)} + 1)\rho P_\ell^{(\pi)}(j)\gamma_\ell^{(\pi)}(j)\right)
$$

$$
+ \frac{1}{n_r}\sum_{i=1}^{n_r}\log_2\left(1 + \frac{1}{(N_k^{(\pi)}+1)n_t}\sum_{\ell\in\mathcal{K}_k^{(\pi)}\cup\{k\}}\sum_{j=1}^{n_t}\sigma_\ell(i,j)\delta_\ell^{(\pi)}(j)\right)
$$

$$
- \frac{1}{n_r}\sum_{\ell\in\mathcal{K}_k^{(\pi)}\cup\{k\}}\sum_{j=1}^{n_t}\gamma_\ell^{(\pi)}(j)\delta_\ell^{(\pi)}(j)\log_2 e
$$

$$
- \frac{1}{n_r}\sum_{\ell\in\mathcal{K}_k^{(\pi)}}\sum_{j=1}^{n_t}\log_2\left(1 + N_k^{(\pi)}\rho P_\ell^{(\pi)}(j)\phi_\ell^{(\pi)}(j)\right)
$$

$$
- \frac{1}{n_r}\sum_{i=1}^{n_r}\log_2\left(1 + \frac{1}{N_k^{(\pi)}n_t}\sum_{\ell\in\mathcal{K}_k^{(\pi)}}\sum_{j=1}^{n_t}\sigma_\ell(i,j)\psi_\ell^{(\pi)}(j)\right)
$$

$$
+ \frac{1}{n_r}\sum_{\ell\in\mathcal{K}_k^{(\pi)}}\sum_{j=1}^{n_r}\phi_\ell^{(\pi)}(j)\psi_\ell^{(\pi)}(j)\log_2 e \qquad\qquad (8.18)
$$

where $N_k^{(\pi)} = |\mathcal{K}_k^{(\pi)}|$ and the parameters $\gamma_k^{(\pi)}(j)$ and $\delta_k^{(\pi)}(j)$ $\forall j \in \{1,\ldots,n_t\}$, $k \in \mathcal{K}$, $\pi \in \mathcal{P}_K$ are the solutions of:

$$
\begin{cases}
\forall j \in \{1,\ldots,n_t\}, \ell \in \mathcal{K}_k^{(\pi)} \cup \{k\} : \\
\gamma_\ell^{(\pi)}(j) = \dfrac{1}{\left(N_k^{(\pi)}+1\right)n_t} \displaystyle\sum_{i=1}^{n_r} \dfrac{\sigma_\ell(i,j)}{1 + \dfrac{1}{\left(N_k^{(\pi)}+1\right)n_t} \displaystyle\sum_{r \in \mathcal{K}_k^{(\pi)} \cup \{k\}} \sum_{m=1}^{n_t} \sigma_r(i,m)\delta_r^{(\pi)}(m)} \\[4ex]
\delta_\ell^{(\pi)}(j) = \dfrac{\left(N_k^{(\pi)}+1\right)\rho P_\ell^{(\pi)}(j)}{1 + \left(N_k^{(\pi)}+1\right)\rho P_\ell^{(\pi)}(j)\gamma_\ell^{(\pi)}(j)},
\end{cases} \tag{8.19}
$$

and $\phi_\ell^{(\pi)}(j)$, $\psi_\ell^{(\pi)}(j)$, $\forall j \in \{1,\ldots,n_t\}$ and $\pi \in \mathcal{P}_K$ are the unique solutions of the following system:

$$
\begin{cases}
\forall j \in \{1,\ldots,n_t\}, \ell \in \mathcal{K}_k^{(\pi)} : \\
\phi_\ell^{(\pi)}(j) = \dfrac{1}{N_k^{(\pi)}n_t} \displaystyle\sum_{i=1}^{n_r} \dfrac{\sigma_\ell(i,j)}{1 + \dfrac{1}{N_k^{(\pi)}n_t} \displaystyle\sum_{r \in \mathcal{K}_k^{(\pi)}} \sum_{m=1}^{n_t} \sigma_r(i,m)\psi_r^{(\pi)}(m)} \\[4ex]
\psi_\ell^{(\pi)}(j) = \dfrac{N_k^{(\pi)}\rho P_\ell^{(\pi)}(j)}{1 + N_k^{(\pi)}\rho P_\ell^{(\pi)}(j)\phi_\ell^{(\pi)}(j)}.
\end{cases} \tag{8.20}
$$

The corresponding water-filling solution is:

$$
P_k^{(\pi),\mathrm{NE}}(j) = \left[\frac{1}{n_r\lambda_k \ln 2} - \frac{1}{N_k^{(\pi)}\rho\gamma_k^{(\pi)}(j)} \right]^+, \tag{8.21}
$$

where $\lambda_k \geq 0$ is the Lagrangian multiplier tuned in order to meet the power constraint:

$$
\sum_{\pi \in \mathcal{P}_K} \sum_{j=1}^{n_t} p_\pi \left[\frac{1}{n_r\lambda_k \ln 2} - \frac{1}{N_k^{(\pi)}\rho\gamma_k^{(\pi)}(j)} \right]^+ = n_t\overline{P}_k.
$$

In order to measure the efficiency of the decentralized network w.r.t. its centralized counterpart, the following quantity is introduced:

$$
\mathrm{SRE} = \frac{R_{\mathrm{sum}}^{\mathrm{NE}}}{C_{\mathrm{sum}}} \leq 1, \tag{8.22}
$$

where SRE stands for sum-rate efficiency; the quantity $R_{\mathrm{sum}}^{\mathrm{NE}}$ represents the sum-rate of the decentralized network at the Nash equilibrium, which is achieved for certain choices of coding and decoding strategies; the quantity C_{sum} corresponds to the sum-capacity of the centralized network, which is reached only if the optimum coding and decoding schemes are known. Note that this is the case for the MAC, but not for other channels, such as the interference channel. Obviously, the efficiency measure used here is strongly connected with the price of anarchy (PoA) introduced in Roughgarden and Tardos (2002). The SRE measures the gap between the sum-rate at the NE and the network sum-rate obtained with the optimal decoding technique, whereas the PoA does not consider the optimal decoding technique. In our context, information theory provides us with fundamental physical limits to the social welfare (network sum-capacity) while in general no such upper bound is available.

We have proved that in both the low and high SNR regimes the SRE tends to 1. This means that in the extreme SNR regimes, the sum-capacity of the fast fading MAC is achieved at the NE point, in spite of the sub-optimal coordination mechanism applied at the receiver. For non-extreme SNR regimes, a closed-form expression of the SRE cannot be found. Numerical results have been provided to assess this optimality gap. When SIC is assumed, for the three power allocation games (TPA, SPA, STPA), the sum-rate efficiency at the NE is close to 1. Quite surprisingly, the NE of the STPA game performs a little worse than its purely spatial counterpart. This highlights another Braess-type paradox as in Sec. 8.2.1.1 (see Belmega et al. (2010b)).

8.3 ON THE CASE OF POWER ALLOCATION GAMES IN PARALLEL MULTIPLE ACCESS CHANNELS

8.3.1 The Uniqueness of the Nash Equilibrium

Parallel multiple access channels (PMAC) consist of a special case of MIMO multiple access channels for which the transfer matrices $\mathbf{H}_k$, $k \in \mathcal{K}$, are diagonal (as argued in Sec. 8.2.1.1). Although the analysis of fast fading PMAC does not call for particular comments, the case of static channels does. Indeed, in static PMAC, the condition for NE uniqueness provided in Sec. 8.2.1.1 is not necessary. In fact, the authors of Mertikopoulos et al. (2011) went even further by proving that the refined sufficient conditions given in Scutari et al. (2009) for parallel interference channels hold with probability zero in static PMAC. By using the fact that Shannon transmission rate-efficient power allocation games in PMAC are exact potential games (when single-user decoding is used), and also the concept of degeneracy introduced in Mertikopoulos et al. (2011), it can be proved that uniqueness holds almost definitely in PMAC. Indeed, as shown in Perlaza et al. (2009a), a potential function for the power allocation game in PMAC is:

$$\phi(\underline{p}) = \sum_{s=1}^{S} \log_2 \left(\sigma_s^2 + \sum_{k=1}^{K} p_{k,s} g_{k,s} \right) \tag{8.23}$$

where $\underline{p} = (\underline{p}_1, \underline{p}_2, \ldots, \underline{p}_K)$, $\underline{p}_k = (p_{k,1}, p_{k,2}, \ldots, p_{k,S})$, K is the number of transmitters, S the number of bands/channels/receivers, $p_{k,s}$ the power allocation by transmitter k to channel s, $g_{k,s} = |h_{k,s}|^2$ the channel gain of link (k,s), and σ_s^2 the noise level in channel s.

8.3.2 Pricing for Channel Selection Games in Parallel Multiple Access Channels

We always consider a parallel multiple access channel with K transmitters, each of them having S static channels at his disposal. The idea is to consider a special case of static PMAC which may not always be realistic but has the advantage of being simple and allows us to better illustrate the Wardrop equilibrium (WE) notion and a pricing technique to improve the WE efficiency. Indeed, as in Belmega et al. (2008a), a simplified system model is assumed:

i. Transmitters are assumed to have the same transmit power $\forall k \in \mathcal{K}, P_k = P$.

ii. Each transmitter can only use one channel $s \in \mathcal{S}$ at a given instant. The problem of distributed power allocation therefore boils down to a channel selection problem, since the decision of a transmitter doesn't consist in sharing his power but in choosing a channel to allocate it to.

iii. The channel gains are assumed to be identical and normalized, i.e., $|h_{k,s}|^2 = 1$.

iv. The number of transmitters (K) is large.

Assumptions (i) and (ii) are very common in the wireless literature and do not call for particular comment. Assumption (iii), which is made in Belmega et al. (2008a), is more questionable. The merit of this assumption is that it allows us to pose the problem and apply the existing concepts in a simple manner. Obviously, an interesting technical extension of Belmega et al. (2008a) would be to remove this assumption. Note that, however, it would make sense in a context where transmitters undergo quite similar propagation effects or channel gains are compensated at the receiver(s). This assumption will be made here to show to the reader how to apply pricing to the channel selection problem. Assumption (iv) is also made for the sake of simplicity and clarity. To better illustrate the concepts used, we assume that the K transmitters are mobile stations and the S channels correspond to base stations operating in non-overlapping frequency bands. The base stations are assumed to always implement single-user decoding. The base station selection problem can therefore be described by a static game for which the action sets are identical and equal to $\mathcal{S} = \{1, \ldots, S\}$ and the utility functions are denoted by v_i. When transmitter k chooses to connect to base station s, his utility function is $v_{i,s}$ and is chosen to be:

$$v_{k,s}(\underline{x}) = \log_2\left[1 + \frac{P}{\sigma_s^2 + Kx_sP}\right] \qquad (8.24)$$

where x_s is the fraction of transmitters who are connected to base station s. The equilibrium point of this game is obtained by applying the indifference principle, used by Wardrop (1952) in the context of congestion games. Knowing that $\sum_{s=1}^{S} x_s = 1$, from Belmega et al. (2008a), we find that the repartition of the selfish users at the equilibrium, denoted by $\underline{x}^* = (x_1^*, \ldots, x_S^*)$, is given by:

$$\forall s \in \mathcal{S}, \; x_s^* = \frac{1}{S} + \frac{1}{S} \sum_{j=1}^{S} \frac{\sigma_j^2 - \sigma_s^2}{KP}. \tag{8.25}$$

In order for the vector of fractions $\underline{x}^*$ to always be in the simplex $\Delta(\mathcal{S})$, it is possible to use the simple iterative optimization algorithm proposed in Belmega et al. (2008a). The indifference principle applied to derive (8.25) can be easily understood by considering a framework similar to the one used for studying learning algorithms in games where transmitters are assumed to know their utilities at any time. If we consider the dynamic game where the transmitters have to choose between different base stations, it can be shown (Milchtaich, 1996a; Roughgarden and Tardos, 2004) that the mobiles are going to react as long as their utility in the system s is not equal to that in system $j \neq s$ where $j \in \{1, \ldots, S\}$. For conventional models of dynamics, this observation–reaction process converges to a stable equilibrium point, from which no user can unilaterally deviate without losing in terms of individual utility. For instance, as the game under investigation is potential, the reinforcement learning algorithm of Mosteller and Bush (1955) and Sastry et al. (1994b) is an example of an algorithm which converges to the mentioned equilibrium point.

The problem with the solution in (8.25) is that the network can be inefficient in terms of sum-rate. As the considered problem is a wireless congestion problem, many tools from the literature on congestion (e.g., see Roughgarden (2005)) can be applied, in particular, pricing techniques and the concept of the price of anarchy (North, 1981; Papadimitriou, 2001). In general, there is no reason why the selfish behavior of the active transmitters should maximize the network sum-rate. A simple intuitive explanation for this can be given in the case where there are two base stations. Let us assume that the utility in base station 2 is much less sensitive to multiuser interference (this could be analyzed more precisely by inspecting the derivatives of the utilities at the equilibrium point) than that of base station 1. Then, by forcing a small fraction of transmitters of base station 1 to move to base station 2, these transmitters will get a very similar utility due to the assumption made above. However, the transmitters who remain connected to base station 1 benefit from a non-negligible gain, which translates into a better sum-rate. As seen in Chapter 7, a way of forcing some transmitters to move from one base station to another is to introduce a pricing mechanism (e.g., see DaSilva (2000), Gupta et el. (1997) and Ramesh et al. (1996)). The technique presented here relies on the idea that the transmitters can either be punished or rewarded by a central authority, in order to deviate the (selfish) transmitters from the spontaneous equilibrium into a stimulated one. It turns out that it is not easy to design such a mechanism for the type of transmitters's utility chosen in this chapter; namely the individual transmission rate. This is because the system owner cannot guarantee a transmitter will have a better rate if this is not feasible or achievable. To circumvent this difficulty, one solution is to define new utilities such that incentives and penalties can be properly defined. Another is to associate a cost with the connection time of a transmitter (Li and Chao, 2004; Li et al., 2002; Yee et al., 2006). Note that the standpoint here is the one of the wireless engineer: we want to maximize the sum-rate of the network instead of maximizing the total revenue generated by the transmitters' activity.

Without loss of generality, assume that each mobile has to transmit a fixed number of bits $n_k \geq 0$. The time that transmitter $k \in \mathcal{K}$ spends to transmit his data if he is connected to base station $s \in \mathcal{S}$ is:

$$\tau_{k,s}(\underline{x}) = \frac{n_k}{\upsilon_{k,s}(\underline{x})}, \tag{8.26}$$

where $\upsilon_{k,s}$ is the transmission rate for transmitter k in base station s. We assume that the price charged by the system owner is a function of the connection time. The price function $p(\tau)$ is a positive, strictly increasing, and linear function (for more details see Daoud et al. (2008), Maillé and Stier-Moses (2009), and Shen and Basar (2004, 2006)). For example $p(\tau) = \sigma\tau$, where $\sigma \geq 0$ represents the price in units per second. It is also assumed that the unitary (virtual or not) price that the mobile pays is the same whatever the base station that they choose to be connected to. The objective of the transmitters is to minimize the cost of transmitting their individual bits. The cost for transmitter k if he decides to connect to base station s is defined as:

$$c_{k,s}(\underline{x}) = p(\tau_{k,s}(\underline{x})) + \beta_s, \tag{8.27}$$

where the parameter β_s is real and can be positive or negative. As a monetary cost is considered instead of the transmission rate for the transmitter's utility, the game is modified. This is why the notations $\widetilde{\underline{x}}$, $\widetilde{\upsilon_{k,s}}$ are used to refer to this modified game. Note that, if for all $s \in \mathcal{S}$, $\beta_s = 0$, and if $p(\tau)$ is positive and strictly increasing, the modified game is equivalent to the original game. As the objective is to maximize the sum-rate, we have to choose the parameters $\beta_s, s \in \mathcal{S}$ to be non-negative according to the following relation:

$$\beta_s = -p(\tau_{k,s})(\widetilde{x}_s^*) + p(\tau_{k,s}(x_s^*)), \tag{8.28}$$

where $\widetilde{\underline{x}}^*$ is the operating point desired by the engineer. In order to measure the social efficiency of the proposed pricing mechanism, the price of anarchy can be used. In our context, it is defined as follows:

$$\mathrm{PoA}(\widetilde{\underline{x}}^*) = \frac{\displaystyle\max_{\underline{x} \in \Delta(\mathcal{S})} \sum_{s=1}^{S} \sum_{i=1}^{K} \upsilon_{k,s}(\underline{x})}{\displaystyle\sum_{s=1}^{S} \sum_{i=1}^{K} \upsilon_{k,s}(\widetilde{\underline{x}}^*)}. \tag{8.29}$$

Figure 8.1 illustrates the corresponding analysis for a PMAC with a large number of transmitters and $S = 2$ channels. In this figure, it is seen how efficient the spontaneous WE is, and what can be gained by introducing a pricing technique to stimulate a new outcome.

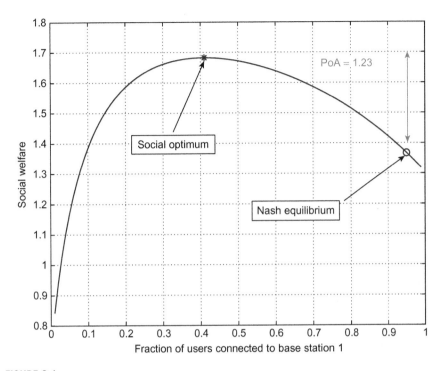

FIGURE 8.1

Performance in terms of utility-sum of the PMAC as a function of the user repartition between the two available channels.

8.4 POWER ALLOCATION GAMES IN PARALLEL INTERFERENCE RELAY CHANNELS

In this section, a different network model is studied: the parallel interference relay channel (Figure 8.2). The Shannon-rate efficient power allocation game for the interference channels has been extensively studied in the literature: Etkin et al. (2007) and Yu et al. (2002) for SISO frequency selective channels, Chung et al. (2003) and Luo and Pang (2006) for the static parallel interference channel and Arslan et al. (2007), Larsson and Jorswieck (2008) and Scutari et al. (2008a,b,c) for the static MIMO channels. The difference in this instance is that the transmitters can exploit the existence of some relaying nodes to improve their communication performance.

As opposed to the previous section, the focus here is on the particular case of static parallel IRC (Belmega et al., 2011).

The system under investigation is composed of two source nodes S_1, S_2, transmitting their private messages to their respective destination nodes D_1, D_2. To this end, each source can exploit Q non-overlapping frequency bands (the notation (q) will be used to refer to band $q \in \{1, \dots, Q\}$) which are assumed to be of normalized

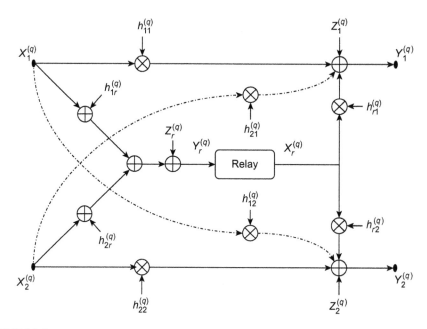

FIGURE 8.2

The parallel interference relay channel model in band (q).

bandwidth. The signals transmitted by S_1 and S_2 in band (q), denoted by $X_1^{(q)}$ and $X_2^{(q)}$ respectively, are assumed to be independent and power constrained:

$$\forall k \in \{1,2\}, \ \sum_{q=1}^{Q} \mathbb{E}|X_k^{(q)}|^2 \leq \overline{P}_k. \tag{8.30}$$

For $k \in \mathcal{K} \triangleq \{1,2\}$, denote by $\theta_k^{(q)}$ the fraction of power that is used by S_k for transmitting in band (q); that is, $\mathbb{E}|X_k^{(q)}|^2 = \theta_k^{(q)}\overline{P}_k$. Therefore, the set of possible power allocation policies for user k can be defined as:

$$\mathcal{A}_k = \left\{ \underline{\theta}_k \in [0,1]^Q \ \middle| \ \sum_{q=1}^{Q} \theta_k^{(q)} \leq 1 \right\} \tag{8.31}$$

Additionally, assume that there exists a multi-band relay $\mathcal{R}$. With these notations, the signals received by $\mathcal{D}_1$, $\mathcal{D}_2$, and $\mathcal{R}$ in band (q) are expressed as:

$$\begin{cases} Y_1^{(q)} = h_{11}^{(q)}X_1^{(q)} + h_{21}^{(q)}X_2^{(q)} + h_{r1}^{(q)}X_r^{(q)} + Z_1^{(q)} \\ Y_2^{(q)} = h_{12}^{(q)}X_1^{(q)} + h_{22}^{(q)}X_2^{(q)} + h_{r2}^{(q)}X_r^{(q)} + Z_2^{(q)} \\ Y_r^{(q)} = h_{1r}^{(q)}X_1^{(q)} + h_{2r}^{(q)}X_2^{(q)} + Z_r^{(q)} \end{cases} \tag{8.32}$$

where $Z_k^{(q)} \sim \mathcal{N}(0, N_k^{(q)})$, $k \in \{1, 2, r\}$, represents the Gaussian complex noise in band (q) and, for all $(k, \ell) \in \mathcal{K}^2$, $h_{k\ell}^{(q)}$ is the channel gain between $\mathcal{S}_k$ and $\mathcal{D}_\ell$, $h_{kr}^{(q)}$ is the channel gain between $\mathcal{S}_k$ and $\mathcal{R}$, and $h_{rk}^{(q)}$ is the channel gain between $\mathcal{R}$ and $\mathcal{D}_k$ in band (q). As far as the channel state information (CSI) is concerned, we always assume coherent communications for each transmitter–receiver pair $(\mathcal{S}_k, \mathcal{D}_k)$, but at the transmitters, the information assumptions will be context-dependent. The single-user decoding (SUD) will always be assumed at $\mathcal{D}_1$ and $\mathcal{D}_2$.

At the relay, the implemented reception scheme will depend on the protocol assumed. The expressions of the signals transmitted by the relay, $X_r^{(q)}, q \in \{1, \dots, Q\}$, will also depend on the relay protocol and will therefore also be explained in the corresponding sections. So far, we have not mentioned any power constraint on the signals $X_r^{(q)}$. We also assume that the relay implements a fixed power allocation policy between the Q available bands $(\mathbb{E}|X_r^{(q)}|^2 = P_r^{(q)}, q \in \{1, \dots, Q\})$. As in Maric et al. (2008) and Sahin and Erkip (2007a,b), the relay is assumed to operate in the full-duplex mode.

In what follows, the existence of an NE solution for the non-cooperative power allocation game where the transmitters are assisted by several relaying nodes is investigated. Three different games are analyzed depending on the relaying protocol assumed, which can be: estimate-and-forward (EF), decode-and-forward (DF) or amplify-and-forward. In general, the multiplicity of the NE points is an intractable problem. However, in the special case of fixed amplification gain AF protocol, the number of NE is characterized and the convergence of the best-response based iterative algorithm is proven.

8.4.1 Decode-and-Forward (DF) Protocol

We start with the decode-and-forward protocol. The basic idea behind this protocol is as follows. From each message intended for the destination, the source builds a coarse and a fine message. On these two messages, the source superimposes two codewords. The rates associated with these codewords (or messages) are such that the relay can reliably decode both of them while the destination can only decode the coarse message. After decoding this message, the destination can subtract the corresponding signal and try to decode the fine message. To help the destination to do so, the relay cooperates with the source by sending some information about the fine message.

Mathematically, this translates as follows. The signal transmitted by $\mathcal{S}_k$ in band (q) is structured as $X_k^{(q)} = X_{k,0}^{(q)} + \sqrt{\frac{\tau_k^{(q)}}{\nu_k^{(q)}} \frac{\theta_k^{(q)} P_k}{P_r^{(q)}}} X_{r,k}^{(q)}$ where the signals $X_{k,0}^{(q)}$ and $X_{r,k}$ are independent, and correspond exactly to the coarse and fine messages respectively; the parameter $\nu_k^{(q)}$ represents the fraction of transmit power the relay allocates to user i, hence we have $\nu_1^{(q)} + \nu_2^{(q)} \leq 1$; the parameter $\tau_k^{(q)}$ represents the fraction of transmit power which $\mathcal{S}_k$ allocates to the cooperation signal (conveying the fine message). The transmitted signal by the relay $\mathcal{R}$ is written as: $X_r^{(q)} = X_{r,1}^{(q)} + X_{r,2}^{(q)}$. From Sahin and Erkip (2007a), and for a given allocation policy $\underline{\theta}_k = \left(\theta_k^{(1)}, \ldots, \theta_k^{(Q)}\right)$, the source–destination pair $(\mathcal{S}_k, \mathcal{D}_k)$ achieves the transmission rate $\sum_{q=1}^{Q} R_k^{(q),\text{DF}}$ where:

$$\begin{cases} R_1^{(q),\text{DF}} = \min\left\{ R_{1,1}^{(q),\text{DF}}, R_{1,2}^{(q),\text{DF}} \right\} \\ R_2^{(q),\text{DF}} = \min\left\{ R_{2,1}^{(q),\text{DF}}, R_{2,2}^{(q),\text{DF}} \right\} \end{cases}, \tag{8.33}$$

$$\begin{cases} R_{1,1}^{(q),\text{DF}} = C\left(\dfrac{\left|h_{1r}^{(q)}\right|^2 \overline{\tau_1^{(q)} \theta_1^{(q)}} P_1}{\left|h_{2r}^{(q)}\right|^2 \overline{\tau_2^{(q)} \theta_2^{(q)}} P_2 + N_r^{(q)}} \right) \\[3ex] R_{2,1}^{(q),\text{DF}} = C\left(\dfrac{\left|h_{2r}^{(q)}\right|^2 \overline{\tau_2^{(q)} \theta_2^{(q)}} P_2}{\left|h_{1r}^{(q)}\right|^2 \overline{\tau_1^{(q)} \theta_1^{(q)}} P_1 + N_r^{(q)}} \right) \\[3ex] R_{1,2}^{(q),\text{DF}} = C\left(\dfrac{\left|h_{11}^{(q)}\right|^2 \theta_1^{(q)} P_1 + \left|h_{r1}^{(q)}\right|^2 \overline{v^{(q)}} P_r^{(q)} + 2\text{Re}\left(h_{11}^{(q)} h_{r1}^{(q),*}\right)\sqrt{\tau_1^{(q)} \theta_1^{(q)} P_1 \overline{v^{(q)}} P_r^{(q)}}}{\left|h_{21}^{(q)}\right|^2 \theta_2^{(q)} P_2 + \left|h_{r1}^{(q)}\right|^2 \overline{v^{(q)}} P_r^{(q)} + 2\text{Re}\left(h_{21}^{(q)} h_{r1}^{(q),*}\right)\sqrt{\tau_2^{(q)} \theta_2^{(q)} P_2 \overline{v^{(q)}} P_r^{(q)}} + N_1^{(q)}} \right) \\[3ex] R_{2,2}^{(q),\text{DF}} = C\left(\dfrac{\left|h_{22}^{(q)}\right|^2 \theta_2^{(q)} P_2 + \left|h_{r2}^{(q)}\right|^2 \overline{v^{(q)}} P_r^{(q)} + 2\text{Re}\left(h_{22}^{(q)} h_{r2}^{(q),*}\right)\sqrt{\tau_2^{(q)} \theta_2^{(q)} P_2 \overline{v^{(q)}} P_r^{(q)}}}{\left|h_{12}^{(q)}\right|^2 \theta_1^{(q)} P_1 + \left|h_{r2}^{(q)}\right|^2 v^{(q)} P_r^{(q)} + 2\text{Re}\left(h_{12}^{(q)} h_{r2}^{(q),*}\right)\sqrt{\tau_1^{(q)} \theta_1^{(q)} P_1 v^{(q)} P_r^{(q)}} + N_2^{(q)}} \right) \end{cases}, \tag{8.34}$$

and $(v^{(q)}, \tau_1^{(q)}, \tau_2^{(q)})$ is a given triple of parameters in $[0,1]^3$, $\tau_1^{(q)} + \tau_2^{(q)} \leq 1$, and $C(x) \triangleq \log_2(1+x)$ denotes the capacity function for complex signals.

The achievable transmission rate of user k is given by:

$$\mu_k^{\text{DF}}(\underline{\theta}_k, \underline{\theta}_{-k}) = \sum_{q=1}^{Q} R_k^{(q),\text{DF}}(\theta_k^{(q)}, \theta_{-k}^{(q)}). \tag{8.35}$$

The one-shot game is defined by the triplet $\mathcal{G}^{\text{DF}} = \left(\mathcal{K}, \{\mathcal{A}_k\}_{k \in \mathcal{K}}, \{\mu_k^{\text{DF}}\}_{k \in \mathcal{K}}\right)$. Although this setup might seem to be demanding in terms of CSI at the source nodes, it turns out that the equilibria predicted in such a framework can be effectively observed in more realistic frameworks where each player observes the strategy played by the other players and reacts accordingly by maximizing his utility iteratively.

The existence theorem for the DF protocol is:

Theorem 189

If the channel gains satisfy the condition $\mathcal{R}e(h_{kk}^{(q)} h_{rk}^{(q)*}) \geq 0$, for all $k \in \mathcal{K}$ and $q \in \{1, \ldots, Q\}$ the game defined by $\mathcal{G}^{\text{DF}} = \left(\mathcal{K}, \{\mathcal{A}_k\}_{k \in \mathcal{K}}, \{\mu_k^{\text{DF}}\}_{k \in \mathcal{K}}\right)$ has always at least one pure-strategy NE.

This theorem shows that, for the pathloss channel model where $h_{k\ell} > 0$, $(k, \ell) \in \{1, 2, r\}^2$, there always exists an equilibrium. As a consequence, if some relays are added into the network, the transmitters will adapt their PA policies accordingly and, whatever the locations of the relays, an equilibrium will be observed. This is a nice property for the system under investigation. Since the PA game with DF is concave, it is tempting to try to verify whether a sufficient condition for uniqueness of Rosen (1965) is met here. It turns out that the diagonally strict concavity condition of Rosen (1965) is not trivial to confirm. Additionally, it is possible that the game has several equilibria, as is proven to be the case for the AF protocol.

In a context of decentralized networks, each source $\mathcal{S}_k$ has to optimize the parameter τ_k in order to maximize its transmission rate R_k. In the rate region above, one can observe that this choice is not independent of the choice of the other source. Therefore, each source finds its optimal strategy by optimizing its rate w.r.t. $\tau_k^*(\tau_\ell)$. In order to do that, each source has to make some assumptions about the value τ_ℓ used by the other source. This is in fact a non-cooperative game where each player makes some assumptions about the other player's behavior and maximizes its own utility. Interestingly, we see that, even in the single-band case, the DF protocol introduces a power allocation game through the parameter τ_k representing the cooperation degree between the source $\mathcal{S}_k$ and the relay. For more details about the game induced by the cooperation degrees the reader is referred to Belmega et al. (2009a).

8.4.2 Estimate-and-Forward (EF) Protocol

The quality of the links from the sources to the relay represents the bottleneck of the DF protocol. In some situations, the presence of the relays may even degrade the performance of the transmission (for example, if the relay is situated far away from the sources such that the destinations have better reception conditions). We will now consider a protocol that always improves the performance of the transmission: the estimate-and-forward. The principle of the EF protocol for the standard relay channel is that the relay sends an approximated version of its observed signal to the receiver. In our setup, we have two different receivers. The relay can either create a single quantized version of its observation, common to both receivers, or two quantized versions, one for each destination (see Djeumou et al. (2009)). Here, we consider that the relay uses two resolution levels to compress its observation signal, each of these levels being adapted to the considered destination; the corresponding version of the EF protocol is called "bi-level compression EF". The same assumptions as in Sec. 8.4.1 concerning the reception schemes and PA policies at the relays are made: each node $\mathcal{R}$, $\mathcal{D}_1$, and $\mathcal{D}_2$ implements single-user decoding, and the PA policy at each relay, i.e., $\underline{v} = \left(v^{(1)}, \ldots, v^{(Q)}\right)$, is fixed. The utility for User $k \in \mathcal{K}$ can be expressed as follows:

$$\mu_k^{\mathrm{EF}}(\underline{\theta}_k, \underline{\theta}_{-k}) = \sum_{q=1}^{Q} R_k^{(q), \mathrm{EF}} \tag{8.36}$$

where:

$$R_1^{(q),\text{EF}} = C\left(\frac{\left(\left|h_{2r}^{(q)}\right|^2\theta_2^{(q)}P_2+N_r^{(q)}+N_{wz,1}^{(q)}\right)\left|h_{11}^{(q)}\right|^2\theta_1^{(q)}P_1+\left(\left|h_{21}^{(q)}\right|^2\theta_2^{(q)}P_2+\left|h_{r1}^{(q)}\right|^2\overline{v^{(q)}}P_r^{(q)}+N_1^{(q)}\right)\left|h_{1r}^{(q)}\right|^2\theta_1^{(q)}P_1}{\left(N_r^{(q)}+N_{wz,1}^{(q)}\right)\left(\left|h_{21}^{(q)}\right|^2\theta_2^{(q)}P_2+\left|h_{r1}^{(q)}\right|^2\overline{v^{(q)}}P_r^{(q)}+N_1^{(q)}\right)+\left|h_{2r}^{(q)}\right|^2\theta_2^{(q)}P_2\left(\left|h_{r1}^{(q)}\right|^2\overline{v^{(q)}}P_r^{(q)}+N_1^{(q)}\right)}\right)$$

$$R_2^{(q),\text{EF}} = C\left(\frac{\left(\left|h_{1r}\right|^2\theta_1^{(q)}P_1+N_r^{(q)}+N_{wz,2}^{(q)}\right)\left|h_{22}^{(q)}\right|^2\theta_2^{(q)}P_2+\left(\left|h_{12}^{(q)}\right|^2\theta_1^{(q)}P_1+\left|h_{r2}^{(q)}\right|^2v^{(q)}P_r^{(q)}+N_2^{(q)}\right)\left|h_{2r}^{(q)}\right|^2\theta_2^{(q)}P_2}{\left(N_r^{(q)}+N_{wz,2}^{(q)}\right)\left(\left|h_{12}^{(q)}\right|^2\theta_1^{(q)}P_1+\left|h_{r2}^{(q)}\right|^2v^{(q)}P_r^{(q)}+N_2^{(q)}\right)+\left|h_{1r}^{(q)}\right|^2\theta_1^{(q)}P_1\left(\left|h_{r2}^{(q)}\right|^2v^{(q)}P_r^{(q)}+N_2^{(q)}\right)}\right)\text{,}$$

$$\text{(8.37)}$$

$$N_{wz,1}^{(q)} = \frac{\left(\left|h_{11}^{(q)}\right|^2\theta_1^{(q)}P_1+\left|h_{21}^{(q)}\right|^2\theta_2^{(q)}P_2+\left|h_{r1}^{(q)}\right|^2\overline{v^{(q)}}P_r^{(q)}+N_1^{(q)}\right)A^{(q)}-\left|A_1^{(q)}\right|^2}{\left|h_{r1}^{(q)}\right|^2v^{(q)}P_r^{(q)}}$$

$$N_{wz,2}^{(q)} = \frac{\left(\left|h_{22}^{(q)}\right|^2\theta_2^{(q)}P_2+\left|h_{12}^{(q)}\right|^2\theta_1^{(q)}P_1+\left|h_{r2}^{(q)}\right|^2v^{(q)}P_r^{(q)}+N_2^{(q)}\right)A^{(q)}-\left|A_2^{(q)}\right|^2}{\left|h_{r2}^{(q)}\right|^2\overline{v^{(q)}}P_r^{(q)}}\text{,}$$

$$\text{(8.38)}$$

$v^{(q)} \in [0,1]$, $A^{(q)} = |h_{1r}^{(q)}|^2\theta_1^{(q)}P_1+|h_{2r}^{(q)}|^2\theta_2^{(q)}P_2+N_r^{(q)}$, $A_1^{(q)} = h_{11}^{(q)}h_{1r}^{(q),*}\theta_1^{(q)}P_1+h_{21}^{(q)}h_{2r}^{(q),*}\theta_2^{(q)}P_2$ and $A_2^{(q)} = h_{12}^{(q)}h_{1r}^{(q),*}\theta_1^{(q)}P_1+h_{22}^{(q)}h_{2r}^{(q),*}\theta_2^{(q)}P_2$. What is interesting with this EF protocol is that, here again, one can prove that the utility function is concave for every user. This is the purpose of the following theorem.

Theorem 190

The game defined by $\mathcal{G}^{\text{EF}} = \left(\mathcal{K},\{A_k\}_{k\in\mathcal{K}},\left\{\mu_k^{\text{EF}}\right\}_{k\in\mathcal{K}}\right)$ has always at least one pure-strategy NE.

The proof is similar to that of Theorem 189. To be able to apply the Rosen existence theorem (Rosen, 1965), we must prove that the utility μ_k^{EF} is concave w.r.t. $\underline{\theta}_k$. The problem is less simple than for DF because the compression noise $N_{wz,k}^{(q)}$, which appears in the denominator of the capacity function in Eq. (8.37), depends on the strategy $\underline{\theta}_k$ of transmitter k. It turns out that it is still possible to prove the desired result as shown in Belmega et al. (2011).

8.4.3 Amplify-and-Forward (AF) Protocol

In this section, we assume that the the relay implements an analog amplifier which does not introduce any delay on the relayed signal. The signal transmitted by the relay is written $X_r = a_r Y_r$ where a_r corresponds to the relay amplification gain. The corresponding protocol is called the zero-delay scalar amplify-and-forward (ZDSAF). One of the nice features of the ZDSAF protocol is that relays are very easy to deploy since they can be used without any change to the existing (non-cooperative) communication system. The amplification gain for the relay on band (q) will be denoted by $a_r^{(q)}$. It is such that the relay exploits all the available power on each band. The achievable transmission rate is given by:

$$\mu_k^{\mathrm{AF}}(\underline{\theta}_k, \underline{\theta}_{-k}) = \sum_{q=1}^{Q} R_k^{(q),\mathrm{AF}}(\theta_k^{(q)}, \theta_{-k}^{(q)}) \tag{8.39}$$

where $R_k^{(q),\mathrm{AF}}$ is the rate which user k obtains by using band (q) when the ZDSAF protocol is used by the relay $\mathcal{R}$.

$$\forall k \in \mathcal{K}, R_k^{(q),\mathrm{AF}} = C \left(\frac{\left| a_r^{(q)} h_{kr}^{(q)} h_{rk}^{(q)} + h_{kk}^{(q)} \right|^2 \theta_k^{(q)} \rho_k}{\left| a_r^{(q)} h_{\ell r} h_{rk} + h_{\ell k} \right|^2 \rho_\ell \theta_\ell^{(q)} \frac{N_\ell^{(q)}}{N_k^{(q)}} + \left(a_r^{(q)} \right)^2 \left| h_{rk}^{(q)} \right|^2 \frac{N_r^{(q)}}{N_k^{(q)}} + 1} \right) \tag{8.40}$$

where $a_r^{(q)} = \tilde{a}_r^{(q)}(\theta_1^{(q)}, \theta_2^{(q)}) \triangleq \sqrt{\dfrac{P_r}{\left| h_{1r}^{(q)} \right|^2 P_1 + |h_{2r}|^2 P_2 + N_r}}$ and $\rho_k^{(q)} = \dfrac{P_k}{N_k^{(q)}}$. Without loss of generality and for the sake of clarity we will assume in Sec. 8.4.3 that $\forall (k,q) \in \{1,2,r\} \times \{1,\ldots,Q\}, N_k^{(q)} = N, P_r^{(q)} = \overline{P}_r$ and we introduce the quantities $\rho_k = \dfrac{\overline{P}_k}{N}$. In this setup the following existence theorem can be proven.

···

Theorem 191

If one of the following conditions is met: i) $\left| a_r^{(q)} h_{kr}^{(q)} h_{rk}^{(q)} \right| \gg \left| h_{kk}^{(q)} \right|$ and $\left| a_r^{(q)} h_{\ell r}^{(q)} h_{rk}^{(q)} \right| \gg \left| h_{\ell k}^{(q)} \right|$ (i.e., the direct links $\mathcal{S}_k - \mathcal{D}_k$ are negligible and the communication is possible only through the relay node), ii) $\left| h_{kk}^{(q)} \right| \gg \left| a_r^{(q)} h_{kr}^{(q)} h_{rk}^{(q)} \right|$ and $\left| h_{\ell k}^{(q)} \right| \gg \left| a_r^{(q)} h_{\ell r}^{(q)} \right| \min \left\{ 1, \left| h_{rk}^{(q)} \right| \right\}$ (i.e., the links $\mathcal{R} - \mathcal{D}_k$ are negligible), iii) $a_r^{(q)} = A_r^{(q)} \in [0, \tilde{a}_r^{(q)}(1,1)]$ (i.e., the amplification gain is constant), there exists at least one pure-strategy NE in the PA game $\mathcal{G}^{\mathrm{AF}} = \left(\mathcal{K}, \{\mathcal{A}_k\}_{k \in \mathcal{K}}, \left\{ \mu_k^{\mathrm{AF}} \right\}_{k \in \mathcal{K}} \right)$.

The proof is similar to that of Theorem 189. The sufficient conditions ensure the concavity of the function $R_k^{(q),\mathrm{AF}}$ w.r.t. $\theta_k^{(q)}$. Notice that, under the last two conditions, the analysis is very similar to that of a parallel IC for which the existence of the NE is always guaranteed (Chung et al., 2003; Luo and Pang, 2006).

We have seen that, under certain sufficient conditions on the channel gains, the existence of the NE point is ensured for all three of the protocols investigated. The multiplicity and the convergence of best-response algorithms is in general not trivial. In Belmega et al. (2011), a particular case of the AF protocol is studied: $Q = 2$ and constant amplification factors $\forall q \in \{1,2\}, a_r^{(q)} = A_r^{(q)} \in [0, \tilde{a}_r(1,1)]$. It turns out that for this particular case, a complete characterization of the number of NE can be made, based on the best-response function analysis. The best-response correspondences are piece-wise affine functions, and thus, the network can have one, two, three or an infinite number of NE. Based on the "Cournot duopoly" (Cournot, 1838), the iterative algorithm based on the best-response functions is guaranteed to converge to one of the NE points.

In Belmega et al. (2011) it is proven that using a time-sharing technique, the existence of an NE can always be ensured, irrespective of the relaying protocol or channel gains. The basic idea is that assuming that the transmitters could be coordinated, the achievable rates become concave by using time-sharing techniques.

A strong motivation for studying IRCs is to be able to introduce relays into a network with non-coordinated and interfering pairs of terminals. For example, relays could be introduced by an operator aiming to improve the performance of the communications of his customers. In such a scenario, the operator acts as a player, and more precisely as a game leader in the sense of von Stackelberg (1952). In this context, the game leader is the operator/engineer/relay who chooses the parameters of the relays. The followers are the adaptive/cognitive transmitters that adapt their PA policy to what they observe. In the preceding sections we have mentioned some of these parameters: the location of each relay; the amplification gain of each relay in the case of AF; and in the case of DF and EF, the power allocation policy between the two cooperative signals at each relay, i.e., the parameter $v^{(q)}$. Therefore, the relay can be thought as a player maximizing its own utility. This utility can be either the individual utility of a given transmitter (picture one WiFi subscriber wanting to increase his downlink throughput by locating his cellular phone somewhere in his apartment while his neighbor can also exploit the same spectral resources) or the network sum-rate (in the case of an operator). The Stackelberg formulation was studied by using numerical results for the particular case of $Q = 2$ and $K = 2$. Several interesting observations were made. For the ZDSAF with constant amplification gain, it is not necessarily optimal to saturate the relay transmit power. For the general ZDSAF, the optimal relay node position w.r.t. the system sum-rate lies exactly on the segment between S_k and $\mathcal{D}_k$ of the better receiver. This is due to the fact that the selfish behavior of the users leads to an operating point where there is no or little interference. Assuming the DF and EF protocols, the optimal power allocation policy at the relay is to allocate all its available power to the better receiver.

8.5 LEARNING DISCRETE POWER ALLOCATION POLICIES IN FAST FADING MIMO MULTIPLE ACCESS CHANNELS

In this section, we present a way of explaining how players may converge to an equilibrium point in terms of power allocation policies. We will mainly present a reinforcement learning algorithm similar to Sastry et al. (1994a). In this framework, the users are simple automata capable of choosing their actions from a finite set. Their choices are based on past results, and feed-back from the environment. Thus, they can improve their performance over time while operating in an almost unknown environment.

More specifically, we study the power allocation game in fast fading MIMO multiple access channels, similarly to Sec. 8.2.2.1. We will restrict our attention to the case where single user decoding is used at the receiver side. The same notations will be used in this section. However, the power allocation game considered here differs from the game in Sec. 8.2.2.1 in two respects. First, the action sets of the players are discrete and finite as opposed to the convex and compact set of positive semi-definite matrices of constrained trace. Second, the channel matrices are restricted to live in compact sets of bounded support: $|\mathbf{H}_k(i,j)| \leq h_{\max} < +\infty$ (no other constraints are made on the channel matrices). The analysis of the Nash equilibrium for the discrete game is completely different to that in Sec. 8.2.2.1. For example, the existence of a pure strategy Nash equilibrium is generally not guaranteed for this type of game. Also, the concavity property of the utility functions and the results in Rosen (1965) cannot be used here. We consider the following action set for user k which is a simple quantized version of $\mathcal{A}_k^{\text{SUD}}$ in (8.2):

$$\mathcal{D}_k = \left\{ \frac{\overline{P}_k}{\ell} \mathbf{Diag}(\underline{e}_\ell) \,\middle|\, \ell \in \{1,\dots,n_t\}, \underline{e}_\ell \in \{0,1\}^{n_t}, \sum_{i=1}^{n_t} e_\ell(i) = \ell \right\}. \tag{8.41}$$

Notice that $\mathcal{D}_k$ represents the set of diagonal matrices resulting from uniformly spreading the available power over a subset of $\ell \in \{1,\dots,n_t\}$ eigenmodes. The components of the game are $\mathcal{G}_{\text{D}}^{\text{SUD}} = \left(\mathcal{K}, \{u_k^{\text{SUD}}\}_{k\in\mathcal{K}}, \{\mathcal{D}_k\}_{k\in\mathcal{K}}\right)$ where the utility function is given by the achievable ergodic rate in (8.11). As discussed in Sec. 8.2.2.1, the discrete game is an exact potential game. The potential function is given by the achievable system sum-rate:

$$V(\mathbf{Q}_1,\dots,\mathbf{Q}_K) = \mathbb{E} \log_2 \left| \mathbf{I}_{n_r} + \rho \sum_{k=1}^{K} \mathbf{H}_k \mathbf{Q}_k \mathbf{H}_k^H \right|. \tag{8.42}$$

Thus, the game $\mathcal{G}_{\text{D}}^{\text{SUD}}$ has at least one pure-strategy Nash equilibrium. However, the uniqueness property of the NE is lost in general. In Belmega et al. (2010a), it is proven that for full-rank channels, the NE is unique, but for unit-rank channels all the strategy profiles are NE points. Knowing that the game is an exact potential game, by the Finite Improvement Property (Monderer and Shapley, 1996), the iterative algorithms based on the best-response dynamics converge to one of the possible pure strategy NE depending on the starting point.

In the remainder of this section, a reinforcement learning algorithm similar to Sastry et al. (1994a) is studied. In contrast to the best-response type algorithm, the users are no longer rational devices but simple automata that know only their own action sets. They start at a completely naive state, randomly choosing their action (following the uniform distribution over the action sets, for example). After the play, each user obtains a certain feed-back from the nature (e.g., the realization of a random variable, the value of its own utility). Based only on this value, each user applies a simple updating rule to its own discrete probability distribution or mixed strategy over its action set. It turns out that, in the long-run, the updating rules converge to some desirable system states. It is important to notice that the transmitters don't even need to know the structure of the game or that a game is played at all. The price to pay for the lack of knowledge and rationality will be reflected in the longer convergence time.

Let us index the possible actions that user k can take as follows: $\mathcal{D}_k = \{\mathbf{D}_k^{(1)}, \ldots, \mathbf{D}_k^{(M_k)}\}$ with $M_k = \mathrm{Card}(\mathcal{D}_k)$ (i.e., the cardinality of $\mathcal{D}_k$). At step $n > 0$ of the iterative process, user k randomly chooses a certain action $\mathbf{Q}_k^{[n]} \in \mathcal{D}_k$ based on the probability distribution $p_k^{[n-1]}$ from the previous iteration. As a consequence, it obtains the realization of a random variable, which is, in our case, the normalized instantaneous mutual information $\mu_k^{[n]} = \dfrac{\mu_k\left(\mathbf{Q}_k^{[n]}, \mathbf{Q}_{-k}^{[n]}\right)}{I_{\max}} \in [0,1]$ such that:

$$\mu_k(\mathbf{Q}_k, \mathbf{Q}_{-k}) = \log_2 \left| \mathbf{I}_{n_r} + \rho \mathbf{H}_k \mathbf{Q}_k \mathbf{H}_k^H + \rho \sum_{\ell \neq k}^{K} \mathbf{H}_\ell \mathbf{Q}_\ell \mathbf{H}_\ell^H \right| - \log_2 \left| \mathbf{I}_{n_r} + \rho \sum_{\ell \neq k} \mathbf{H}_\ell \mathbf{Q}_\ell \mathbf{H}_\ell^H \right|$$

and $I_{\max}$ is the maximum value of the mutual information (under the assumption that the channel takes values in a compact set: $I_{\max} < +\infty$). Based on this feed-back, user k updates its own probability distribution as follows:

$$p_{k,j_k}^{[n]} = p_{k,j_k}^{[n-1]} - \gamma^{[n]} \mu_k^{[n]} p_{k,j_k}^{[n-1]} + \gamma^{[n]} \mu_k^{[n]} \mathbb{1}_{\left(\mathbf{Q}_k^{[n]} = \mathbf{D}_k^{(j_k)}\right)} \tag{8.43}$$

where $0 < \gamma^{[n]} < 1$ is the step size and $p_{k,j_k}^{[n]}$ represents the probability that user k chooses $\mathbf{D}_k^{(j_k)}$ at iteration n. Notice that the assumption on the channel matrices w.r.t. their bounded support is required for the normalization of the mutual information. In order to make sure that the discrete probability distribution in (8.43) is well defined, the random utility must be bounded as follows: $0 \leq \mu_k^{[n]} \leq 1$. An intuition behind this updating rule is that the frequency with which an action is played depends on its score (i.e., the utility obtained by playing this action) such that the actions which perform well are played more frequently.

In order to study the asymptotic behavior and convergence properties of the discrete stochastic process in (8.43), the objective is to prove that it can be approximated with the solution of the following deterministic ODE:

$$\frac{dp_{k,j_k}}{dt} = p_{k,j_k} \sum_{i_k=1}^{M_k} p_{k,i_k} [h_{k,j_k}(\underline{p}_{-k}) - h_{k,i_k}(\underline{p}_{-k})], \tag{8.44}$$

where:

$$h_{k,j_k}(\underline{p}_{-k}) = \sum_{\underline{i}_{-k}} u_k^{(\mathrm{SUD})} \left(D_k^{(j_k)}, D_{-k}^{(i_{-k})}\right) \prod_{\ell \neq k} p_{\ell, i_\ell}.$$

The idea is to apply the stochastic approximation results and, in particular, to check whether the conditions [H1]–[H5] in Appendix A.2 are satisfied in our context.

The vector field of the ODE in (8.44) is $f(\underline{p}) = \left\{ f_{k,j_k}(\underline{p}) \right\}_{k \in \mathcal{K}, j_k \in \{1,\dots,M_k\}}$ with:

$$f_{k,j_k}(\underline{p}) = p_{k,j_k} \sum_{i_k=1}^{M_k} p_{k,i_k} [h_{k,j_k}(\underline{p}_{-k}) - h_{k,i_k}(\underline{p}_{-k})],$$

where $\underline{p}$ represents the concatenation vector of the discrete probability distributions of all the players. It is easy to see that $f(p)$ is a class C^1 vector field (i.e., differentiable with continuous partial derivatives). This implies that $f(p)$ is Lipschitz continuous ([H2]). Since $\underline{p} \in \prod_k \Delta(\mathcal{D}_k)$, the condition [H3] is straightforward. The noise process is given by:

$$Z_{k,j_k}^{[n]} = -\mu_k^{[n]} p_{k,j_k}^{[n-1]} + \mu_k^{[n]} \mathbb{1}_{\left(Q_k^{[n]} = D_k^{(j_k)} \right)} - f_{k,j_k}(\underline{p}^{[n-1]}).$$

We can observe that $\left\{ \underline{p}^{[n]}, (Q^{[n-1]}, \underline{\mu}^{[n-1]}) \right\}_n$ is a Markov process where $Q^{[n-1]} = \{Q_k^{[n-1]}\}_{k \in \mathcal{K}}$ and $\underline{\mu}^{[n-1]} = \{\mu_k^{[n-1]}\}_{k \in \mathcal{K}}$. Also, for time invariant distributions $\underline{p}^{[n-1]} = \underline{p}$, the sequence $(Q^{[n]}, s^{[n]})_n$ is an i.i.d. sequence which implies that condition [H4] is met. Also, condition [H5] can be verified easily since $\underline{p}^{[n-1]} \in \prod_k \Delta(\mathcal{D}_k)$ and $\mu_k^{[n]} \in [0,1]$. Therefore, depending on the choice of the step size (either constant or verifying [H1]) the almost sure or weak convergence is guaranteed.

In order to study the stochastic process $p_{k,j_k}^{[n]}$, we can therefore focus on the study of the deterministic ODE that captures its average behavior. Notice that ODE (8.44) is similar to the replicator dynamics (Hofbauer and Sigmund, 2003). The mixed and pure-strategy NE are rest points of this kind of dynamics. However, all the pure-strategy profiles, even those which are not NE, are also rest points. Moreover, the border of the domain where the vector of probability distributions p lives is an invariant set. Using the fact that the game is an exact potential game, the pure-strategy NE points can be proved to be stable in the sense of Lyapunov.

8.5.1 Numerical Results

We consider the following scenario: $K = 2$, $n_r = n_t = 2$; the entries of the channel matrices are drawn independently following a truncated complex Gaussian distribution, i.e., $|\mathbf{H}_k(i,j)| \leq h_{\max} = 1$ and $\rho = 10$ dB, $\bar{P}_1 = \bar{P}_2 = 1$ W. The actions the users can take are: $\mathbf{D}_k^{(1)} = \bar{P}_k \mathbf{diag}(0,1)$, $\mathbf{D}_k^{(2)} = \bar{P}_k \mathbf{diag}(1,0)$, $\mathbf{D}_k^{(3)} = \frac{\bar{P}_k}{2} \mathbf{diag}(1,1)$. The beam-forming strategies are identical in terms of utility, and the users can be considered as having only two strategies: beam-forming (BF) (either $\mathbf{D}_k^{(1)}$ or $\mathbf{D}_k^{(2)}$) and uniform power allocation (UPA) $\left(\mathbf{D}_k^{(3)} \right)$. The utility matrix for user 1 is given by its ergodic achievable rate:

$$\mathbf{ER}_1 = \begin{pmatrix} 2.6643 & 1.9271 \\ 3.0699 & 2.2146 \end{pmatrix}$$

and since the example is symmetric we have $\mathbf{ER}_2 = \mathbf{ER}_1^T$. The elements of these matrices correspond to the utilities of the two users in the following situations: $\mathbf{ER}_k(1,1)$ when both players choose BF; $\mathbf{ER}_k(1,2)$ when user k chooses BF while its opponent plays UPA; $\mathbf{ER}_k(2,1)$ when user k chooses UPA while its opponent plays BF; $\mathbf{ER}_k(2,2)$ when both players choose UPA. In this case, we observe that the unique NE consists in playing the uniform power allocation by both users.

We apply the reinforcement algorithm by considering $I_{\max} = 8.7846$ bpcu (i.e., the maximum single-user instantaneous mutual information under the assumption that $h_{max} = 1$). In Fig. 8.3, the expected utility is represented, depending on the probability distribution over the action sets at every iteration for user 1 in Fig. 8.3(a) and for user 2 in Fig. 8.3(b) assuming $\overline{P}_1 = \overline{P}_2 = 5$ W. We observe that the users converge to the Nash equilibrium after approximatively 1.3×10^4 iterations, whereas using a best-response algorithm convergence is almost instantaneous (only 2 or 3 iterations). However, in the latter case, more assumptions are required in terms of user knowledge of the game structure. As already mentioned, one of the strengths of reinforcement learning is that it can be used in various communications contexts (in terms of channel models, utility functions, etc.) but at the price of slower convergence rates.

8.6 LEARNING DISCRETE POWER ALLOCATION POLICIES IN SLOW FADING SINGLE-USER MIMO CHANNELS

We have seen that the method for using reinforcement learning consists of modeling a transmitter by an automaton interacting with its environment. Therefore, a given learning algorithm can be applied in a single-user channel or any multiuser channel (e.g., multiple access or interference channels). However, even if the structure of the interactive situations is not known to the transmitters, it has an influence on the dynamics of the system (e.g., in terms of convergence). In this section, the dynamic aspects such as convergence are not studied (and for the sake of clarity), we focus on single-user MIMO channels. The case of interest is the one where the utility function is the goodput (Goodman and Mandayam, 2000), which is proportional to the packet success rate. When slow fading is assumed, the mutual information is a random variable, varying from block to block, and thus it is not possible to guarantee that it is above a certain threshold. In this case, the achievable transmission rate in the sense of Shannon is zero. A more suitable performance metric is the probability of an outage for a given transmission rate target (Ozarow et al., 1994). This metric allows one to quantify the probability that the rate target is not reached by using a good channel coding scheme, and is defined as follows:

$$P_{\text{out}}(\mathbf{Q}, R) = \Pr[\mu(\mathbf{Q}) < R]. \qquad (8.45)$$

Here, $\mu(\mathbf{Q})$ denotes the instantaneous mutual information:

$$\mu(\mathbf{Q}) = \log_2 \left| \mathbf{I}_{n_r} + \rho \mathbf{HQH}^H \right|. \qquad (8.46)$$

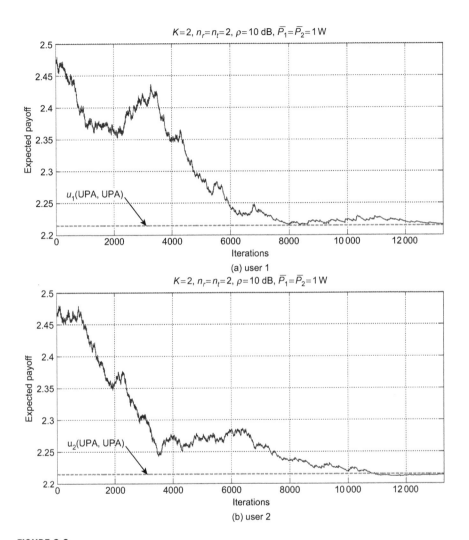

FIGURE 8.3

Expected utility vs. iteration number for $K = 2$ users.

The idea is to implement a reinforcement algorithm that allows the transmitter to compute its best precoding matrix minimizing the outage probability. We start with the same simple action set given in Sec. 8.5:

$$\mathcal{D} = \left\{ \frac{\overline{P}}{\ell} \mathbf{Diag}(\underline{e}_\ell) \, \middle| \, \ell \in \{1, \ldots, n_t\}, \underline{e}_\ell \in \{0, 1\}^{n_t}, \sum_{i=1}^{n_t} e_\ell(i) = \ell \right\}. \qquad (8.47)$$

The choice of this set is motivated by two factors:

i. for Rayleigh fading, the optimal covariance matrix is diagonal;
ii. Telatar (1999) conjectured that the optimal covariance matrix is to uniformly allocate the power on a subset of antennas.

We assume that the only feed-back the transmitter receives at iteration n is an ACK/NACK bit denoted $s^{[n]}$, i.e., the realization of the following random variable: $S = 0$ if $\mu(\mathbf{Q}) \leq R$ otherwise $S = 1$. Therefore, if an outage occurs at time n, the receiver feeds back $s^{[n]} = 0$, otherwise $s^{[n]} = 1$. Notice that the random variable S follows a Bernoulli distribution of parameter $q = 1 - P_{\text{out}}(\mathbf{Q},R)$. Its expected value is equal to $1 - P_{\text{out}}(\mathbf{Q},R)$ and, therefore, if the instantaneous utility is $s^{[n]}$, then its expected utility is exactly the success probability: $1 - P_{\text{out}}(\mathbf{Q},R)$.

Based only on $s^{[n]}$, the user applies a simple updating rule over its own probability distribution over the action space. Let us index the elements of $\mathcal{D} = \{\mathbf{D}^{(1)}, \ldots, \mathbf{D}^{(M)}\}$ with $M = \text{Card}(\mathcal{D})$ (i.e., the cardinality of $\mathcal{D}$). We want to find out whether using a simple reinforcement learning algorithm will allow us to solve the open problem:

$$u_{\max} = \max_{j \in \{1,\ldots,M\}} u(\mathbf{D}^{(j)}), \qquad (8.48)$$

where $u(\mathbf{D}^{(j)}) = 1 - P_{\text{out}}(\mathbf{D}^{(j)},R)$ represents the success probability. At step $n > 0$ of the iterative process, the transmitter randomly chooses a certain action $\mathbf{Q}^{[n]} \in \mathcal{D}$ based on the probability distribution $p^{[n-1]}$ from the previous iteration. As a consequence, it obtains the realization of a random variable, which is, in our case, $s^{[n]}$. Based on this value, the transmitter updates its own probability distribution as follows:

$$p_j^{[n]} = p_j^{[n-1]} - \gamma^{[n]} s^{[n]} p_j^{[n-1]} + \gamma^{[n]} s^{[n]} \mathbb{1}_{(\mathbf{Q}^{[n]} = \mathbf{D}^{(j)})} \qquad (8.49)$$

where $0 < \gamma^{[n]} < 1$ is a step size and $p_j^{[n]}$ represents the probability that the transmitter chooses $\mathbf{D}^{(j)}$ at iteration n. Notice that, as opposed to Sec. 8.5, no assumptions on the channel matrix have to be made, since $s^{[n]} \in \{0,1\}$.

The sequence $p^{[n]}$ can be approximated in the asymptotic regime with the solution of the following ODE:

$$\frac{dp_j}{dt} = p_j \left[u(\mathbf{D}^{(j)}) - \sum_{i=1}^{M} p_i u(\mathbf{D}^{(i)}) \right], \qquad (8.50)$$

for all $j \in \{1,\ldots,M\}$.

Similarly to the previous section, the conditions [H2]–[H5] can be proved to be satisfied. First, we observe that the vector field $f(p)$, with the components given by $f_j(\underline{p}) = p_j \left[u(\mathbf{D}^{(j)}) - \sum_{i=1}^{M} p_i u(\mathbf{D}^{(i)}) \right]$ for all $j \in \{1,\ldots,M\}$, is a class $\mathcal{C}^1$ function. This implies that $f(p)$ is Lipschitz continuous and [H2] is met. Since the updated process corresponds to the discrete probability distribution, $\underline{p}^{[n]} \in \Delta(\mathcal{D})$, it is always bounded

([H3]). The noise is given by:

$$Z_j^{[n]} = -s^{[n]}p_j^{[n-1]} + s^{[n]}\mathbb{1}_{(Q^{[n]}=D^{(j)})} - f_j(\underline{p}^{[n-1]}). \qquad (8.51)$$

By construction, we have that $\left\{\underline{p}^{[n]},(Q^{[n-1]},s^{[n-1]})\right\}_n$ is a Markov process and for a time invariant distribution $\underline{p}^{[n-1]} = \underline{p}$ $(Q^{[n-1]},s^{[n-1]})_n$ is an i.i.d. sequence, which implies that condition [H4] is met. Also, condition [H5] can be easily verified.

We can observe that only the corner points of the simplex $\Delta(\mathcal{D})$ (i.e., the pure-actions) are stationary states of the ODE in (8.50). It turns out that the only stable states are solutions to the optimization problem in (8.48).

8.6.1 Numerical Results

Consider the simple case of an i.i.d. channel matrix of complex standard Gaussian entries, $n_t = n_r = 2$, $R = 1$ bpcu, $\overline{P} = 0.1$ W, $\rho = 10$ dB. In this case, the user can choose between beam-forming and the uniform power allocation: $\mathbf{D}^{(1)} = \overline{P}\mathbf{diag}(0,1)$, $\mathbf{D}^{(2)} = \overline{P}\mathbf{diag}(1,0)$, $\mathbf{D}^{(3)} = \frac{\overline{P}}{2}\mathbf{diag}(1,1)$. The success probability is given by $u(\mathbf{D}^{(1)}) = u(\mathbf{D}^{(2)}) = 0.7359$, $u(\mathbf{D}^{(3)}) = 0.8841$. Notice that the positions of the active antennas do not matter, and only the number of active modes has an influence on the success probability. Figure 8.4 represents the expected utility $\sum_{j=1}^{M} p_j^{[n]} u(\mathbf{D}^{(j)})$ as a function of the iterations. We assume $\gamma^{[n]} = \gamma = 0.01$ (constant step-size) and observe that the optimal solution is reached after 2554 iterations. However, the performance of the algorithm depends on the choice of the learning parameter: a larger

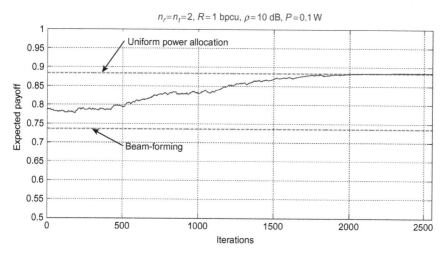

FIGURE 8.4

Average utility vs. number of iterations.

Table 8.1 Trade-off between the Convergence Time and the Convergence to the Optimal Point

γ	Time [nb. iterations]	Convergence to Optimum [%]
0.001	3755	100
0.1	261	71
0.5	27	45
0.9	9	39

γ gives a smaller convergence time. The problem with large steps is that the algorithm may converge to a corner of the simplex which is not a maximizer of the success probability. This scenario is investigated in Table 8.1. Here, we summarize the results of the reinforcement algorithm obtained after 1000 experiments in terms of average number of iterations and convergence to the maximum point. We observe that there is a trade-off between the convergence time and the convergence to the optimal point. This trade-off can be controlled by tuning the learning step-size γ.

8.7 LEARNING CONTINUOUS POWER ALLOCATION POLICIES IN STATIC PARALLEL MULTIPLE ACCESS CHANNELS

Reinforcement learning algorithms rely on the assumption of discrete action sets for the players. Here, we briefly present a learning technique which is designed for continuous action spaces (Mertikopoulos et al., 2011). One of the reasons for this is that it bridges the gap between best-response dynamics (iterative water-filling algorithms) presented at the beginning of this chapter and reinforcement learning algorithms. A second reason is to present a short convergence analysis of the proposed learning algorithm that is complementary to the preceding sections. The scenario is always the same: there are K transmitters which can share their power between S bands/channels/receivers. An interesting observation is that rational transmitters will necessarily use all their transmit power when maximizing their own Shannon transmission rates. As a consequence, the power constraint is saturated for every transmitter. The action spaces therefore have the structure of a simplex, just like conventional probability distributions. This makes the design or algorithms which learn continuous policies easier.

Having determined the properties of the game's unique equilibrium point in Sec. 8.2.1.1, our objective is now to present a decentralized learning scheme which allows the users to converge to this equilibrium point (and to estimate the speed of this convergence). Given that the structure of the power allocation game does not adhere to the (multilinear) setting of Nash (1950), the usual theory of evolutionary "random-matching" games does not apply either. This leaves us in a rather unclear position regarding how to proceed, but since the players will invariably want to increase their rewards and an increase in utility is equivalent to a decrease in potential, we will begin by considering the directional derivatives of the potential function ϕ in (8.23):

$$v_{k,s}(\underline{p}) \equiv -\frac{\partial \phi}{\partial p_{k,s}} = \frac{g_{k,s}}{\sigma_s^2 + \sum_\ell g_{\ell,s} p_{\ell,s}}. \tag{8.52}$$

Clearly, if a player transmits with positive power on channel s, then he will be able to calculate the gradient $v_{k,s}(\underline{p})$ in terms of the observable $p_{k,s}$ (the user's power allocation), $g_{k,s}$ (his channel gain coefficients), and the spectral efficiency $u_{k,s}(\underline{p}) = \log\left(1 + g_{k,s} p_{k,s}/\left(\sigma_s^2 + \sum_{\ell \neq k} g_{\ell,s} p_{\ell,s}\right)\right)$ which user k observes on channel s.[1] As a result, any learning scheme which relies only on the $v_{k,s}$'s will be inherently distributed in the sense that it only requires information that is readily obtainable by the individual players. With all this in mind, a particularly simple scheme to follow is that of the replicator dynamics (Weibull, 1997) associated with the "marginal utilities" $v_{k,s}$. More specifically, this means that the players will update their power allocations according to the differential equation:

$$\frac{\mathrm{d}p_{k,s}}{\mathrm{d}t} = p_{k,s}\big(v_{k,s}(\underline{p}(t)) - v_k(\underline{p}(t))\big), \tag{8.53}$$

where v_k is just the user average $v_k(\underline{p}) = P_k^{-1} \sum_{s'} p_{k,s'} v_{k,s'}(\underline{p})$.

As usual, the rest points of (8.53) are characterized by the (waterfilling) property; that is, for every pair of nodes $s, s' \in \mathrm{supp}(\underline{p})$ to which user k allocates positive power, we have $v_{k,s}(\underline{p}) = v_{k,s'}(\underline{p})$. Hence, comparing this to the corresponding KKT conditions, we immediately see that the Nash equilibria of the game are stationary in the replicator equation (8.53). This result is well-known in finite Nash games with multilinear payoffs (Fudenberg and Levine, 1998) and in continuous population games (Sandholm, 2001), but the converse does not hold: for instance, every vertex of the action space (which is a simplex, as already pointed out) is stationary in (8.53), so the stationarity of (8.53) does not imply equilibrium.

Nevertheless, *only* Nash equilibria can be attracting points, and, in fact, they attract almost every replicator solution orbit:

Theorem 192

Let $\underline{q}$ be the unique (a.s.) equilibrium of the power allocation game. Then, every solution orbit of the replicator dynamics (8.53) which begins at finite Kullback-Leibler entropy from $\underline{q}$ will converge to it. Furthermore, even if the game does not admit a unique equilibrium, every interior trajectory still converges to a Nash equilibrium (and not merely to the Nash set of the game).

---

[1]Note that this is different from gradient techniques applied to the utility functions themselves, a practice which requires the utility functions to be known.

Remark Recall that the *Kullback-Leibler divergence* (or *relative entropy*) of $\underline{p}$ with respect to $\underline{q}$ is:

$$H_{\underline{q}}(\underline{p}) = \sum_k H_{q_k}(p_k) = \sum_{k,s} q_{k,s} \log\left(\frac{q_{k,s}}{p_{k,s}}\right). \tag{8.54}$$

Clearly, $H_q(p)$ is finite if and only if p_k allocates positive power $p_{k,s} > 0$ to all nodes $s \in \text{supp}(\underline{q})$ which are present in q_k; more succinctly, the domain of $H_{\underline{q}}$ consists of all power allocations which are absolutely continuous w.r.t. $\underline{q}$.

This convergence result is extremely powerful because it shows that the network's users will eventually settle down to a stable state which discourages unilateral deviations, even though they only have local information at their disposal. The only case that is left open in the above theorem is the case where the initial K-L entropy of the solution orbit is infinite, i.e., if the users' initial power allocation does not support all of the nodes which are present in equilibrium. If this is the case, then the face-invariance property of the replicator dynamics ($p_{k,s}(t) = 0$ if and only if $p_{k,s} = 0$) will prevent the users from settling down to a Nash equilibrium. However, an easy analysis shows that if one considers the reduced game where each user only has access to the nodes to which he initially allocates positive power, then the users will actually converge to an equilibrium of this reduced game (Mertikopoulos et al., 2011).

Medium Access Control Games

9.1 INTRODUCTION

Medium access control protocols are mechanisms that allow several users or transmitters to access a common medium or channel. They have played, and still play, an important role in the development of both wired and wireless networks. In particular, distributed protocols facilitate network deployment and allow networks to accommodate more users/transmitters without any change. An important class of protocols with contention (namely, for which there is a risk of collision) is random medium access protocols (for a classification of protocols see for example Kumar et al. (2006)). The Aloha, slotted Aloha, and CSMA (Carrier Sense Multiple Access) protocols are much used in communication networks. In the classical Aloha protocol, users can access the medium at any time, but if two or more users transmit at the same time, a collision occurs and all the packets are lost. The slotted Aloha is a variant of the classical Aloha in which users are synchronized, in the sense that they can only start transmitting at the beginning of a time-slot. In the CSMA protocol, each transmitter first tests the medium to check whether it is idle or busy. If the medium is busy, the transmitter defers its own transmission to prevent a collision with the existing signal. Otherwise, the user transmits while continuing to monitor the medium.

CSMA is much used in wired networks, but it cannot directly be applied to wireless networks. Due to the path loss effects, some undesirable effects such as hidden-terminal or exposed-terminal problems occur (see Kleinrock and Tobagi (1975) and Poojary et al. (2001) for more details on spatial reuse based CSMA protocols). Therefore, although CSMA protocols based on a perfect sensing mechanism offer better throughputs (or sum-rates) than the Aloha protocols, the latter have the advantage of being easier to implement in wireless networks and more robust to effects such as those mentioned. This is one of the reasons why we will mainly focus on Aloha-type protocols. The CSMA protocol will be considered only in Sec. 9.5. As in Chapters 2 and 3, transmitters interact because of multiuser interference; but, in contrast with models called physical layer models in the Open System Interconnection Reference – OSI – structure (Zimmermann, 1980), in this chapter a receiver cannot accommodate two users at the same time on the same channel. Indeed, the classical Aloha protocol does not allow two or more transmitters to transmit simultaneously. We will also use a refined model for which simultaneous transmissions are possible under some conditions on the transmit powers chosen by the terminals.

This allows one to partially bridge the gap between the models used in Chapters 2 and 3 and the standard Aloha-based system models.

We can see that wireless networks based on Aloha or CSMA protocols are decentralized or distributed by design. Since transmitters are autonomous decision-makers who interact (through collisions), the use of game theory is completely relevant in this context. The goal of this chapter is to show how some important game-theoretic concepts presented in Part I can be exploited to analyze the medium access control problem in wireless networks. In Sec. 9.2, the network model considered is a multiple access channel with K transmitters and one receiver. The game model is a static (or one-shot) game and the following concepts are emphasized: Pareto optimal points, Nash equilibrium, Stackelberg solution, correlated equilibria, evolutionarily stable strategy, and satisfaction solution and strong equilibrium. In Sec. 9.3, the model under study is a multiple access channel with K transmitters and S receivers. We give two examples of parallel multiple access channels. The receivers are assumed to use orthogonal channels, which means that there is no collision when two users transmit at the same time but on different channels. Nash equilibria and Pareto optimal solutions are discussed in the static game as well as in the unconstrained repeated game. This analysis is extended to the case where channels associated with the receivers are non-orthogonal (Sec. 9.3.3); for this purpose a certain spatial structure is assumed, which allows one to characterize the interference between the signals transmitted over the different channels. In Sec. 9.4 dynamic games are considered, in order to study the multi-stage medium access control problem, in which users can adjust their probability of using the medium over time. In Sec. 9.5, stochastic games are considered. The motivation for this is that the environment, the medium, the receiver, or the transmitters, etc. can take some states, e.g., the backoff state, the transmitter battery state. Finally, note that in the whole chapter, except in Secs. 9.4 and 9.3.3, transmitters will be assumed to operate in the saturated regime for which each of them always has a packet to send.

9.2 ACCESS CONTROL GAMES FOR MULTIPLE ACCESS COLLISION CHANNELS

In this section, we consider a wireless network model which is commonly called the (slotted) multiple access collision channel. Packets or time-slots of fixed length can be transmitted over the channel provided that they are synchronized. The destination of all the packets is a single common receiver. When two or more users send a packet on the same time-slot, these packets collide, and, consequently, they are lost (i.e., the receiver cannot determine the packet contents and retransmission is necessary). The study of collision channels, also called random access channels, started with Abramson's Aloha protocol (Abramson, 1970), which is a non-slotted protocol and uses only binary feedbacks (collision/no collision). In this case, collisions give some information: all the users, including those who are not transmitting, are informed immediately of the multiplicity of the collision. This type of feedback mechanism is not assumed in this section. As introduced above, random access games study the problem of medium access. We assume that the channels are perfectly reliable for packet transmissions, and that all the errors (or erasures) are due to collisions. A user decides either to transmit a packet with some power level, or not to transmit (null power level) to the receiver when they are within the transmission range of each other. Interference occurs as in the Aloha protocol where the power control is introduced: there is a collision if more than one neighbor of the receiver transmits a packet with a power level which is greater than the corresponding power of the transmitter at the same time-slot.

The game considered in this section is a static game for which the strategic form is given by the triplet:

$$\mathcal{G}_{K,m} = \left(\mathcal{K}, \{\mathcal{P}_i\}_{i \in \mathcal{K}}, \{u_i\}_{i \in \mathcal{K}}\right) \tag{9.1}$$

where: $\mathcal{K} = \{1, \ldots, K\}$ is the set of players (namely the transmitters); $\mathcal{P}_i = \{P^0, \ldots, P^m\}$ is the set of actions (or pure strategies) of transmitter i, that is, the possible levels for his transmit power; and u_i is the utility function of the transmitter i, which will be defined in the different subsections of this part. The power levels are ordered according to $P^0 < P^1 < \cdots < P^m$, where by convention $P^0 = 0$ and P^m is the maximum transmit power. As a comment, note that we have implicitly assumed that the transmitters have the same set of actions ($\mathcal{P}_i = \mathcal{P}$).

9.2.1 Aloha-Based Protocol Medium Access Control Games

The utility function of the transmitter $i \in \mathcal{K}$ is defined as follows:

$$u_i : \left| \begin{array}{ccc} \mathcal{P}^K & \to & [0, 1] \\ (a_1, \ldots, a_K) & \mapsto & \mathbb{1}_{\{a_i > \max_{j \neq i} a_j\}} \end{array} \right. \tag{9.2}$$

where $\mathbb{1}_{\{.\}}$ is the indicator function, which is equal to 1 if $a_i > \max_{j \neq i} a_j$ and 0 otherwise. This model means that the transmission of user i is successful only if the power level he chooses is strictly greater than those chosen by other transmitters. We consider now the mixed extension (see Chapter 1 for more details) of $\mathcal{G}_{K,m}$, which allows us to address the more general situation in which transmitters can randomize over their power levels. The extended game is denoted by $\tilde{\mathcal{G}}_{K,m}$ and is given by the following triplet:

$$\tilde{\mathcal{G}}_{K,m} = \left(\mathcal{K}, \Delta(\mathcal{P}), \{\tilde{u}_i\}_{i \in \mathcal{K}}\right) \tag{9.3}$$

where $\Delta(\mathcal{P})$ is the unit simplex of $\mathbb{R}_+^{m+1}$, defined by:

$$\Delta(\mathcal{P}) = \left\{ (\lambda_0, \ldots, \lambda_m) \in \mathbb{R}_+^{m+1} \; \middle| \; \sum_{\ell=0}^{m} \lambda_\ell = 1, \right\} \tag{9.4}$$

and the utility functions $\{\tilde{u}_i\}_i$ of the transmitters are given by:

$$\forall i \in \mathcal{K}, \; \tilde{u}_i(\underline{\pi}_1, \ldots, \underline{\pi}_K) = \sum_{\underline{a} \in \mathcal{P}^K} u_i(\underline{a}) \prod_{j=1}^{K} \pi_j(a_j) \tag{9.5}$$

where $\underline{\pi}_i = \left(\pi_i(P^0), \ldots, \pi_i(P^m)\right)$ is the mixed strategy chosen by transmitter $i \in \mathcal{K}$, i.e., for a given action, and $a_i \in \mathcal{P}$, $\pi_i(a_i)$ is the probability that the transmitter i assigns to the action a_i. Quite often in this chapter we will study the specific case where the transmitters' action sets have only two elements, i.e. $m = 1$. Therefore, the actions P^0 and P^1 will also be called Wait and Transmit respectively. The two-action case has the advantage of being simple to describe, and the generalization to the $m + 1$-action case, with $m + 1 > 2$, is doable for most of the solution concepts studied here. To clearly distinguish between the two cases we use a notation that is specific to the two-action case. The probability that transmitter i chooses the action Transmit (resp. Wait) will be denoted by x_i (resp. $1 - x_i$). Thus, the utility function for the two-action case will be denoted by:

$$\tilde{v}_i : \begin{vmatrix} [0,1]^K & \to & [0,1] \\ \underline{x} & \mapsto & x_i \prod_{j \neq i} (1 - x_j) \end{vmatrix} \tag{9.6}$$

with $\underline{x} = (x_1, \ldots, x_K)$. In the sequel, a mixed strategy profile will be said to be an *interior strategy* if $0 < x_i < 1$ for all $i \in \mathcal{K}$. These notations being defined, it is now possible to apply some of the solution concepts presented in Chapter 1 to the medium access control game under study.

9.2.1.1 *Pareto Optimal Solutions*

In this part, consider the Pareto optimal solutions of $\mathcal{G}_{K,1}$. The Pareto optimal strategy profiles correspond to operating points for which the utility of any of the users cannot be improved without harming another user. An allocation of utilities is Pareto optimal if there is no other feasible utility profile which Pareto dominates this allocation (see Chapter 1). One of the reasons to be interested in Pareto optimal solutions for access control problems is that they are often fair in some sense (maxmin fairness (Keshav, 1997)). More generally, as pointed out in Chapter 1, Pareto optimality is a minimum requirement which is often expected for an efficient network operating state.

Let us determine the Pareto optimal solutions of the game $\mathcal{G}_{K,1}$. Consider a configuration in which only one user transmits and the other users stay quiet. Then, the user who transmits has a successful transmission (his utility is 1, i.e., the maximum utility) and the utility of the others is zero. There is no other action profile at which the utility of this user can be improved, and it is impossible to make another user better off without necessarily making this user worse off. This statement holds for any other configuration where only one transmitter is active. This leads to K possible Pareto solutions in pure strategies and the corresponding solution profiles are the elements of the canonical basis of $\mathbb{R}^K$, which are denoted by $\underline{e}_1, \ldots, \underline{e}_K$.

We now focus on the feasible utilities in mixed strategies. The set of feasible utilities in mixed strategies is defined by:

$$\mathcal{F} = \left\{ \left(\tilde{v}_1(\underline{x}), \ldots, \tilde{v}_K(\underline{x})\right) \in [0,1]^K, \ \underline{x} \in [0,1]^K \right\}. \tag{9.7}$$

The set of utility profiles which are Pareto optimal is called the Pareto frontier of $\mathcal{F}$. The mixed strategy profile parameterized by $(\frac{1}{K}, \ldots, \frac{1}{K})$ belongs to the Pareto frontier and the corresponding utility is $\frac{1}{K}(1-\frac{1}{K})^{K-1}$. The sum of utilities or social welfare is $(1-\frac{1}{K})^{K-1}$, which goes to $\frac{1}{e}$ when K goes to the infinity. For this reason, the bound $\frac{1}{e}$ is called the asymptotic capacity of the system. It can be checked that all results given so far hold in the $m+1$-action case when $m+1>2$.

9.2.1.2 *Nash Equilibria*

First, note that from the Nash existence theorem (see Chapter 2), the existence of at least one mixed Nash equilibrium is guaranteed, since the game $\mathcal{G}_{K,m}$ is finite. In fact, the number of mixed Nash equilibria is infinite: all mixed strategy profiles for which one component is 1 and the others are arbitrary mixed strategies $(x_i, 1-x_i)$ are mixed Nash equilibria. The game $\mathcal{G}_{K,m}$ has also pure Nash equilibria. Indeed, all the pure strategy Pareto optimal solutions derived in the previous section are pure Nash equilibria. In Table 9.1 the 2-transmitter 2-action access control game is represented in a bi-matrix form. Pareto optimal solutions and pure Nash equilibria are highlighted.

9.2.1.3 *Stackelberg Solutions*

In this section we consider a Stackelberg formulation (see Chapter 2) of the two-transmitter two-action access control game when only pure strategies are allowed. The game is known to have at least one Stackelberg solution in pure strategies. More precisely, we assume that transmitter 1 is the leader and transmitter 2 is the follower, that is, transmitter 2 can observe (without observation costs and manipulation of the information by the leader) the action played by transmitter 1. As in Chapter 7, the leader is assumed to know that his action is observed. The corresponding game is called $\mathcal{G}_{2,1}^{S}$. If the action of transmitter 1 is Wait, then transmitter 2 prefers to transmit to get 1; transmitter 1 gets 0. If the action of transmitter 1 is Transmit then, transmitter 2 is indifferent between Transmit and Wait because in both cases, he gets 0. Therefore, there are two pure Stackelberg solutions: a pessimistic one and an optimistic one, which are respectively: $\underline{p}^{\text{WSS}}=(P^1,P^1)$ (worse Stackelberg solution) and $\underline{p}^{\text{BSS}}=(P^1,P^0)$ (best Stackelberg solution). Notice that these two action profiles are also Nash equilibria of the game $\mathcal{G}_{2,1}$. This equilibrium analysis can be easily

Table 9.1 The Access Control Game with Two Transmitters and Two Actions: Transmitter 1 Chooses a Row, Transmitter 2 Chooses a Column. The Superscript ♯ Indicates a Pareto Optimal Solution and ∗ Indicates a Pure Nash Equilibrium

	P^1	P^0
P^1	$(0,0)^*$	$(1,0)^{\sharp,*}$
P^0	$(0,1)^{\sharp,*}$	$(0,0)$

generalized to the game $\mathcal{G}_{K,m}$ with $m+1 > 2$. Concerning the efficiency of the pure Stackelberg solutions, it can be noticed that, even with zero observation cost, the hierarchical game solutions are not necessarily Pareto optimal.

9.2.1.4 *Correlated Equilibria*

Correlated equilibria (see Chapter 2) can be obtained very easily for $\mathcal{G}_{K,1}$. If all the transmitters can observe a signal which recommends to one user to transmit and the others to stay quiet, any unilateral deviation from such recommendation is non-profitable. Any convex combinations of pure Nash equilibria are correlated equilibria, and all the distributions under the form:

$$\Pr[\underline{a} \notin \{P^1\underline{e}_1,\ldots,P^1\underline{e}_K\}] = 0, \ \Pr[\underline{a} = P^1\underline{e}_i] = \lambda_i \ \text{with} \ \sum_{i=1}^{K}\lambda_i = 1 \qquad (9.8)$$

are correlated equilibria, where $\underline{a} \in \mathcal{P}^K$ always represents the possible action profiles. The corresponding sum-rate or throughput is 1, which means there is no loss with respect to the global optimum (centralized solution). Note that this result does not hold if some transmitters operate in a non-saturated regime in which they do not always have packets to transmit. In such cases, the performance of correlated equilibria is bounded by the probability that a user has a packet to transmit.

9.2.1.5 *Evolutionarily Stable Strategies*

Since the game $\mathcal{G}_{K,1}$ is finite and symmetric, there exists a symmetric Nash equilibrium (see Chapter 1). The one-shot access control game has no interior Nash equilibrium. The reason for this is that, for any transmitter $i \in \mathcal{K}$, the best response of i to an interior profile $\underline{a}_{-i}$ is to transmit. Therefore the only candidate to be an evolutionarily stable strategy is Transmit. This can be checked very easily. Let $0 < \epsilon < 1$ and define the convex combination $x_\epsilon = (1-\epsilon) + \epsilon y$ with $0 \le y < 1$. We have that:

$$\forall i \in \mathcal{K}, \ \forall x' \in [0,1[, \ \tilde{v}_i(1,x_\epsilon,\ldots,x_\epsilon) > \tilde{v}_i(x',x_\epsilon,\ldots,x_\epsilon), \qquad (9.9)$$

which proves that the action P^1 is evolutionarily stable. As a consequence, the utilities are zero for all transmitters, which is the worse case in terms of sum utility. This result indicates that the evolutionarily stable strategy of Aloha-like systems is inefficient in term of system throughput. It is hence natural to ask whether it is possible to obtain an unstable but more efficient solution. In the following subsection, we present a satisfaction solution for access control games. The idea of such a solution is not to maximize the utility in the strict sense but to satisfy the users' demands.

9.2.1.6 *Satisfaction Solutions*

Assume that, in the game $\widetilde{\mathcal{G}}_{K,1}$, each user has a demand $\rho_j \ge 0$ and is satisfied if his utility is greater than ρ_j. Clearly, in such a scenario, transmitters do not necessarily want to maximize their utility and every action profile such that all transmitters are satisfied is a suitable solution to the problem. This requires a feasibility condition: the demand profile should be in the set of feasible utilities (in mixed strategies).

A satisfaction solution is a solution of the following system of inequalities:

$$\begin{cases} x_1 \displaystyle\prod_{j\neq 1}(1-x_j) \geq \rho_1 \\[2mm] x_2 \displaystyle\prod_{j\neq 2}(1-x_j) \geq \rho_2 \\[1mm] \quad\quad\quad \vdots \\[1mm] x_K \displaystyle\prod_{j\neq K}(1-x_j) \geq \rho_K \end{cases} \tag{9.10}$$

It can be easily verified that the utility profile $(\frac{1}{K},\ldots,\frac{1}{K})$ is not achievable. Therefore, there is no satisfaction solution corresponding to the demand profile $(\frac{1}{K},\ldots,\frac{1}{K})$. This proves that the set $\mathcal{F}$ is not convex. In addition, the set $\mathcal{F}$ is a non-empty interior whenever there exists a feasible utility profile which strictly dominates (component-wise) the demand profile. However, from Sec. 9.2.1, we know that the demand profile:

$$\underline{\rho}^\star = \left(\frac{1}{K}\left(1-\frac{1}{K}\right)^{K-1}, \ldots, \frac{1}{K}\left(1-\frac{1}{K}\right)^{K-1} \right) \tag{9.11}$$

is feasible and the symmetric strategy profile:

$$\underline{x}^\star = \left(\frac{1}{K}, \ldots, \frac{1}{K} \right) \tag{9.12}$$

for which users transmit with probability $\frac{1}{K}$ is a satisfaction solution. Moreover, we know from Sec. 9.2.1.1 that this solution is Pareto optimal and maxmin fair, but it is not a Nash equilibrium. Obviously, when the set of satisfaction solutions is not empty, there is no reason why the solution should be unique. In fact, there are generally multiple solutions. The question of selection then arises. A relevant assumption is that users will try to minimize their effort to reach a satisfaction solution. In the context of the access control game under investigation, this consists of minimizing the individual transmit probability. We refer to such a solution as a power-efficient satisfaction solution. For instance, in the 2-transmitter 2-action access control game, a sufficient condition for the existence of a satisfaction solution is given by $0 \leq \rho_1 + \rho_2 \leq \frac{1}{4}$. An *efficient satisfaction solution* is a solution to the following system:

$$\begin{cases} x_1(1-x_2) = \rho_1 \\ x_2(1-x_1) = \rho_2 \\ 0 \leq x_1 \leq 1 \\ 0 \leq x_2 \leq 1 \end{cases}. \tag{9.13}$$

For feasible demand profiles, the system (9.13) has two solutions which correspond to the roots of $x_1^2 - (1+\rho_1-\rho_2)x_1 + \rho_1 = 0$, x_2 follows from (9.13). By inspecting the two solutions, it is possible to select the most efficient one (e.g., in terms of sum of efforts). Interestingly, both for this game and the more general game

$\tilde{\mathcal{G}}_{K,1}$, $K > 2$, there exists a dynamic procedure which converges to satisfaction solutions. The best response adjustment is such a dynamic. Assume that the game is played a large number of times (stages) and at each stage t, the transmitters adjust/update their distribution over their actions, which are denoted by $x_i(t)$, $i \in \mathcal{K}$. For this purpose, the following public signal is available:

$$\omega(t) = \prod_{i=1}^{K}[1 - x_i(t)]. \tag{9.14}$$

As a comment, note that we deliberately name the public signal $\omega(t)$ to establish a connection with Chapters 2 and 3. Indeed, in Chapter 2, the public signal was the sum $\sigma^2 + \sum_{i=1}^{K}|h_i|^2 p_i$ where p_i was the transmit power of user i. This signal was sufficient to reach an efficient cooperation plan in the repeated power control game. In Chapter 3, the same type of signal represented the potential of the power allocation game knowledge, which is sufficient to implement the alternate/best response dynamics[1]. Here, in (9.14), we have a product of mixed strategies. These three examples indicate that in some wireless games, the knowledge of an aggregate signal can be sufficient to implement the desired solution. This class of game is called *aggregative games*, where the aggregative can be additive, multiplicative, logarithmic, and so on. For a given transmitter, knowledge of $\omega(t)$ amounts to knowing his instantaneous utility since:

$$\forall i \in \mathcal{K}, \forall \underline{x}(t), \forall x_i(t) \neq 1, \ \tilde{v}_i(\underline{x}(t)) = x_i(t)\prod_{j \neq i}[1 - x_j(t)] = \omega(t)\frac{x_i(t)}{1 - x_i(t)}. \tag{9.15}$$

For any initial point $\underline{x}(0) = (x_1(0),\dots,x_K(0))$, the best response dynamics for slotted Aloha is given by:

$$x_i(t+1) = \min\left(1, \frac{\rho_i[1 - x_i(t)]}{\omega(t)}\right) \tag{9.16}$$

and converges to a satisfaction solution.

In scenarios where the public signal $\omega(t)$ is not available, other solutions have to be found. For example, in Chen et al. (2008), each transmitter infers the contention of the wireless network by observing a contention measure δ_i, which is a function of the users' medium access probabilities. The authors of Chen et al. (2008) propose general utility functions for random access games and give sufficient conditions under which random access games are supermodular games. Convergence to equilibria as well as selection of efficient equilibria can also be found in Chen et al. (2008).

---

[1] but the game studied here is not a potential game because the 2-transmitters, 3-receivers game does not have a pure Nash equilibrium.

9.2.2 Aloha-Based Protocol Medium Access Control Games with Pricing

As mentioned in Chapters 2 and 3, pricing is way of improving the efficiency of the outcome of a game. In this case, there is usually an additional player (e.g., the network owner or base station) who wants to maximize a certain global performance metric. But pricing can also be used to model the cost of transmission, in which case the cost is evaluated by the transmitter himself. The one-shot access control game of Sec. 9.2 becomes a new game:

$$\mathcal{G}_{K,m}^c = \left(\mathcal{K}, \mathcal{P}, \{u_i^c\}_{i \in \mathcal{K}} \right) \tag{9.17}$$

with:

$$u_i^c : \left| \begin{array}{ccc} \mathcal{P}^K & \rightarrow & [0,1] \\ (a_1, \ldots, a_K) & \mapsto & \mathbb{1}_{\{a_i > \max_{j \neq i} a_j\}} - c(a_i) \end{array} \right. \tag{9.18}$$

and $c : \mathbb{R}_+ \rightarrow \mathbb{R}_+$ is the cost or pricing function of transmitter i (assumed to be identical for all transmitters), which only depends on the action of transmitter i and not on the other played actions. The cost function is assumed to be strictly increasing, that is:

$$c(P^0) = 0 < c(P^1) < c(P^2) \ldots < c(P^m) \leq 1. \tag{9.19}$$

In the following, some different solution concepts of this game are analyzed.

9.2.2.1 Nash Equilibria

In order to determine the mixed Nash equilibria, which are the fixed points of the best response correspondence, we first determine the best response correspondence of the mixed extension of game $\mathcal{G}_{K,1}^c$, which is denoted by $\widetilde{\mathcal{G}}_{K,1}^c$, following the convention adopted in the whole book. The utility function $\widetilde{v}_i$ of transmitter i is written:

$$\widetilde{v}_i : \left| \begin{array}{ccc} [0,1]^K & \rightarrow & [0,1] \\ \underline{x} & \mapsto & x_i \left[-(1 - \omega_{-i})c(P^1) + (1 - c(P^1))\omega_{-i} \right] \end{array} \right. \tag{9.20}$$

where $c(P^1) > 0$, and $\omega_{-i} = \prod_{j \neq i}(1 - x_j)$. The best response ($\mathrm{BR}_i$) of transmitter i to the profile $\underline{x}_{-i} = (x_1, \ldots, x_{i-1}, x_{i+1}, \ldots, x_K)$ is given by:

$$\mathrm{BR}_i(x_1, \ldots, x_{i-1}, x_{i+1}, \ldots, x_K) = \left| \begin{array}{ll} \{1\} & \text{if } \omega_{-i} > c(P^1) \\ \{0\} & \text{if } \omega_{-i} < c(P^1) \\ [0,1] & \text{if } \omega_{-i} = c(P^1) \end{array} \right. . \tag{9.21}$$

It can be verified that this game has a unique interior strategy equilibrium which can be obtained by noticing that the quantity $\omega = \prod_{j=1}^K (1 - x_j)$ is independent of the user's index, that is $(i,j) \in \mathcal{K}^2$, $\frac{\prod_{k \neq i}(1-x_k)}{\prod_{k' \neq j}(1-x_{k'})} = 1 = \frac{1-x_j}{1-x_i}$. This implies that $x_1 = \cdots = x_K$. The interior Nash equilibrium follows:

$$(x_1^*, \ldots, x_K^*) = 1 - c(P^1)^{\frac{1}{K-1}} \times (1, 1, \ldots, 1). \tag{9.22}$$

The determination of the pure, mixed, and partially mixed Nash equilibria is fairly straightforward. Here, we summarize the main remarks that can be made about the corresponding analysis.

- If only one user transmits and the others stay quiet, then the user who transmits gets the utility $1 - c(P^1)$ and the others receive nothing and have no cost. This configuration is an equilibrium.
- There are exactly K pure Nash equilibria, and all of them are Pareto optimal. They are also strong equilibria and global optima.
- There are exactly $\sum_{k=1}^{K-2} \binom{K}{k} = 2^K - (K+2)$ partially mixed Nash equilibria. These equilibria correspond to the situations where k (with $1 \leq k < K - 1$) of the K transmitters choose to stay quiet, and the others are active and play the optimal mixed strategy in the game $\widetilde{\mathcal{G}}^c_{K-k}$; that is, $x_i = 1 - c(P^1)^{\frac{1}{K-k-1}}$ for the active users and x_i for the silent users.
- The game has a unique strictly mixed Nash equilibrium given by:

$$(x_1^*, \dots, x_K^*) = 1 - c(P^1)^{\frac{1}{K-1}} \times (1, 1, \dots, 1) \tag{9.23}$$

as pointed out previously.

- The partially mixed Nash equilibria and strictly mixed Nash equilibria are not Pareto optimal.

These results cannot be generalized to the multi-action case with $m + 1 \geq 3$. Indeed, for more than three actions, the game may have no pure equilibrium, as shown in Table 9.2, which is the bi-matrix representation of $\mathcal{G}^c_{K,2}$. This implies that the game $\mathcal{G}^c_{K,2}$ is not a potential game.

9.2.2.2 Stackelberg Solutions
In this subsection, we study the Stackelberg formulation of the game $\mathcal{G}^c_{K,m}$ where there is one leader, say transmitter 1, and $K - 1$ followers. The optimal action of the leader is to anticipate the reaction of the followers by playing his maximum power P^m. The followers react by playing their best response which is P^0 for all the followers. The

Table 9.2 Medium Access Control Game with Two Transmitters and Three Actions: Transmitter 1 Chooses a Row, Transmitter 2 Chooses a Column. There is No Pure Equilibrium. The Symbol $\sharp$ Indicates a Pareto Optimal Solution

	P^2	P^1	P^0
P^2	$(-c(P^2), -c(P^2))$	$(1 - c(P^2), -c(P^1))$	$(1 - c(P^2), 0)$
P^1	$(-c(P^1), 1 - c(P^2))$	$(-c(P^1), -c(P^1))$	$(1 - c(P^1), 0)^\sharp$
P^0	$(0, 1 - c(P^2))$	$(0, 1 - c(P^1))^\sharp$	$(0, 0)$

action profile:

$$\underline{a}^S = (P^m, P^0, \ldots, P^0) \tag{9.24}$$

is therefore a pure Stackelberg equilibrium. This solution is not Pareto optimal since it is Pareto-dominated by all profiles under the form $(P^\ell, P^0, \ldots, P^0)$ for any $\ell < m$ since $c(P^\ell) > c(P^m)$.

9.2.2.3 *Evolutionarily Stable Strategies*

It can be shown that the random access game $\mathcal{G}^c_{K,1}$ has a unique evolutionarily stable strategy (ESS) which is given by:

$$\underline{x}^{\text{ESS}} = 1 - c(P^1)^{\frac{1}{K-1}} \times (1, 1, \ldots, 1), \tag{9.25}$$

which is the unique interior equilibrium of the game.

9.2.2.4 *Correlated Equilibria*

The efficient correlated equilibria of the game $\mathcal{G}^0_{K,1} \equiv \mathcal{G}_{K,1}$ described in Sec. 9.2.1.4 are correlated equilibria of the game $\mathcal{G}^c_{K,1}$.

9.2.3 Heterogeneous Medium Access Control Games

The network is assumed to be heterogeneous in the sense that the transmitters evaluate the risk/cost of collision differently. To simplify the analysis, we assume $m = 1$. For every user $i \in \mathcal{K}$, the utility function of user i in the heterogeneous random medium access game is denoted by $v_i^h : \mathcal{P}^K \to \mathbb{R}$ and given by:

$$v_i^h(\underline{a}) = \begin{vmatrix} 1 - \alpha_i & \text{if} & a_i = P^1 \text{ and } \forall j \neq i, a_j = P^0 \\ 0 & \text{if} & a_i = P^0 \\ -\alpha_i & \text{if} & a_i = P^1 \text{ and } \sharp\{j \in \mathcal{K} \mid a_j = P^1\} \geq 2 \end{vmatrix} \tag{9.26}$$

with $\underline{a} \in \mathcal{P}^K$ and $\alpha_i = c_i(P^1)$. Under feasibility conditions, the mixed extension of this game has a mixed equilibrium point given by:

$$\forall i \in \mathcal{K}, \, x_i^* = 1 - \frac{1}{\alpha_i} \left(\prod_{j=1}^K \alpha_j \right)^{\frac{1}{K-1}}. \tag{9.27}$$

Let us show how to determine this equilibrium. To this end, let us rewrite the utility of user i in the mixed extension of the game as:

$$\widetilde{v}_i(\underline{x}) = x_i \times (\omega_{-i} - \alpha_i) \tag{9.28}$$

always with $\omega_{-i} = \prod_{j \neq i}(1 - x_j)$. The best response of transmitter i to $\underline{x}_{-i}$ is:

$$\text{BR}_i(\underline{x}_{-i}) = \begin{vmatrix} \{1\} & \text{if } \omega_{-i} > \alpha_i \\ \{0\} & \text{if } \omega_{-i} < \alpha_i \\ [0, 1] & \text{if } \omega_{-i} = \alpha_i \end{vmatrix} . \tag{9.29}$$

Every mixed equilibrium point $\underline{x}^*$ must satisfy:

$$\prod_{j \neq i}(1 - x_j^*) = \alpha_i. \tag{9.30}$$

Let $\omega^* := \prod_{j=1}^{K}(1 - x_j^*)$. By using (9.30), one has that:

$$\frac{\omega_{-i}}{\omega_{-k}} = \frac{1 - x_k^*}{1 - x_i^*}. \tag{9.31}$$

This implies that $(1 - x_i^*)\alpha_i = \text{constant} = \omega^*$. Therefore $\prod_{j=1}^{K}\alpha_j = (\omega^*)^{K-1}$ and thus:

$$x_i^* = 1 - \frac{\omega^*}{\alpha_i} = 1 - \frac{1}{\alpha_i}\left(\prod_{j=1}^{K}\alpha_j\right)^{\frac{1}{K-1}}. \tag{9.32}$$

9.2.4 Medium Access Control Games with Collision Costs

In this section, we study the random access game proposed in Inaltekin and Wicker (2006, 2008). We assume that $m = 1$ and that the transmitters always have a packet to transmit (saturated case), and that a transmission fails if there are more than one simultaneous transmissions. The cost of an unsuccessful transmission by any node is $\alpha_i \in]0, 1[$. If a transmission is successful, the node which successfully transmitted its packet gets a utility 1 (instead of $1 - \alpha_i$.). Note that the cost incurred by transmitter i does not only depend on his own action but also on the actions played by the others, which is different from the analysis on the game with pricing (Sec. 9.2.2). For every user $i \in \mathcal{K}$, the utility function of user i in the random medium access game with collision cost is given by $u_i : \mathcal{P}^K \to \mathbb{R}$

$$u_i(a) = \begin{vmatrix} 1 & \text{if} & a_i = P^1 \text{ and } a_j = P^0, \ \forall j \neq i \\ 0 & \text{if} & a_i = P^0 \\ -\alpha_i & \text{if} & a_i = P^1 \text{ and } \sharp\{j \in \mathcal{K} \mid a_j = P^1\} \geq 2 \end{vmatrix} \tag{9.33}$$

with $\alpha_i = c_i(P^1)$.

9.2.4.1 Symmetric Cost Functions

Assume that the cost α_i is identical and equal to γ. Then, the one-shot game has a total number of $2^K - 1$ Nash equilibria; K of them are optimal in the Pareto sense and unique ESS:

- If only one node *transmits* and the others *stay quiet*, then the transmitting node gets the utility 1 and the others receive nothing and have no cost. This configuration is a pure Nash equilibrium.
- There are exactly K pure equilibria, and all these pure equilibria are Pareto optimal.

- k $(1 \leq k < K - 1)$ of the K nodes choose to stay quiet and the $K - k$ others are active and play the optimal mixed strategy in the game with $K - k$ users: $\left(1 - \beta^{\frac{1}{K-k-1}}, \beta^{\frac{1}{K-k-1}}\right)$ where $\beta := \frac{\gamma}{1+\gamma}$. Thus, there are exactly $\sum_{k=1}^{K-2} \binom{K}{k} = 2^K - (K+2)$ partially mixed Nash equilibria.
- The game has a unique strictly mixed Nash equilibrium given by $\left(1 - \beta^{\frac{1}{K-1}}, \beta^{\frac{1}{K-1}}\right)$.
- The allocation of utility obtained in these (partially or completely) mixed strategies is not Pareto optimal.
- The ratio of the worse Nash equilibria to the global optimum is zero (the counterpart of the price of anarchy for Nash equilibria).
- The ratio of the best Nash equilibrium to the global optimum is 1 (the counterpart of the price of stability of Nash equilibria).
- The ratio of the best correlated Nash equilibria to the global optimum is 1 (the counterpart of the price of stability for correlated equilibria).

9.2.5 Medium Access Control Games with Regret Costs

The work in Tembine et al. (2008a) generalizes the case of collision costs by adding a regret cost. This allows one to model the cost related to backoff delay (the cost to retransmit the same packet when it is backlogged). Again, for the sake of simplicity, we only study the case of K users and two actions. By denoting by δ the transmission cost, γ the collision cost, and κ the regret cost, the utility function becomes $v_i^r :$ $\mathcal{P}^K \to \mathbb{R}$ with:

$$
v_i^r(\underline{q}) = \begin{vmatrix} 1 - \delta & \text{if} & a_i = P^1 \text{ and } a_j = P^0, \ \forall j \neq i \\ 0 & \text{if} & a_i = P^0 \text{ and } \sharp\{j \in \mathcal{K} \mid a_j = P^1\} \geq 1 \\ -\gamma - \delta & \text{if} & a_i = P^1 \text{ and } \sharp\{j \in \mathcal{K} \mid a_j = P^1\} \geq 2 \\ -\kappa & \text{if} & a_j = P^0, \ \forall j \in \mathcal{K} \end{vmatrix}. \tag{9.34}
$$

The game has a unique ESS $s^*(1,\ldots,1)$ with $s^* = 1 - \beta_r^{\frac{1}{K-1}}$ and $\beta_r = \frac{\gamma+\delta}{1+\gamma+\kappa}$. The other results are similar to the random access game without regret costs: the factor β becomes β_r:

- If only one node *transmits* and the others *stay quiet*, then the node which transmits gets the utility 1 and the others receive nothing and incur no cost. This configuration is a Nash equilibrium.
- There are exactly K pure Nash equilibria, and all of them are Pareto optimal.
- k $(1 \leq k < K - 1)$ of the K nodes choose to stay quiet and the $K - k$ others are active and play the optimal mixed strategy in the game with $K - k$ users: $\left(1 - \beta_r^{\frac{1}{K-k-1}}, \beta_r^{\frac{1}{K-k-1}}\right)$. Thus, there are exactly $\sum_{k=1}^{K-2} \binom{K}{k} = 2^K - (K+2)$ partially mixed Nash equilibria.

- The game has a unique strictly mixed Nash equilibrium given by
$$\left(1 - \beta_r^{\frac{1}{K-1}}, \beta_r^{\frac{1}{K-1}}\right).$$
- The allocation of utility obtained in these (partially or completely) mixed strategies is not Pareto optimal.

9.3 ACCESS CONTROL GAMES FOR MULTI-RECEIVER NETWORKS

Consider a wireless communication network with K transmitters and with S receivers. As a first step, receivers are assumed to use orthogonal channels: two packets sent over two different channels do not collide. As a second step, a network model for which inter-channel interference is possible is studied. The utility functions are defined as in Sec. 9.2 and the game is denoted by $\Gamma_{K,1,S}$.

9.3.1 Two-User MAC Game

In this subsection, we look at the case where the number of receivers is greater than the number of users. Table 9.3 gives the example of two transmitters and three receivers (resources). The convention used is as follows: the action r_j means that transmitter $i \in \{1,2\}$ has chosen receiver or channel $j \in \{1,2,3\}$.

9.3.2 Three-User MAC Game

9.3.2.1 The One-Shot Game (Table 9.4)

Denote the action spaces by $\mathcal{A}_i = \{r_1, r_2\}$, $i \in \{1,2,3\}$. The one-shot game $\Gamma(3,1,2)$ has 6 pure equilibria, an infinite number of partially mixed equilibria and one completely mixed equilibria. Let $\underline{\alpha} = (\alpha_1, \alpha_2, \alpha_3) \in \Delta(\mathcal{A}_1) \times \Delta(\mathcal{A}_2) \times \Delta(\mathcal{A}_3)$, be a

Table 9.3 Two Users for Three Resources. Transmitter 1 Chooses a Row. Transmitter 2 Chooses a Column. The Symbol* Represents Pareto-Nash Configurations. These Configurations are Global Optima

		Transmitter 2		
		r_1	r_2	r_3
Transmitter 1	r_1	(0,0)	(1,1)*	(1,1)*
	r_2	(1,1)*	(0,0)	(1,1)*
	r_3	(1,1)*	(1,1)*	(0,0)

mixed strategy of the one-shot game. Denote by $\alpha_1(r_1) = x, \alpha_2(r_1) = y, \alpha_3(r_1) = z$. Thus, a mixed strategy profile of the one-shot game is parameterized by (x, y, z). We have that:

$$\begin{cases} \tilde{u}_1(\alpha) = (1-x)yz + x(1-y)(1-z) \\ \tilde{u}_2(\alpha) = x(1-y)z + (1-x)y(1-z) \\ \tilde{u}_3(\alpha) = xy(1-z) + (1-x)(1-y)z \end{cases} \tag{9.35}$$

The utility $\tilde{u}_i(\alpha)$ is exactly the probability of user i to have sole use of a channel, i.e., the probability of success. Denote by NEU the set of Nash equilibrium utilities of the one-shot game $\Gamma(3, 1, 2)$. The pure Nash equilibria of this game are $(r_1, r_2, r_1), (r_2, r_1, r_1),\ (r_2, r_2, r_1),\ (r_1, r_1, r_2),\ (r_1, r_2, r_2),\ (r_2, r_1, r_2)$ with utilities $(0, 1, 0), (1, 0, 0), (0, 0, 1),\ (0, 0, 1), (1, 0, 0), (0, 1, 0)$. Denote by e_{r_1} the vector $(1, 0)$. e_{r_2} denotes $(0, 1, 0)$. The partially mixed equilibria are obtained when two of the three users choose different resources and the third chooses a mixed action (he gets zero in both cases) $(xe_{r_1} + (1-x)e_{r_2}, e_{r_1}, e_{r_2}), (xe_{r_1} + (1-x)e_{r_2}, e_{r_2},$ $e_{r_1}), (e_{r_1}, xe_{r_1} + (1-x)e_{r_2}, e_{r_2}), (e_{r_2}, xe_{r_1} + (1-x)e_{r_2}, e_{r_1}), (e_{r_1}, e_{r_2}, xe_{r_1} + (1-x)e_{r_2}),$ $(e_{r_2}, e_{r_1}, xe_{r_1} + (1-x)e_{r_2})$ with utility $(0, x, 1-x), x \in [0, 1]$ and all possible permutations of the coordinates. The one-shot game $\Gamma_{3,1,2}$ has a unique interior mixed equilibrium given by $\left(\frac{1}{2}e_{r_1} + \frac{1}{2}e_{r_2}, \frac{1}{2}e_{r_1} + \frac{1}{2}e_{r_2}, \frac{1}{2}e_{r_1} + \frac{1}{2}e_{r_2}\right)$, with the utility $\frac{1}{4}$ for each user. Hence, the set of Nash equilibrium utilities is:

$$\text{NEU} = \{(0, x, 1-x), x \in [0, 1]\} \bigcup \{(x, 0, 1-x), x \in [0, 1]\} \tag{9.36}$$

$$\bigcup \{(x, 1-x, 0), x \in [0, 1]\} \bigcup \left\{ \left(\frac{1}{4}, \frac{1}{4}, \frac{1}{4}\right) \right\}$$

The convex hull of the set NEU is denoted by co(NEU):

$$\text{co(NEU)} = \text{co}\left\{ (0, 1, 0), (1, 0, 0), (0, 0, 1), \left(\frac{1}{4}, \frac{1}{4}, \frac{1}{4}\right) \right\}. \tag{9.37}$$

Table 9.4 Medium Access Control with Three Users. User 3 Chooses the Matrix T_1 or T_2

		Transmitter 2				Transmitter 2	
		r_1	r_2			r_1	r_2
Transmitter 1	r_1	(0,0,0)	(0,1,0)*	Transmitter 1	r_1	(0,0,1)*	(1,0,0)*
	r_2	(1,0,0)*	(0,0,1)*		r_2	(0,1,0)*	(0,0,0)
		T_1				T_2	

9.3.2.2 *Unconstrained Repeated Games with Perfect Monitoring*

We focus on the repeated game analysis of the 3-transmitter 2-action game with perfect monitoring (see Chapter 3). In order to characterize the set of utilities of the discounted repeated game, we determine the punishment levels. The vector of punishment level (PL) utilities is $\underline{u}_{PL} = (0,0,0)$, and the set of feasible and individually rational utilities of the one-shot game $\Gamma_{3,1,2}$ is:

$$V = \text{co}\{(0,0,0),(1,0,0),(0,1,0),(0,0,1)\} \tag{9.38}$$

$$= \{(v_1,v_2,v_3) \in \mathbb{R}^3 : \forall i \in \{1,2,3\}, v_i \geq 0, \ v_1 + v_2 + v_3 \leq 1\}. \tag{9.39}$$

The dimension of V is three, i.e., the number of users in interaction. Thus, the full dimensionality condition is satisfied (the interior of V is non-empty). The standard Folk theorem with complete information (see Chapter 3) can be applied here. If E_λ denotes the set of Nash equilibrium utilities of the dynamic discounted game with discount factor $0 \leq \lambda < 1$ then:

$$\lim_{\lambda \to 0} E_\lambda = V. \tag{9.40}$$

9.3.2.3 *Unconstrained Repeated Games with a Public Signal*

In this section, the assumption of perfect monitoring is relaxed, and only a public signal which indicates the last crowd receiver is assumed to be available to the transmitters. Compared to unconstrained repeated games with perfect monitoring, for which the action profile is observed by all the three transmitters at the end of each time-slot, we now assume that, at the end the each time-slot, the public signal:

$$\omega(t) \in \Omega = \{r_1, r_2\} \tag{9.41}$$

is observed by each user. The public signal is the crowd receiver. Our objective is to characterize the set of equilibrium utilities of the discounted repeated game with the public signal (the crowd receiver). The next realization of the public signal is drawn according to the crowd receiver at the last time-slot. If an action profile $\underline{a}(t-1)$ is played at time $t-1$ then:

$$\omega_t = \begin{vmatrix} r_1 & \text{if } \sharp\{j \in \mathcal{K},\ a_{j,t-1} = r_1\} > \sharp\{i \in \mathcal{K},\ a_{i,t-1} = r_2\} \\ r_2 & \text{otherwise} \end{vmatrix} \tag{9.42}$$

action	$r_1r_1r_1, r_1r_2r_1,$ $r_2r_1r_1, r_1r_1r_2$	$r_1r_2r_2, r_2r_1r_2,$ $r_2r_2r_2, r_2r_2r_1$
announced signal	r_1	r_2

The probability that the signal $\omega \in \Omega$ is observed when $\underline{a}(t-1)$ is played is denoted by $\xi(\omega|\underline{a})$. We have that:

$$\xi(r_1|\underline{a}) = \begin{vmatrix} 1 & \text{for } \underline{a} \in \{r_1r_1r_1, r_1r_2r_1, r_2r_1r_1, r_1r_1r_2\} =: \Omega_{r_1} \\ 0 & \text{for } \underline{a} \in \{r_2r_2r_2, r_2r_1r_2, r_1r_2r_2, r_2r_2r_1\} =: \Omega_{r_2} \end{vmatrix},$$

$$\xi(r_2|\underline{a}) = \begin{vmatrix} 0 & \text{for } \underline{a} \in \Omega_{r_1} \\ 1 & \text{for } \underline{a} \in \Omega_{r_2} \end{vmatrix}$$

Define:

$$\xi(\omega|\underline{\alpha}) = \sum_{\underline{a} \in A} \xi(\omega|\underline{a})\alpha(\underline{a}) \text{ with } \alpha(\underline{a}) = \alpha_1(a_1)\alpha_2(a_2)\alpha_3(a_3), \ a_i \in \{r_1, r_2\}. \quad (9.43)$$

with $A = A_1 \times A_2 \times A_3$. It can be shown that, for all Nash equilibrium action profiles of the static game, the following inequalities hold:

$$\begin{aligned} 2\tilde{u}_1(\underline{\alpha}) + \tilde{u}_2(\underline{\alpha}) + \tilde{u}_3(\underline{\alpha}) &\geq 1 \\ \tilde{u}_1(\underline{\alpha}) + 2\tilde{u}_2(\underline{\alpha}) + \tilde{u}_3(\underline{\alpha}) &\geq 1 \ . \\ \tilde{u}_1(\underline{\alpha}) + \tilde{u}_2(\underline{\alpha}) + 2\tilde{u}_3(\underline{\alpha}) &\geq 1 \end{aligned} \quad (9.44)$$

These inequalities characterize the convex hull of the set of equilibrium utilities. These inequalities can be extended to the class of strategies that induce the same public signal.

We now describe the structure of public signals induced by a mixed strategy. The probability that the public signal will be r_1 when $\underline{\alpha}$ is played is $\xi(r_1|\underline{\alpha})$:

$$\xi(r_1|\underline{\alpha}) = xyz + x(1-y)z + xy(1-z) + (1-x)yz, \quad (9.45)$$

which can be rewritten as:

$$\xi(r_1|\underline{\alpha}) = xy + z[x(1-y) + (1-x)y] \quad (9.46)$$

$$= yz + x[(1-y)z + y(1-z)] \quad (9.47)$$

$$= zx + y[x(1-z) + (1-x)z]. \quad (9.48)$$

Define:

$$\mathcal{D}(\underline{\alpha}) := \{\beta = (\beta_1, \beta_2, \beta_3) \in [0,1]^3 \mid \xi(r_1|\underline{\alpha}_{-1}, \beta_1) = \xi(r_1|\underline{\alpha}_{-2}, \beta_2) = \xi(r_1|\underline{\alpha}_{-3}, \beta_3)\}. \quad (9.49)$$

The set $\mathcal{D}(\underline{\alpha})$ represents the set of possible deviations from the mixed action profile $\underline{\alpha}$ that gives the same public signal. Then, for a given time-slot, if the recommendation is to follow $\underline{\alpha}$ and:

- user 1 plays β_1,
- user 2 uses β_2 instead of $(y, 1-y)$,
- user 3 uses β_3 instead of $(z, 1-z)$

and if $\beta \in \mathcal{D}(\underline{\alpha})$, then these three deviations will induce the same public signal, and hence the same continuation utility. For any mixed strategy $\underline{\alpha}$ parameterized by

$(x, y, z) \in [0, 1]^3$, there exists $\beta \in \mathcal{D}(\underline{\alpha}) = \{\beta = (\beta_1, \beta_2, \beta_3) \in [0, 1]^3 \mid \xi(r_1 | \underline{\alpha}_{-1}, \beta_1) = \xi(r_1 | \underline{\alpha}_{-2}, \beta_2) = \xi(r_1 | \underline{\alpha}_{-3}, \beta_3)\}$ such that:

$$
\begin{cases}
2\tilde{u}_1(\underline{\alpha}_{-1}, \beta_1) + \tilde{u}_2(\underline{\alpha}_{-2}, \beta_2) + \tilde{u}_3(\underline{\alpha}_{-3}, \beta_3) \geq 1 \\
\tilde{u}_1(\underline{\alpha}_{-1}, \beta_1) + 2\tilde{u}_2(\underline{\alpha}_{-2}, \beta_2) + \tilde{u}_3(\underline{\alpha}_{-3}, \beta_3) \geq 1 \\
\tilde{u}_1(\underline{\alpha}_{-1}, \beta_1) + \tilde{u}_2(\underline{\alpha}_{-2}, \beta_2) + 2\tilde{u}_3(\underline{\alpha}_{-3}, \beta_3) \geq 1
\end{cases}
\tag{9.50}
$$

where the probabilities to observe r_1 for the public signal realization are:

$$
\begin{cases}
\xi(r_1 | \underline{\alpha}_{-1}, \beta_1) = yz + \beta_1 [(1 - y)z + y(1 - z)] \\
\xi(r_1 | \underline{\alpha}_{-2}, \beta_2) = zx + \beta_2 [x(1 - z) + (1 - x)z] \\
\xi(r_1 | \underline{\alpha}_{-3}, \beta_3)\} = xy + \beta_3 [x(1 - y) + (1 - x)y]
\end{cases}
\tag{9.51}
$$

Therefore, the set of subgame perfect equilibria (Chapter 3) of the 3-transmitter 2-action repeated game with a public signal is the convex hull of the set of Nash equilibrium utilities of the one-shot game when the discount factor goes to zero. In other words, if E'_λ denotes the set of perfect public equilibrium of the repeated game, then the following holds:

$$
\lim_{\lambda \to 0} E'_\lambda = \mathrm{co}(\mathrm{NEU}).
\tag{9.52}
$$

In particular, the global optimum of the one-shot game can be approximated by a perfect public equilibrium of the repeated game with small discount factor. Another important feature of the repeated game studied here is that the set of perfect public equilibrium utilities when the discount factor goes to zero coincides with V.

9.3.3 Spatial Random Access Games

Based on the two examples studied in Secs. 9.3.1 and 9.3.2, a spatial random access game is formulated here. The motivation for this is to take into account the spatial structure of the network, namely the locations of the transmitters and receivers in the game formulation. All transmitters have the same action set. The number of transmitters with which a given transmitter interacts can vary from one transmitter to another. The reciprocity property is not required here: if transmitter Tx_1 interacts with transmitter Tx_2, we do not require the converse to be true. Local interactions between transmitters are considered. For each transmitter there corresponds a receiver. We shall say that a transmitter Tx_1 is subject to the influence (interference) of transmitter Tx_2 if the transmission from Tx_2 overlaps the one from Tx_1, provided that the receiver of the transmission from Tx_1 is within interference range of transmitter Tx_2. Consider the example represented in Figure 9.1. The network comprises 3 receivers (represented by squares) with range $\mathcal{N}_i$, $i \in \{1, 2, 3\}$. A signal from transmitter i, within a distance r of the receiver Rx_1, causes interference to any signal from transmitter $j \neq i$ if the latter is located in the area covered by Tx_1. We see that the transmitters in the area $A1 := \mathcal{N}_1 \backslash (A12 \cup A13 \cup A123)$ cause no interference

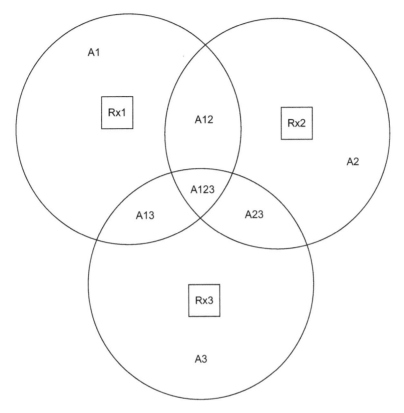

FIGURE 9.1

Non-reciprocal interference control.

to any transmitter from $A12 \cup A13 \cup A123$ transmitting to Rx_2 or Tx_3. The transmitters located in area $A1$ cause no interference to any transmitter located in areas $A2$ or $A3$ but the transmission from transmitters in $A1$ suffers from interference from sources located in $A12$, $A13$, $A123$. The transmitters in $A1$ interfere with each other at their common destination Rx_1. Thus, the interference effect is not reciprocal. In term of games, a transmitter in areas $A12 \cup A13 \cup A23 \cup A123$ plays several one-shot games simultaneously with two actions (transmit or not transmit), and receives a vector of utilities. Only the utility obtained at the receiver to which he transmits can be non-zero.

9.3.3.1 *Analysis of Equilibria and Optimality*

An arbitrary number of transmitters in each area is assumed. Let n_{ij} be the cardinality of Aij and n_{123} be the cardinality of $A123$. Denote by $N_1 := n_1 + n_{12} + n_{13} + n_{123}$ the number of transmitters located in $\mathcal{N}_1 = A12 \cup A13 \cup A1 \cup A123$ (the total number of transmitters covered by receiver Rx$_1$). Similarly, $N_2 := n_2 + n_{12} + n_{23} + n_{123}$, $N_3 := n_3 + n_{13} + n_{23} + n_{123}$.

9.3.3.2 *Equilibria and Pareto Optimal Solutions*
Consider the following configurations:

(C3) Equilibria with three successful transmissions at the same slot:
- From each area $A1, A2, A3$, only one transmitter transmits: only one transmitter from $A1$, one from $A2$, and one from $A3$ transmit, and the others stay quiet. These configurations are equilibria. There are three successful transmissions. Any permutation of the n_1 transmitters in $A1$, n_2 transmitters in $A2$, and n_3 transmitters in $A3$ for which the other transmitters in areas $A12, A13, A23$, and $A123$ stay quiet, lead a pure equilibrium which are also Pareto optimal. They are global optima (social welfare) in the sense that the total sum of utilities is maximized (one can have at most three successful transmissions at the slot).

(C2) Equilibria with two successful transmissions at the same slot:
- Only one transmitter located in $A12$ transmits, one transmitter from $A3$ transmits, and the others stay quiet.
- Only one transmitter located in $A13$ transmits, one transmitter from $A2$ transmits, and the others stay quiet. There are two successful transmissions.
- Only one transmitter located in $A23$ transmits, one transmitter from $A1$ transmits, and the others stay quiet.

(C1) Equilibria with one successful transmission at the same slot:
- Only one transmitter located in $A123$ transmits and all the other transmitters stay quiet. There is one successful transmission in all of the three receivers. Since this transmitter is located at the intersection between $\mathcal{N}_1$, $\mathcal{N}_2$, and $\mathcal{N}_3$, there is interference to any transmission at any receiver. By permutation and selection of one of the n_{123} transmitters located in $A123$, we obtain the other equilibria at this configuration. These equilibria are not global optima.

(C0) At least two transmitters from each neighborhood $\mathcal{N}_j$, $j \in \{1,2,3\}$ transmit, and the others play any mixed strategies. Hence, there is an infinite number of equilibria of this type. These equilibria are not Pareto optimum because each transmitter gets zero, which is Pareto dominated by the configurations of type $C1 - C3$.

Note that if there is no pairwise intersection between the neighborhoods $\mathcal{N}_j$, $j \in \{1,2,3\}$, (the numbers $n_{12}, n_{123}, n_{23}, n_{13}$ equals zero) then the interactions become reciprocal (if transmitter i interferes on the signal of transmitter j then transmitter j interferes on the signal of transmitter i) and the games in $\mathcal{N}_1, \mathcal{N}_2, \mathcal{N}_3$ can be analyzed separately. Finally, note that this model can easily be extended to finitely many intersecting neighborhoods.

9.4 MULTI-STAGE ACCESS CONTROL GAMES WITH ENERGY CONSTRAINTS

In this section, the problem of medium access control, in which (K) transmitters take into account some energy constraints, is studied. It is assumed in a realistic manner that the energy consumed by a given transmitter only depends on his own actions; always for the sake of simplicity only the two-action case is considered, i.e., $\mathcal{A}_i = \{\text{Transmit}, \text{Wait}\}$. However, the performance in terms of transmission rate for each transmitter in a multiuser system depends strongly on the actions taken by the other users. We have so far essentially focused on static games; hence the natural framework for studying control games with energy constraints is to move to multi-stage games. A stage corresponds to one unit time namely, the block, frame, packet, or time-slot duration in the context of wireless communications. The time horizon of the game is given by the number of stages T. The probability of transmission of a terminal in time-slot t depends on the amount of energy he has left prior to this slot. The probability that transmitter i sends a packet is denoted by $x_i(t)$. In order to define the energy consumed by a transmitter, we first need to define its utility and strategy in the context of the multi-stage under investigation. The private history of transmitter i at stage $t \in \{1, \ldots, T\}$ is defined by:

$$\underline{h}_{i,t} = (x_i(1), \ldots, x_i(t-1)).\tag{9.53}$$

Assuming perfect recall, we know from Kuhn's theorem (see Chapter 3) that restricting the transmitters' strategies to behavioral strategies instead of mixed strategies over T-action sequences does not induce any loss of optimality. A behavioral strategy for transmitter i is therefore defined by:

$$\sigma_{i,t} : \begin{vmatrix} \mathcal{H}_{i,t} & \to & \Delta(\mathcal{A}_i) \\ \underline{h}_{i,t} & \mapsto & x_i(t) \end{vmatrix}\tag{9.54}$$

where $\mathcal{H}_{i,t}$ is the set of possible private histories at time t. As in Chapter 2, the notation σ_i is used to refer to the sequence of causal functions which define the behavioral strategy of transmitter i. The utility of transmitter i is defined by the Cesaro mean of the instantaneous utilities over the whole duration of the multi-stage (MS) game, that is:

$$\widetilde{v}_i^{\text{MS}}(\sigma_1, \ldots, \sigma_K) = \sum_{t=1}^{T} x_i(t) \prod_{j \neq i} \left[1 - x_j(t)\right]\tag{9.55}$$

where $(x_1(t), \ldots, x_K(t))$ is the transmission probability profile induced by the joint behavior strategy $(\sigma_1, \ldots, \sigma_K)$ at stage t. Now we can define the energy consumed by transmitter i to play the T-stage game:

$$E_i(\sigma_i) = \sum_{t=1}^{T} x_i(t).\tag{9.56}$$

The available energy for transmitter $i \in \mathcal{K}$ is denoted by $\mathcal{E}_i$, and each transmission requires 1 unit of energy. The objective of a transmitter is to maximize his utility over his strategy under the constraint $E_i \leq \mathcal{E}_i$. Each terminal has therefore to solve the following problem:

$$\max_{\sigma_i} \tilde{v}_i^{MS}(\sigma_1, \ldots, \sigma_K) \text{ s.t } E_i(\sigma_i) \leq \mathcal{E}_i \qquad (9.57)$$

having complete information about the game and knowing his private history. As this optimization problem depends on what the other transmitters do and know, this problem is a game. Let us analyze a few of its solution concepts.

Recall that a joint behavior strategy is a generalized Nash equilibrium (Debreu, 1952) if it satisfies the energy constraint and if no transmitter can benefit from deviating unilaterally under the constraints. More precisely, if only transmitter i deviates then either his utility does not increase, or his energy constraints are violated. A strategy $\underline{\sigma} = (\sigma_1, \ldots, \sigma_K)$ is a constrained Pareto efficient strategy if it is feasible and there is no other strategy $\underline{\tau} = (\tau_1, \ldots, \tau_K)$ which satisfies the energy constraints and verifies $\tilde{v}_i(\underline{\tau}) \geq \tilde{v}_i(\underline{\sigma})$ for all i with strict inequality for at least one terminal i. A strategy $\underline{\sigma}$ is a Nash–Pareto equilibrium if it is a constrained Nash equilibrium and is Pareto efficient. A Nash–Pareto policy may not exist in general, but we will see that such an equilibrium does exist in the constrained multi-stage random access game under consideration. Another feature of the game under study is that it also possesses strong Nash equilibria. From Chapter 1, we know that a strategy is a k-strong Nash equilibrium if no coalition of transmitters of size k can improve the number of successful transmissions of each of its members. A strategy is a strong equilibrium if it is a k-strong equilibrium for any size $k \in \{1, \ldots, K\}$. A strong equilibrium is therefore a strategy from which no coalition (of any size) can deviate and improve the utility of every member of the coalition, while possibly lowering the utilities of transmitters outside the coalition. This notion of strong equilibria is very attractive because it is robust to the formation of coalitions of transmitters. Most of the games do not possess strong equilibria but, very interestingly, this particular constrained random access game has an exponential number of constrained strong equilibria.

Note that if $T \geq \sum_{i=1}^{K} \mathcal{E}_i$ then it is possible to assign K disjoint sets of slots to the K transmitters such that each transmitter has sufficiently many slots to use his energy budget. Such a strategy is obviously a constrained Nash equilibrium and is Pareto efficient. This is why we only study the case where $T < \sum_{i=1}^{K} \mathcal{E}_i$. For this analysis, the notion of a *Time Division Multiplexing* (TDM) strategy is important.

Definition 193: TDM Policy

A TDM policy $\Pi_{\mathcal{T}}$ is a partition of the set $\mathcal{T} = \{1, \ldots, T\}$ into K disjoint sets $\Pi_1, \ldots, \Pi_K$, such that at each time-slot, $t \in \mathcal{T}$, the transmitter i for which $t \in \Pi_i$ transmits a packet with probability 1 and all the others do not transmit. The policy $\Pi_{\mathcal{T}}$ is said to be feasible if the individual energy constraint is met for every transmitter.

The main results obtained by Altman et al. (2009b) can be summarized as follows:

- Any feasible TDM policy is a Nash equilibrium and is Pareto efficient for the constrained multi-stage game if $(\mathcal{E}_i)_i \in \mathbb{N}^K$.
- The number of TDM strategies is exactly the Stirling number of the second kind $S(T,K)$ which is the number of partitions of a set of size T into K blocks ($K \leq T$). $S(T,K)$ satisfies the recursive formula:

$$
\begin{aligned}
S(T,K) &= \binom{K}{1}S(T-1,K) + S(T-1,K-1) \\
&= KS(T-1,K) + S(T-1,K-1)
\end{aligned}
\tag{9.58}
$$

with $S(T,1) = S(T,T) = 1$. The Stirling numbers of the second kind are given by the explicit formula:

$$
S(T,K) = \frac{1}{K!}\sum_{i=0}^{K}(-1)^{K-i}i^T\binom{K}{i}.
\tag{9.59}
$$

- The total number of Nash–Pareto equilibria of the constrained multi-stage game is bounded by $\sum_{i=1}^{K} iT!S(T,i)$.
- If the budgets are identical:
  - if $\mathcal{E}_i = \mathcal{E} < T < K\mathcal{E}$, then the joint behavior strategy for which each transmitter always chooses $x_i(t) = \frac{\mathcal{E}}{T}$ is a Nash equilibrium;
  - if $\mathcal{E} > T$, then there is no symmetric interior equilibrium.
- The feasible TDM strategies are *constrained strong equilibria*.
- These constrained strong equilibria are also energy-efficient equilibria in the sense that the constrained Nash equilibrium is realized with a minimum energy investment for each transmitter and it is a global optimum, i.e., it solves the maximization problem:

$$
\max_{\underline{\sigma}} \ \mathcal{V}(\underline{\sigma}) = \sum_{i=1}^{K}\tilde{v}_i(\underline{\sigma}) \ \text{ subject to}
$$

$$
\begin{cases}
E_i(\sigma_i) \leq \mathcal{E}_i, \\
i \in \{1,\dots,K\}.
\end{cases}
\tag{9.60}
$$

  This means that, for the dynamic random access game with the number of successful transmissions as the utility, the *strong price of anarchy* is one.
- In the non-integer constraints case, the TDM strategies can be suboptimal and infeasible.

Detailed proofs of these statements can be found in Altman et al. (2009b).

9.5 STOCHASTIC ACCESS CONTROL GAMES

Stochastic games, introduced by Shapley in 1953 (Shapley, 1953b), are in the class of dynamic games with probabilistic transitions played by one or more players. They generalize standard repeated games (Chapter 2) which correspond to the special case where the game always stays in the same state. They also generalize Markov decision processes which correspond to the special case where there is only one player. In Shapley (1953b), the game is played in a sequence of stages. At the beginning of each stage, the game is in a given state. The players select actions depending on their state and history, and each player receives an instantaneous utility which depends on the current state and the chosen action profile. The game then moves to a new (random) state whose distribution depends on the previous state and the actions chosen by all the players. The procedure is repeated for the new state, and play continues for a finite or infinite number of stages. The total utility for a player is often taken to be the discounted sum of the stage utilities or the limit inferior of the averages of the stage utilities. For more details on stochastic games, we refer the reader to Chapter 2. In this section, a stochastic formulation of the access control problem is presented, in order to model situations where the game has a state. In access control, a state can correspond to the medium state, the server state, the backoff state, or the environment state. The transition between the states depends on the transmission probabilities for all the active users (those which have a packet to transmit at the corresponding time-slot). The instantaneous utility corresponds to the success probability at that time, possibly minus an energy consumption cost. Note that the non-saturated regime assumption is relaxed here, since transmitters do not always have packets to send.

9.5.1 Medium-State Dependent Access Control Games with Partial Information

We first study a simple stochastic competitive game between two transmitters, for which the state of the game is given by the medium state (available or not available). This game is particularly interesting because it shows the non-existence of quasi-stationary (which depends only on the last move) optimal strategies. However, the stochastic game has an optimal strategy in two-step history dependent strategies and the equilibrium strategies and the value of the stochastic game can be determined explicitly.

Consider two users who compete by sending packets over a single channel. Because of constraints imposed by the receiver (say a server), only one user can transmit over a given time-slot. The server requires a transmitter to send a request before sending a packet and there is no ACK after transmission. At every time-slot, each user can send a request, send a packet, or do nothing (for example, if the transmitter is not expected to have a packet). If both transmitters send a request at the same time-slot, then there is a collision, which is not known by the users, and the server does not transmit a packet on the following time-slot. If only one user sends a request, and that user sends a packet at the subsequent time-slot, the server does transmit this packet. Otherwise the request is offset, and the transmitter who made the request incurs a cost c. The server then becomes free again, the transmitters have this information, and the server waits for another request. Although the transmitters know when the server becomes free, its state becomes unknown after one time-slot: if transmitter 1 sends a request, transmitter 1 does not know whether the server is free or is waiting for transmitter 1's packet; if transmitter 1 does nothing or sends a packet, the server may either be free or waiting for transmitter 2's packet. As the transmitters compete among themselves in a variable environment, the analysis requires the use of stochastic games. Since the state is not completely known by both transmitters, this dynamic game is in the class of stochastic games with partial information.

The stochastic game is played as follows. At every slot each transmitter has three available actions:

- send a request: SR;
- send a packet: TD;
- stay quiet or wait: W.

The server remains free as long as one of the transmitters sends a request. It then waits for a packet from the requesting transmitter. If a packet arrives, it is forwarded to the network. Otherwise, the requesting transmitter has a cost c. To analyze the competitive behavior of the transmitters, we assign a utility value to each outcome. We focus on competitive situations where each transmitter aims to maximize his long run average utility. Figure 9.2 represents the transitions between the three states.

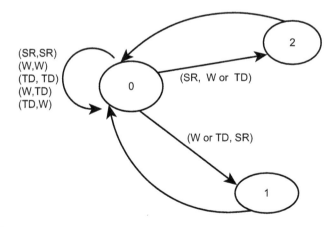

FIGURE 9.2

Structure of the Markov chain.

Let "0" be the state when the channel is free. The utilities at this state are given by:

	TD	SR	W
TD	$O_{(1,0,0)}$	$O_{(0,1,0)}$	$O_{(0,1,0)}$
SR	$O_{(0,0,1)}$	$O_{(1,0,0)}$	$O_{(0,1,0)}$
W	$O_{(0,0,1)}$	$O_{(0,0,1)}$	$O_{(1,0,0)}$

Denote by "1" the state where the channel is a waiting a transmission from transmitter "1". The utilities at state "1" are given by:

	TD	SR	W
TD	$C_{(1,0,0)}$	$C_{(1,0,0)}$	$C_{(1,0,0)}$
SR	$-C_{(1,0,0)}$	$-C_{(1,0,0)}$	$-C_{(1,0,0)}$
W	$-C_{(1,0,0)}$	$-C_{(1,0,0)}$	$-C_{(1,0,0)}$

The utilities at state "2" are given by:

	TD	SR	W
TD	$-C_{(1,0,0)}$	$C_{(1,0,0)}$	$C_{(1,0,0)}$
SR	$-C_{(1,0,0)}$	$C_{(1,0,0)}$	$C_{(1,0,0)}$
W	$-C_{(1,0,0)}$	$C_{(1,0,0)}$	$C_{(1,0,0)}$

Transitions are represented by 3-dimensional vector. From state 0 to 1, the transition is $(0,1,0)$ if the action (TD, SR) is played. Transitions at state "1": after the transmitters choose their actions, packets are forwarded to the network, and the server moves back to 0. When the server becomes free, a new game starts from this time-slot on. Therefore, it is sufficient to study the game until the first time the server restarts (see Figure 9.2).

The strategy of transmitter i in this stochastic game is defined by:

$$\sigma_{i,t}^{SG} : \left| \begin{array}{ccc} \{TD, SR, W\}^{t-1} & \to & \Delta(\mathcal{A}_i) \\ (a_i(1), \ldots, a_i(t-1)) & \mapsto & \pi_i(t) \end{array} \right. \tag{9.61}$$

where $\pi(t)$ is the distribution played at stage t. This stochastic game (SG) with partial information has the following properties.

- The strategies which assigns W at each time-slot are weakly dominated by TD.
- If the value exists, the value is zero.
- The game has a value.
- No transmitter has a quasi-stationary (which depends only on the last move) optimal strategy.
- The stochastic game has an optimal strategy in two-step history dependent strategies.
- The equilibrium strategies modulo events with zero measure are given by: for $t = 1, \pi_i^*(1) = \left(\frac{1}{2}, \frac{2}{3}, 0\right)$, and for $t \geq 2$

$$\pi_i^*(t) = \begin{vmatrix} (1,0,0) & \text{if} & a_i(t-2) = a_i(t-1) = \text{SR} \\ \left(\frac{1}{2}, \frac{1}{2}, 0\right) & \text{if} & a_i(t-1) = \text{SR}, a_i(t-2) = \text{TD} \\ \left(\frac{1}{3}, \frac{2}{3}, 0\right) & \text{if} & a_i(t-1) = \text{TD} \end{vmatrix}. \qquad (9.62)$$

9.5.2 Battery-State Dependent Access Control

In this section, the state of the game is given by the battery levels of the transmitters. Each user stays permanently in the system. From time to time, his battery state changes. Each user is thus faced with a Markov decision process (MDP) in which he wants to maximize the expected average successful transmission probability under a varying energy constraint. Each transmitter only knows the state of his own MDP, and does not know the state of the other transmitters. The transition probabilities of a transmitter's MDP are only controlled by that transmitter. Terminals are always assumed to transmit packets in a sporadic manner (non-saturated regime). Occasionally, their destination may simultaneously receive a transmission from another terminal, which results in a collision. It is assumed, however, that even when packets collide, one of the packets can be received correctly if transmitted at the highest power. The transmitters' utility is the probability of successful transmission. By allowing the transmitter's battery to be reloaded (say with solar power), the batteries of transmitters may have a very large lifetime, and a relevant performance criterion is the Cesaro-mean of the instantaneous utility. Assume that each battery has $L+1$ energy states $\mathcal{L} = \{0, 1, \ldots, L\}$. The state 0 corresponds to the state Empty and the state L is the state Full for the battery. The other states $1, \ldots, L-1$ are intermediate states. In the state 0, the action set is $\{P^0\}$.

We associate with each transmitter an MDP (which represents the transition probabilities between energy levels). Let $L_{i,t}$ be the energy level of battery i at time t. Given a strategy of the transmitter, the transition probability of the energy level of the battery is described by the (first order, time-homogeneous) Markov process $(L_{i,t})$ where the transition probability law q_i is given by:

$$q_i(\ell_i'|\underline{a}, \ell_i) = q_i(\ell_i'|a_i, \ell_i) \qquad (9.63)$$

with $\underline{a} = (a_1, \ldots, a_K)$ and $a_i \in \{P^0, \ldots, P^m\}$. This gives action dependent transition probabilities, which will change if the power consumption changes. To keep the same property as for the battery, we assume that q_i decreases to state $\ell_i - 1$ if a_i increases. Based on this stochastic modeling for the remaining energy of the battery for each user, the author in Tembine (2009a,b) shows the existence of equilibria and conditions for evolutionarily stable strategies under good and bad weather conditions. These situations are represented by the absorbing or non-absorbing state of the Markov decision process under fixed behavioral strategies. In the irreducible case, for example in presence of cost-free renewable energy during the whole game, one can show that the fully aggressive strategy P^m, which consists of transmitting at maximum power in each state $\ell \neq 0$, is an evolutionarily stable strategy. However, in the absorbing state case, transmitting with maximum power may not be an equilibrium, such as when a solar-powered system operates in cloudy weather (no sunlight) over a long period (cloudy sky, raining, or due to the season). Power storage elements, such as supercapacitors (finite in practice), are therefore proposed in order to have energy available for later use. Because of limited capacity (and hence energy of the battery), the aggressive terminals (which use the most power) will rapidly reach state 0 (the lower state of the battery), which becomes an absorbing state. Those transmitters with empty batteries need an alternative solution, such as external recharger or a new battery, so there is an additional cost in this situation.

If the maximum lifetime of the battery (for example using the power "P^0") is finite, then power control in cloudy weather can be modeled by the stochastic game, with its utility being total successful transmission before reaching the absorbing state 0. Under these conditions, the total utility is the sum over the states of the expected successful transmission times associated with the expected sojourn times spent in this state. It is easy to see that the strategy "stay quiet" (plays P^0 in each state of the battery) cannot be an equilibrium. The work in Tembine (2009a,b) derived sufficient conditions for stationary strategies to be evolutionarily stable. Thanks to renewable energy techniques, designing autonomous transmitter and consumer embedded electronics that exploit energy deriving from the environment is becoming a feasible option. However, the design of such devices requires careful selection of components, such as power consumption and energy storage elements, according to the working environment and features of the application.

9.5.3 Backoff State-Dependent Access Control

Up to this point we have considered saturated access control games. In this section we formulate the non-saturated case as a dynamic game with partial information (see Basar and Olsder (1982) and the references therein). The state of the game is given by the states of the packet buffers at the transmitters. A user only interacts with the active users (the users that have a packet to send). This defines a random set $\mathcal{B}(t)$ describing the set of active users. Each user has its own backoff state which is described by a Markov decision process controlled by all the active users. The backoff stage increases if a collision occurs. The packet is then retransmitted up to a certain threshold, denoted by L. To illustrate this stochastic modeling, we consider a simple example based on CSMA with power control. We consider a wireless LAN with K stations employing the IEEE 802.11 DCF (Bianchi, 2000). When it has a packet, each transmitter sends it at some power level, with distribution x. Each transmitter runs a back-off timer. When a transmitter i is in the state $\ell \in \mathcal{L} = \{0, 1, \ldots, L\}$, it makes a transmission attempt in the next slot with a probability x_s^K and uses a power level among $m_i + 1$ levels in $\mathcal{P}_i = \{P_i^0, P_i^1, \ldots, P_i^{m_i}\}$ where $P_i^0 < \ldots < P_i^{m_i}$. The back-off stage only is reset to 0 after a successful transmission. All nodes that made a successful transmission increment their states by 1. We assume that a successful transmission occurs when the received signal to interference plus noise (SINR) from the sending transmitter is larger than some threshold γ, and its state returns to 0. Otherwise, all other transmitters for which the SINR is under the required threshold lose their packets (collision or not admitted), and their state is incremented by 1 (if the state s is different than L). This class of policy is called SINR-based admission control policy. When a transmitter i makes an attempt at time t, it interacts with a random set of transmitters $\mathcal{B}(t) \subset \mathcal{K}$. It receives a utility u_j that depends on the action of the transmitters involved in the interaction:

$$u_j^{\mathcal{B}(t)}((s_i, a_i)_{i \in \mathcal{B}(t)}) = -c_j(a_j) + \mathbb{1}_{\{\text{SINR}_{j,t} \geq \gamma, \, j \in \mathcal{B}(t)\}}. \tag{9.64}$$

where $c_i(a_i)$ is the instantaneous cost (energy consumption) and $\text{SINR}_{j,t}$ is the signal to noise plus interference ratio of user i at time t when he is an active user. This defines a dynamic game in which each transmitter knows only his own state ℓ_i (its backoff stage). The transition between the states depends on the actions chosen by the other users, but also on the states of the other transmitters which are unknown. This dynamic game is in the class of stochastic games with partial monitoring (only the individual state is observed at each stage). Discussion of the existence of an equilibrium for such a class of stochastic games under average utilities and discounted utilities can be found in Basar and Olsder (1982) and Tembine (2009a,b).

PART

Appendices, References, and Index

IV

Appendices

A.1 REVIEW OF GENERAL MATHEMATICAL NOTIONS

Definition 194: Relation

A relation is any subset of a Cartesian product.

Definition 195: Binary Relation

A binary relation from the set $\mathcal{X}$ to the set $\mathcal{Y}$ is a collection of ordered pairs $(x,y) \in \mathcal{X} \times \mathcal{Y}$. One often writes $x\mathcal{R}y$ to show x is related to y.

Definition 196: Equivalence Relation

An equivalence relation on a set $\mathcal{X}$ is a subset of $\mathcal{X}^2$; that is, a collection $\mathcal{R}$ of ordered pairs of elements of $\mathcal{X}$ which satisfies the following three properties:

1. Reflexiveness: $\forall x \in \mathcal{X}$, $x\mathcal{R}x$.
2. Symmetry: $\forall (x,y) \in \mathcal{X}^2$, $x\mathcal{R}y \Leftrightarrow y\mathcal{R}x$.
3. Transitivity: $\forall (x,y,z) \in \mathcal{X}^3$, $x\mathcal{R}y$ and $y\mathcal{R}z$ implies that $x\mathcal{R}z$.

Definition 197: Countable Set

A set $\mathcal{X}$ is said to be countable if there exists an injective function from $\mathcal{X}$ to the set of natural numbers $\mathbb{N} = \{0,1,2,3,\ldots\}$.

Definition 198: Correspondence

A correspondence is a set-valued function: $f : \mathcal{E} \to \mathcal{F}$ where $\mathcal{F}$ is a set (different from a vector-valued function).

Definition 199: Unit Simplex

Let $\mathcal{X} = \{x_1,\ldots,x_N\}$ be a set. Then the unit simplex associated with $\mathcal{X}$ is given by $\Delta(\mathcal{X}) = \left\{ (q_1,\ldots,q_N) \in \mathbb{R}_+^N : \sum_{i=1}^N q_i = 1 \right\}$.

Definition 200: Connectedness

The set $\mathcal{X}$ is said to be disconnected if it is the union of two disjoint nonempty open sets. Otherwise, $\mathcal{X}$ is said to be connected.

Definition 201: Open Set

A set X is said to be open if $\forall x \in X \subset \mathcal{E}$ (where $\mathcal{E}$ is a normed vector space), $\exists \epsilon > 0$, $\mathcal{B}_o(x, \epsilon) \subset X$, where $\mathcal{B}_o(x, \epsilon)$ is the open ball with center x and radius ϵ.

Definition 202: Interior

If X is a subset of a Euclidean space, then x is an interior point of X if there exists an open set centered at x which is contained in X.

Definition 203: Compactness (General Definition)

A subset X of a normed vector space is compact if every sequence in X has a subsequence that converges in X.

Definition 204: Compactness (Characterization in Finite Dimensional Spaces)

A subset X is compact if and only if it is closed and bounded.

Definition 205: Convex Set

A set X is said to be convex if $\forall (x, y) \in X^2$, $\forall \mu \in [0,1]$, $\mu x + (1 - \mu) y \in X$.

Definition 206: Convex Function

A function $f : X \to Y$ is convex if $\forall (x, x') \in X^2$, $\forall \mu \in [0,1]$:

$$f(\mu x + (1 - \mu) x') \leq \mu f(x) + (1 - \mu) f(x'). \tag{A.1}$$

Definition 207: Concave Function

A function $f : X \to Y$ is concave if $-f$ is convex.

Definition 208: Quasi-Concavity

A function f is quasi-concave on a convex set X if, for all $\alpha \in \mathbb{R}$, the upper contour set $\mathcal{U}_\alpha = \{x \in X, f(x) \geq \alpha\}$ is convex.

Definition 209: Lower/Upper Semi-Continuity

A function f is lower (resp. upper) semi-continuous if the lower (resp. upper) contour set $\mathcal{L}_\alpha = \{x \in X, f(x) \leq \alpha\}$ (resp. $\mathcal{U}_\alpha$) is closed.

A.2 REVIEW OF CONCEPTS RELATED TO DYNAMICAL SYSTEMS AND STOCHASTIC APPROXIMATION

A.2.1 Appendix to Chapters 5 and 6

In Chapters 5 and 6 we have assumed that the stochastic process $\{(\mathbf{x}_{j,t}, y_{j,t})\}_t$ satisfies the conditions in Benaïm (1999). The definitions which are needed to understand these conditions are given here. The Robbins–Monro algorithm is also described, since some results in Chapter 5 are based on it. The following concepts are taken from Benaïm (1999). Let $\mathcal{A}$ be a nonempty subset of $\mathbb{R}^n$ with the Euclidean metric d.

> **Definition 210: Semi-Flow**
>
> A semi-flow is a continuous map $\Phi : \mathcal{A} \times \mathbb{R}_+ \longrightarrow \mathcal{A}$ such that $\Phi(a,0) := \Phi_0(a) = a$ and $\Phi_t(\Phi_{t'}(x)) = \Phi_{t+t'}(a)$, for every $t,t' \in \mathbb{R}_+, a \in \mathcal{A}$.

We extend Φ to be a subset of $\mathcal{A}$ by defining $\Phi_t(A_1) = \{\Phi_t(a), \ a \in A_1\}$ for $A_1 \subseteq \mathcal{A}$.

> **Definition 211: Invariant Subset**
>
> We say that a subset A_1 of $\mathcal{A}$ is invariant if $\Phi_t(A_1) = A_1$ for all $t \in \mathbb{R}_+$.

> **Definition 212: Attracting Set**
>
> We say that A_1 is an attracting subset if A_1 is nonempty and compact, and there exists some open neighborhood $\mathcal{O}_1$ of A_1 such that $\lim_{t \longrightarrow \infty} d(\Phi_t(x), A_1) = 0$ uniformly in $\mathcal{O}_1$. An attractor is an invariant attracting set. $t \in \mathbb{R}_+$.

Every attracting set contains an attractor. A globally attracting set is an attracting set such that $\lim_{t \longrightarrow \infty} d(\Phi_t(a), A_1) = 0$, $\forall a \in \mathcal{A}$.

> **Definition 213: Omega-Limit Set**
>
> The ω-limit set of a point a is the set:
>
> $$\omega(a) = \{a' \in \mathcal{A} \mid \Phi_{t_k}(a) \longrightarrow a', \text{ for some } t_k \longrightarrow +\infty.\}$$

> **Definition 214: Internally Chain-Transitive**
>
> For a nonempty invariant set A_1, a (δ,T)−pseudo-orbit from a to a' in A_1 is a finite series of partial trajectories, $\{\Phi_t(a_k), \ 0 \leq t \leq t_k\}, k \in \{1,2,\dots,l-1\}, t_k \geq T$ such that:
>
> - $a_k \in A_1, \forall k$;
> - $d(a_0, a) < \delta, d(\Phi_{t_k}(a_k), a_{k+1}) < \delta, \ \forall k \in \{1,2,\dots,l-1\}$ and $x_l = a'$.
>
> A nonempty compact, invariant set, A_1, is internally chain-transitive if, for every $a, a' \in_1$ and every $\delta > 0, T > 0$, there is a (δ,T)−pseudo-orbit in A_1 from a to a'.

> **Definition 215**
>
> A subset of a topological space $\mathcal{A}$ is precompact in $\mathcal{A}$ if its closure in $\mathcal{A}$ is compact.

Every internally chain-transitive set is connected. Moreover, a nonempty compact invariant set is internally chain-transitive if and only if the set has no proper attracting set. Internally chain-transitive sets provide a characterization of the possible long-run behavior of the system; Benaïm has shown that this particular concept of long-run behavior is useful when working with stochastic approximations. The stochastic approximation gives an algorithm for a discrete-time stochastic process, whose step size decreases with time, so that asymptotically the system converges to its deterministic continuous-time limit. Early work on stochastic approximation was done by Robbins and Monro (1951) and Kiefer and Wolfowitz (1952), and has since been applied and extended by a number of authors. For our most relevant purposes, the results are the following:

> **Theorem 216**
>
> Consider the discrete time process on a non-empty convex subset $\mathcal{A}$ of $\mathbb{R}^n$, defined by the recursion:
>
> $$q_{t+1} = q_t + \frac{1}{t+1}\left[f(q_t) + \epsilon_{t+1}^1 + \epsilon_t^2\right] \tag{A.2}$$
>
> and the corresponding continuous time semi-flow Φ induced by the system of ordinary differential equations (ODEs):
>
> $$\dot{x}(t) = f(x(t)) \tag{A.3}$$
>
> where:
>
> - $f: \mathcal{A} \longrightarrow \mathbb{R}^n$ is continuous and the ODE (A.3) is globally integrable;
> - the $\{\epsilon_t^j\}$, $j \in \{1,2\}$ are stochastic processes adapted to filtration $\mathcal{F}_t$, i.e., for each time slot $t \in \mathbb{N}$, ϵ_t^1 and ϵ_t^2 are random variables that are measurable with respect to $\mathcal{F}_t$, where $\mathcal{F}_t$ is the σ-algebra corresponding to the history of the system up to the end of period t;
> - $\mathbb{E}\left(\epsilon_{t+1}^1 | \mathcal{F}_t\right) = 0$ almost surely (a.s), and $\mathbb{E}\left(\|\epsilon_{t+1}^1\|^2\right) < \infty$, $\lim_{t \longrightarrow \infty} \epsilon_t^2 = 0$ almost surely;
> - the set $\{q_t, t \geq 0\}$ is precompact in $\mathcal{A}$ almost surely.
>
> Then, with probability 1, every ω-limit of the process $\{q_t\}_t$ lies in an internally chain-transitive set for Φ.

If $\mathcal{A}$ is compact and f is Lipschitz continuous, then ODE is globally integrable and $\{q_t\}_t$ is precompact.

Recall that rest point x^* of the ODE (A.2) (i.e., $f(x^*) = 0$) is linearly unstable if the Jacobian matrix of f at x^*, $Df(x^*)$, has some eigenvalue with a positive real part. Let $\mathbb{R}^m = E_+ \oplus E_-$, where $E+$ and E_- are the generalized eigenspaces of $Df(x^*)$ corresponding to eigenvalues with positive and nonpositive real parts, respectively. If x^* is a linearly unstable rest point, then E_+ has at least one dimension. We are now ready for the result of Brandiére and Duflo (1996).

Theorem 217: Brandiére and Duflo (1996)

Consider the stochastic process (A.2) on a nonempty open subset $\mathcal{A}$ of $\mathbb{R}^n$. Let x^* be a linearly unstable rest point of f and $\epsilon^1_{t,+}$ be the projection of ϵ^1_t on E_+ in the direction of E_-. Assume that:

- f is continuously differentiable and its derivative is Lipschitz continuous on a neighborhood of x^*;
- $\{\epsilon^j_t\}$, $j \in \{1,2\}$ are stochastic processes adapted to filtration $\mathcal{F}_t$, i.e., for each time slot $t \in \mathbb{N}$, ϵ^1_t and ϵ^2_t are random variables that are measurable with respect to $\mathcal{F}_t$, where $\mathcal{F}_t$ is the σ-algebra corresponding to the history of the system up to the end of period t;
- $\mathbb{E}\left(\epsilon^1_{t+1}|\mathcal{F}_t\right) = 0$ almost surely (a.s), and $\limsup_{t \longrightarrow \infty} \mathbb{E}\left(\|\epsilon^1_{t+1}\|^2\right) < \infty$, $\liminf_{t \longrightarrow \infty} \mathbb{E}\left(\|\epsilon^1_{t+1,+}\|^2 \mid \mathcal{F}_t\right) > 0$ almost surely;
- $\sum_{t \geq 1}\|\epsilon^2_t\|^2 < \infty$ almost surely;

Then, $\lim_{t \longrightarrow \infty} q_t = x^*$ with probability 0.

A.2.2 Appendix to Chapter 8

In Chapter 8, we mentioned several assumptions (called [H1]–[H5]) under which the reinforcement learning algorithms almost certainly converge. Here we provide some reviews of dynamical systems and then state the conditions which lead us to the desired stochastic approximation results.

Dynamical systems are mathematical formalizations that are used to capture the evolution of systems over time. In general, they consist of three components:

- *System state*, $\underline{x}(t)$, representing the parameters which characterize the system at time t;
- *State space*, $\mathcal{U}$, representing all possible states of the system;
- *State transition function* $\Phi : \underline{U} \times [0, +\infty) \to \underline{U}$, which is a flow defining the change in the system state from one moment to the other (i.e., $\Phi_\tau(\underline{x_0}) = \underline{x}$ translates the fact that the system is in state $\underline{x}$ at time $t = \tau$ starting from $\underline{x_0}$ at $t = 0$). If this function is differentiable, then it can be characterized by the solution of an autonomous ordinary differential equation (ODE):

$$\frac{d\underline{x}}{dt} = v(\underline{x}(t)), \tag{A.4}$$

where $v(\underline{x}(t))$ is the vector field describing the speed of evolution of the system state and which does not depend explicitly on time. The existence of a uniqueness solution of this ODE, for any initial state $\underline{x_0} \in \mathcal{U}$ is ensured if the vector field is Lipschitz continuous, i.e., there exists a constant $L > 0$ such that, for all $\underline{x}, \underline{y}$: $\|v(\underline{x}) - v(\underline{y})\| \leq L\|\underline{x} - \underline{y}\|$.

In this chapter, we will be interested in characterizing the asymptotic behavior of dynamical systems. To this end, we define the following notions: stationary states, stable states, and asymptotically stable states.

Definition 218

A system state $\tilde{x} \in \mathcal{U}$, such that $v(\tilde{x}) = 0$, is called a *stationary state* or *equilibrium state*. The equilibrium states correspond to fixed points of the flow: $\Phi_t(\tilde{x}) = \tilde{x}$.

However, the equilibrium regime of dynamical systems does not necessarily consist of isolated points, and can consist of a whole subspace of $\mathcal{U}$ (e.g., cycles, periodic orbits). The general concept is that of the invariant set which is defined below.

Definition 219

Let $\mathcal{W}$ be a sub-space of $\mathcal{U}$. $\mathcal{W}$ is called an invariant (respectively positive invariant) set, if, for all $t \in \mathbb{R}$ (respectively in $\mathbb{R}_+$), $\Phi_t(\mathcal{W}) \subseteq \mathcal{W}$. It is said to be "internally chain transitive" if, for any $x, y \in \mathcal{W}$ and any $\varepsilon > 0$, $T > 0$, there exists $n \geq 1$ and $x^{[0]} = x, x^{[1]}, \ldots, x^{[n]} = y$ in $\mathcal{W}$ such that the trajectory in (A.4), initiated at $x^{[m]}$, meets with the ε-neighbourhood of $x^{[m+1]}$ for $0 \leq m \leq n$ after a time $\geq T$. Furthermore, if $x = y$, the set is said to be "internally chain recurrent".

The stability issue of the equilibrium regime answers questions such as: If the system is moved away from an equilibrium state, will it return to this equilibrium? Can a small deviation, which slightly moves the system away from an equilibrium state, have important consequences and be amplified in time?

Definition 220

An equilibrium point $\tilde{x}$ is stable in the sense of Lyapunov if, for any $\varepsilon > 0$, there exists $\eta > 0$ such that, for all $y \in \mathcal{U}$ verifying $\|y - \tilde{x}\| \leq \eta$, then $\|\Phi_t(y) - \tilde{x}\| \leq \varepsilon$ for all $t > 0$.

Definition 221

An equilibrium point $\tilde{x}$, is asymptotically stable in the sense of Lyapunov, if it is stable in the sense of Lyapunov and, for any $y \in \mathcal{U}$ sufficiently close to $\tilde{x}$, $\lim_{t \to +\infty} \Phi_t(y) = \tilde{x}$.

Sufficient conditions that ensure the stability or asymptotic stability are given in the following theorem.

Theorem 222

If $\tilde{x}$ is an equilibrium point, and if there exists a differentiable function $V : \mathcal{U} \to \mathbb{R}_+$ with a continuous derivative such that: $V(\tilde{x}) = 0$, $V(y) > 0$ for all $y \neq \tilde{x}$ and $\frac{dV}{dt} \leq 0$ (i.e., V decreases along all trajectories), then, $\tilde{x}$ is stable in the sense of Lyapunov. If $\frac{dV}{dt} < 0$ for all $y \neq \tilde{x}$, then $\tilde{x}$ is asymptotically stable in the sense of Lyapunov. Furthermore, if $V(x)$ goes to infinity when x approaches infinity, then all trajectories tend to $\tilde{x}$. In this case, $\tilde{x}$ is called the globally asymptotically stable state.

For further details on dynamical systems and their associated asymptotic behavior, the reader is referred to Bonnans and Rouchon (2005), Scheinerman (1995) and Teschl (2010).

The stochastic approximation theory is used to study discrete-time stochastic processes that can be written as:

$$\underline{X}^{[n+1]} = \underline{X}^{[n]} + \gamma^{[n+1]} \left(f\left(\underline{X}^{[n]}\right) + \underline{Z}^{[n+1]} \right), \qquad (A.5)$$

where $\underline{X}^{[n]}$ is a vector in an Euclidean space which is updated based on a noisy observation, $f(\cdot)$ is a deterministic vector field, and $\underline{Z}^{[n+1]}$ is a random noise. The idea is to approximate $\underline{X}^{[n]}$ with a continuous-time interpolation process, e.g., the following piecewise linear function given by:

$$\underline{\widehat{X}}(t) = \underline{X}^{[n]} + \frac{t - \tau^{[n]}}{\tau^{[n+1]} - \tau^{[n]}} (\underline{X}^{[n+1]} - \underline{X}^{[n]}), \quad \text{if} \quad t \in [\tau^{[n]}, \tau^{[n+1]}), \qquad (A.6)$$

where $\tau^{[0]} = 0$ and $\tau^{[n]} = \sum_{m=0}^{n} \gamma^{[m]}$ (i.e., the parameter that covers the time axis).

In the asymptotic regime, i.e., $n \to +\infty$, and under certain conditions on the quantization step $\gamma^{[n]}$, on the vector field $f(\cdot)$ and on the noise process $\underline{Z}^{[n]}$, this interpolated process follows the solution of the deterministic ODE:

$$\frac{dx}{dt} = f(\underline{x}(t)). \qquad (A.7)$$

Therefore, using the stochastic approximation approach, the study of the iterative process $\underline{X}^{[n]}$ amounts to the study of the deterministic continuous-time ODE in (A.7). Notice that the discrete process described by (A.5) can be viewed as a noisy Euler scheme for numerically approximating the trajectory of the ODE in (A.7).

Two different approaches can be distinguished as a function of the step size: i) variable step-size ($\gamma^{[n]}$ changes at each iteration); ii) constant step-size ($\gamma^{[n]} = \varepsilon, \forall n$). For the variable step size case, the main convergence result is given in Benaïm (1999) and a clear proof is presented in Borkar (2008). The author of Benaïm (1999) proved the almost certain convergence (convergence with a probability of one) in the asymptotic regime ($n \to +\infty$) under the following conditions:

[H1] The learning steps satisfy:

$$\begin{cases} \gamma^{[n]} & \geq 0, \forall n \\ \lim\limits_{n \to +\infty} \gamma^{[n]} & = 0 \\ \lim\limits_{n \to +\infty} \sum\limits_{m=0}^{n} \gamma^{[m]} & = +\infty \\ \lim\limits_{n \to +\infty} \sum\limits_{m=0}^{n} \left(\gamma^{[m]}\right)^2 & < +\infty \end{cases} \qquad (A.8)$$

The parameters $\{\gamma^{[n]}\}$ correspond to the quantization steps and, thus, small values are desirable to suppress quantization errors. Too small values imply a long convergence time for the algorithm. In conclusion, relatively large values of $\gamma^{[n]}$ are desirable for the initial steps ($n = 1, 2, \ldots$) but as n grows large, $\gamma^{[n]}$ should become very small. Since a discrete process will be approximated by the continuous-time ODE, the discrete steps must cover the entire time axis, $\sum_{n\geq 0}\gamma^{[n]} = +\infty$. The errors introduced by the noise must also be asymptotically suppressed. The condition $\sum_{n\geq 0}\left(\gamma^{[n]}\right)^2 < +\infty$ is necessary to this purpose.

[H2] The vector field $f(\cdot)$ is Lipschitz continuous.

[H3] The discrete process remains bounded with a probability of one, i.e., $\sup_n \|\underline{X}^{[n]}\| < +\infty$.

[H4] The noise sequence $\{\underline{Z}^{[n]}\}$ is a Martingale difference sequence such that with probability one $\mathbb{E}\left[\underline{Z}^{[n+1]}\,\middle|\,\underline{Z}^{[m]}, m \leq n\right] = \underline{0}$.

[H5] The noise sequence $\{\underline{Z}^{[n]}\}$ is square integrable, i.e., $\mathbb{E}\left[\left\|\underline{Z}^{[n+1]}\right\|^2\,\middle|\,\underline{Z}^{[m]}, m \leq n\right] < +\infty$. This condition, together with $\sum_{n\geq 0}\left(\gamma^{[n]}\right)^2 < +\infty$, asymptotically suppresses the overall contribution of the noise. Thus, the discrete process converges asymptotically to the mean behavior given by the deterministic ODE (A.7).

Before stating the main results, we define $\underline{X}^s(t), s \leq t$ as the trajectory that starts at time $s > 0$ in the point $\widehat{\underline{X}}(s)$:

$$\begin{cases} \frac{d\underline{X}^s}{dt}(t) = f(\underline{X}^s(t)) \\ \underline{X}^s(s) = \widehat{\underline{X}}(s) \end{cases} \tag{A.9}$$

Also, $\underline{X}_s(t), s \geq t$ is the trajectory that ends at time $s > 0$ in the point $\widehat{\underline{X}}(s)$ defined as:

$$\begin{cases} \frac{d\underline{X}_s}{dt}(t) = f(\underline{X}_s(t)) \\ \underline{X}_s(s) = \widehat{\underline{X}}(s) \end{cases} \tag{A.10}$$

The following convergence results for stochastic approximation are due to Benaïm (1999).

Theorem 223

Assuming that the conditions [H1]–[H5] are met (Benaïm, 1999), we have that, for all $T > 0$:

$$\begin{cases} \Pr\left[\lim_{s\to+\infty} \sup_{t\in[s,s+T]} \|\widehat{\underline{X}}(t) - \underline{X}^s(t)\| = 0\right] = 1 \\ \Pr\left[\lim_{s\to+\infty} \sup_{t\in[s-T,s]} \|\widehat{\underline{X}}(t) - \underline{X}_s(t)\| = 0\right] = 1. \end{cases} \tag{A.11}$$

Furthermore, the discrete process $\{\underline{X}^{[n]}\}$ in (A.5) converges almost surely, when $n \to +\infty$, to a (possibly path dependent) compact connected internally chain transitive set of the ODE (A.7).

Intuitively this means that, in the asymptotic regime (i.e., $n \to +\infty$), the interpolation process converges almost certainly to the solution of the deterministic ODE in (A.7).

For the constant-step size case, $\gamma^{[n]} = \gamma$ for all n, only weak convergence results (convergence in distribution) can be proved in the asymptotic regime ($\gamma \to 0$ and $n \to +\infty$) (Kushner and Yin, 1997). The corresponding theorems can be found in Kushner and Yin (1997). Sufficient conditions for ensuring the weak convergence are less restrictive than the conditions for guaranteeing almost sure convergence. Thus, if the conditions [H2]–[H5] are satisfied, certain weak convergence is guaranteed. For a more detailed discussion, the reader is referred to the specialized books (Borkar, 2008; Kushner and Yin, 1997).

In conclusion, the asymptotic study of the iterative process $\underline{X}^{[n]}$ is reduced to the asymptotical study of the ordinal differential equation (A.7).

A.3 REVIEW OF INFORMATION-THEORETIC NOTIONS

Throughout the book, some information-theoretic notions and results concerning multiuser channels are used. Here we review some of them for the reader who is not familiar with communication theory.

A.3.1 General Notions

....

Definition 224: Entropy of a Discrete Random Variable

Let X be a random variable lying in $\mathcal{X} = \{x_1,\ldots,x_M\}$ with $M = |\mathcal{X}|$. The entropy of X is defined by:

$$H(X) = - \sum_{x \in \mathcal{X}} \Pr[X = x] \log \Pr[X = x]. \qquad (A.12)$$

....

Definition 225: Entropy of a Continuous Random Variable

Let X be a random variable whose probability density function is f_X. Its (differential) entropy is:

$$H_d(X) = - \int_{x \in \mathbb{R}} f_X(x) \log f_X(x) \mathrm{d}x. \qquad (A.13)$$

In order to avoid too much redundancy, we only provide the following two definitions for discrete random variables.

....

Definition 226: Conditional Entropy

The uncertainty on X given Y is defined by:

$$H(X|Y) = - \sum_{(x,y) \in \mathcal{X} \times \mathcal{Y}} \Pr[(X,Y) = (x,y)] \log \Pr[X = x | Y = y]. \qquad (A.14)$$

....

Definition 227: Mutual Information

The mutual information between X and Y is defined by:

$$I(X;Y) = \sum_{(x,y) \in \mathcal{X} \times \mathcal{Y}} \Pr[(X,Y) = (x,y)] \log \left[\frac{\Pr[(X,Y) = (x,y)]}{\Pr[X = x]\Pr[Y = y]} \right]. \qquad (A.15)$$

A.3.2 A Few Results Concerning Single-User Channels

The Shannon capacity of a discrete-input discrete-output memory-less single-user channel $(\mathcal{X}, \mathcal{Y}, p(y|x))$ is given by:

$$C = \max_{p(x) \in \Delta(\mathcal{X})} I(X;Y). \qquad (A.16)$$

The Shannon capacity of an additive white Gaussian channel $Y = X + Z$ with $\mathbb{E}|X|^2 \leq P$ and $Z \sim \mathcal{CN}(0, N)$ is given by:

$$C = \log_2\left(1 + \frac{P}{N}\right).$$ (A.17)

In the case of multi-dimensional channel inputs and outputs, $\underline{y} = \mathbf{H}\underline{x} + \underline{z}$, the Shannon capacity becomes:

$$C = \max_{\mathbf{Q}} \log_2\left|\mathbf{I}_r + \frac{1}{N}\mathbf{H}\mathbf{Q}\mathbf{H}^H\right|$$ (A.18)

where r is the dimension of the channel output, $\mathbf{Q} = E(\underline{x}\underline{x}^H)$, $\mathrm{Tr}(\mathbf{Q}) \leq P$, and $|\mathbf{M}| = \mathrm{Det}(\mathbf{M})$.

A.3.3 A Few Results Concerning Multi-Terminal Channels

The capacity region of a discrete-input, discrete-output memory-less multiple access channel $(\mathcal{X}_1, \mathcal{X}_2, \mathcal{Y}, p(y|x_1, x_2))$ is given by:

$$\begin{cases} R_1 & \leq I(X_1; Y|X_2) \\ R_2 & \leq I(X_2; Y|X_1) \\ R_1 + R_2 & \leq I(X_1, X_2; Y) \end{cases}$$ (A.19)

In the Gaussian case (always with complex quantities), this region becomes:

$$\begin{cases} R_1 & \leq \log_2\left(1 + \frac{P_1}{N}\right) \\ R_2 & \leq \log_2\left(1 + \frac{P_2}{N}\right) \\ R_1 + R_2 & \leq \log_2\left(1 + \frac{P_1 + P_2}{N}\right) \end{cases}$$ (A.20)

where $E|X_i|^2 \leq P_i$.

The capacity region for interference relay channels is still unknown. In this book, only achievable rate regions are considered.

References

Abramson, A., 1970. The Aloha system – another alternative for computer communications. AFIPS Conf. Proc. 36, 295–298.

Akyildiz, I., Lee, W., Vuran, M., Mohanty, S., 2006. Next generation/dynamic spectrum access/cognitive radio wireless networks: a survey. Comput. Networks, 50 (12), 2127–2159.

Akyildiz, I.F., Su, W., Sankarasubramaniam, Y., Cayirci, E., 2002. Wireless sensor networks: a survey. Comput. Networks, 38 (4), 393–422.

Al Daoud, A., Alpcan, T., Agarwal, S., Alanyali, M., 2008. A Stackelberg game for pricing uplink power in wide-band cognitive radio networks. In: IEEE Proceedings of the 47th Conference on Decision and Control, pp. 1422–1427.

Alpcan, T., Basar, T., Srikant, R., Altman, E., 2002. CDMA uplink power control as a noncooperative game. Springer Wireless Networks J. 8, 659–670.

Altman, E., Altman, Z., 2003. S-modular games and power control in wireless networks. IEEE Trans. Automat. Control 48 (5), 839–842.

Altman, E., Avrachenkov, K., Cottatellucci, L., Debbah, M., He, G., Suarez, A., 2009a. Equilibrium selection in multiuser communication. In: Proceedings of the 28th IEEE Conference on Computer Communications (INFOCOM).

Altman, E., Basar, T., Menache, I., Tembine, H., 2009b. A dynamic random access game with energy constraints. In: Proceedings of the 7th International Symposium on Modeling and Optimization in Mobile, Ad Hoc, and Wireless Networks (WiOpt'09, June 25–27).

Altman, E., Shwartz, A., 2000. Constrained Markov games: nash equilibria. Ann. Dyn. Games, 5, 213–221.

Altman, E., Silva, A., Bernhard, P., 2008. The mathematics of routing in massively dense ad hoc networks. In: International Conference on Ad-Hoc Networks and Wireless. Sophia Antipolis, France.

Altman, E., Solan, E., 2009. Games with constraints with network applications. IEEE Trans. Automat. Control 54, 2435–2440.

Arora, J.S., 2004. Introduction to Optimum Design, second ed. Academic Press.

Arrow, K.J., 1963. Social Choice and Individual Values. Yale University Press.

Arslan, G., Demirkol, M.F., Song, Y., 2007. Equilibrium efficiency improvement in MIMO interference systems: a decentralized stream control approach. IEEE Trans. Wireless Commun. 6, 2984–2993.

Arthur, W.B., 1993. On designing economic agents that behave like human agents. J. Evol. Econ. 3, 1–22.

Arthur, W.B., 1994. Inductive reasoning and bounded rationality. Am. Econ. Rev. 84.

Athan, T.W., Papalambros, P.Y., 1996. A note on weighted criteria methods for compromise solutions in multi-objective optimization. Eng. Opt. 27, 155–176.

Aubin, J.-P., 2002. Viability Kernels and Capture Basins. Lecture Notes, Universidad Politécnica de Cartagena.

Aumann, R.J., 1959. Chapter Acceptable points in general cooperative n-person games. In: Contributions to the Theory of Games IV. Princeton Univ. Press, Princeton, NJ, pp. 287–324.

Aumann, R.J., 1960. Acceptable points in games of perfect information. Pac. J. Math. 10, 381–417.

Aumann, R.J., 1961. Mixed and behavior strategies in infinite extensive games. In: Econometric Research Program, n032; appeared also in 1964 in Advances in Game Theory.

Aumann, R.J., 1974. Subjectivity and correlation in randomized strategies. J. Math. Econ. 1, 67–96.

Aumann, R.J., 1976. Agreeing to disagree. Ann. Stat. 4, 1236–1239.

Aumann, R.J., 1981. Survey of repeated games. In: Wissenschaftsverlag, D. (Ed.), Essays in Game Theory and Mathematical Economics in Honor of Oskar Morgenstern. Bibliographisches Institut, Mannheim, Wien, Zurich, pp. 11–42.

Aumann, R.J., 1985. Game theory in the Talmud. Technical report, Research Bulletin Series on Jewish Law and Economics.

Aumann, R.J., 1997. Rationality and bounded rationality. Games Econ. Behav. 21, 2–14.

Aumann, R.J., Maschler, M., 1966. Game theoretic aspects of gradual disarmament. Technical report. In: Report of the U.S. Arms Control and Disarmament Agency. Mathematica Inc., Princeton, NJ, ST-80, pp. 1–55 (Chapter V).

Aumann, R.J., Maschler, M., 1985. Game theoretic analysis of a bankruptcy problem from the Talmud. J. Econ. Theory, 36, 195–213.

Aumann, R.J., Peleg, B., 1960. Von Neumann-Morgenstern solutions to cooperative games without side payments. Bull. Am. Math. Soc. 66 (3), 173–179.

Aumann, R.J., Shapley, L., 1976. Long term competition. In: A Game Theoretic Analysis. Mimeo, Hebrew University.

Bardi, M., 2009. On differential games with long-time-average cost. In: Pourtallier, O., Gaitsgory, V., Bernhard, P. (Eds.), Advances in Dynamic Games and Their Applications. Annals of the International Society of Dynamic Games, vol. 10. Birkhäuser, Boston, pp. 1–16.

Basar, T., 1974. A counterexample in linear-quadratic games: existence of nonlinear Nash strategies. J. Optim. Theory Appl. 14, 425–430.

Basar, T., 1975. Nash strategies for m-person differential games with mixed information structures. Automatica. 11, 547–551.

Basar, T., 1977. Informationally non-unique equilibrium solutions in differential games. SIAM J. Control Optim. 15, 636–660.

Basar, T., Olsder, G.J., 1982. Dynamic noncooperative game theory. In: Classics in Applied Mathematics, first ed. SIAM, Philadelphia.

Basar, T., Srikant, R., 2002. Revenue-maximizing pricing and capacity expansion in a many-users regime. In: Proceedings of the 21st IEEE Conference on Computer Communications (INFOCOM). NYC.

Basu, K., Weibull, J., 1991. Strategy subsets closed under rational behavior. Econ. Lett. 36, 140–146.

Beckmann, M., McGuire, C.B., Winston, C.B., 1956. Studies in the economics of transportation. Yale University Press.

Bellman, R., 1952. On the theory of dynamic programming. Proc. Natl. Acad. Sci. USA. 38, 716–719.

Belmega, E.V., Djeumou, B., Lasaulce, S., 2009a. Resource allocation games in interference relay channels. In: IEEE International Conference on Game Theory for Networks (GAMENETS). Istanbul, Turkey, pp. 575–584.

Belmega, E.V., Djeumou, B., Lasaulce, S., 2009b. What happens when cognitive terminals compete for a relay node? In: IEEE International Conference on Acoustics, Speech and Signal Processing (ICASSP). Taipei, Taiwan, pp. 2609–2612.

Belmega, E.V., Djeumou, B., Lasaulce, S., 2011. Power allocation games in interference relay channels: existence analysis of Nash equilibria. EURASIP J. Wireless Commun. Netw. (JWCN).

Belmega, E.V., Jungers, M., Lasaulce, S., 2011. A generalization of a trace inequality for positive definite matrices. Aust. J. Math. Anal. Appl. (AJMAA), 7 (2), 1–5.

Belmega, E.V., Lasaulce, S., 2009. An information-theoretic look at MIMO energy-efficient communications. In: ACM Proceedings of the International Conference on Performance Evaluation Methodologies and Tools (VALUETOOLS). Pisa, Italy.

Belmega, E.V., Lasaulce, S., Debbah, M., 2008a. Decentralized handovers in cellular networks with cognitive terminals. In: IEEE Proceedings of the 3rd International Symposium on Communications, Control and Signal Processing (ISCCSP). St Julians, Malta.

Belmega, E.V., Lasaulce, S., Debbah, M., 2008b. Power control in distributed multiple access channels with coordination. In: IEEE/ACM Proceedings of the International Symposium on Modeling and Optimization in Mobile, Ad Hoc, and Wireless Networks and Workshops (WIOPT). Berlin, Germany.

Belmega, E.V., Lasaulce, S., Debbah, M., 2009a. A trace inequality for Hermitian positive matrices. J. Inequalities Pure Appl. Math. (JIPAM), 10 (1), 1–4.

Belmega, E.V., Lasaulce, S., Debbah, M., 2009b. Power allocation games for MIMO multiple access channels with coordination. IEEE Trans. Wireless Commun. 8 (6), 3182–3192.

Belmega, E.V., Lasaulce, S., Debbah, M., Hjørungnes, A., 2010a. Learning distributed power allocation policies in MIMO channels. In: European Signal Processing Conference (EUSIPCO). Aalborg, Denmark.

Belmega, E.V., Lasaulce, S., Debbah, M., Jungers, M., Dumont, J., 2010b. Power allocation games in wireless networks of multi-antenna terminals. Springer Telecommun. Syst. J. doi:10.1007/s11235-010-9305-3.

Benaïm, M., 1999. Dynamics of stochastic approximation algorithms. Séminaire de probabilités (Strasbourg), 3, 1–68.

Benaïm, M., Faure, M., 2010. Stochastic approximations, cooperative dynamics and supermodular games. Preprint available at http://members.unine.ch/michel.benaim/perso/papers1.html.

Benaim, M., Hofbauer, J., Sorin, S., 2006. Stochastic approximations and differential inclusions. Part II. Appl. Math. Oper. Res. 31 (4), 673–695.

Benoît, J.-P., Krishna, V., 1985. Finitely repeated games. Econometrica, 53 (4), 905–22.

Benoît, J.-P., Krishna, V., 1987. Nash equilibria of finitely repeated games. Int. J. Game Theory 16 (3), 197–204.

Berkowitz, L.D., 1964. A variational approach to differential games. In: Advances in Game Theory., Ann. Math. Stud., p. 127.

Bertrand, J., 1883. Théorie mathématique de la richesse sociale. Journal des savants.

Bertsekas, D.P., 1995. Nonlinear Programming. Athena Scientific.

Bianchi, G., 2000. Performance analysis of the IEEE 802.11 distributed coordination function. IEEE J. Sel. Areas Telecommun. Wireless Ser. 18, 535–547.

Biglieri, E., Taricco, G., Tulino, A., 2002. How far is infinity? Using asymptotic analyses in multiple-antennas systems. In: Proceedings of the International Symposium on Software Testing and Analysis (ISSTA), vol. 1. Rome, Italy, pp. 1–6.

Blackwell, D., 1968. Discounted dynamic programming. Ann. Math. Stat. 39, 226–235.

Boche, H., Schubert, M., Vucic, N., Naik, S., 2007. Non-symmetric Nash bargaining solution for resource allocation in wireless networks and connection to interference calculus. In: Proceedings of the 15th European Signal Processing Conference (EUSIPCO). Poznan, Poland.

Bonnans, F., Rouchon, P., 2005. Commande et optimisation de systèmes dynamiques. Les Éditions de L'École Polytechnique.

Bonneau, N., Altman, E., Debbah, M., 2006. Correlated equilibrium in access control for wireless communications. In: Networking. Coimbra, Portugal, pp. 173–183.

Bonneau, N., Debbah, M., Altman, E., Hjørungnes, A., 2008. Non-atomic games for multi-user system. IEEE J. Sel. Areas Commun. 26 (7), 1047–1058.

Border, K.C., 1985. Fixed Point Theorems with Applications to Economics and Game Theory. Cambridge University Press.

Borel, E., 1921. La théorie du jeu et les équations intégrales á noyau symétrique gauche. Comptes Rendus Hebdomadaires de l'Académie des Sciences, vol. 173, pp. 1304–1308 (in French). [English translation by Savage, L., 1953. The theory of play and integral equations with skew symmetric kernels. Econometrica. 21, 97–100.]

Borgers, T., Sarin, R., 1993. Learning Through Reinforcement and Replicator Dynamics. Mimeo, University College London.

Borkar, V.S., 1997. Stochastic approximation with two timescales. Syst. Control Lett. 29, 291–294.

Borkar, V.S., 2008. Stochastic Approximation: A Dynamical System Viewpoint. Cambridge University Press.

Bournez, O., Cohen, J., 2009. Learning equilibria in games by stochastic distributed algorithms. Available at http://arxiv.org/abs/0907.1916.

Bowley, A.L., 1924. The Mathematical Groundwork of Economics: An Introductory Treatise. Clarendon Press.

Braess, D., 1969. Über ein paradoxon aus der verkehrsplanung. Unternehmens Forsch. 24 (5), 258–268.

Brandiére, O., Duflo, M., 1996. Les algorithmes stochastiques contournent-ils les piéges? Ann. de Inst. Henri Poincaré (B) Probabiliés et Statistiques, 32, 395–427.

Brouwer, L.E.J., 1912. Ueber abbildung von mannigfaltigkeiten. Math. Ann. 71, 97–115.

Brown, G.W., 1949. Some notes on computation of games solutions. Report P-78, The Rand Corporation.

Brown, G.W., 1951. Iterative solutions of games by fictitious play, in activity analysis of production and allocation. In: Koopmans, T.C. (Ed.), Wiley, New York, 13 (1), 374–376.

Brown, G., von Neumann, J., 1950. Solutions of games by differential equations. In: Kuhn, H., Tucker, A. (Eds.), Contributions to the Theory of Games. Ann. Math. Stud., vol. 24. Princeton University Press, Princeton.

Bush, R., Mosteller, F., 1955. Stochastic Models of Learning. Wiley, New York.

Cardaliaguet, P., 2007. Differential games with asymmetric information. SIAM J. Control Optim. (SICON) 46, 816–838.

Carleial, A.B., 1975. A case where interference does not reduce capacity. IEEE Trans. Inf. Theory 21 (5), 569–570.

Carleial, A.B., 1978. Interference channels. IEEE Trans. Inf. Theory 24 (1), 60–70.

Case, J.H., 1967. Equilibrium points of N-person differential games, PhD thesis, University of Michigan, Ann Arbor, MI. Department of Industrial Engineering, Tech. report no. 1967-1.

Chen, L., Cui, T., Low, S.H., Doyle, J.C., 2008. A game-theoretic model for medium access control. IEEE J. Sel. Areas Commun. 26 (7).

Chung, S.T., Kim, S.J., Lee, J., Cioffi, J.M., 2003. A game theoretic approach to power allocation in frequency-selective Gaussian interference channels. In: Proceedings IEEE International Symposium on Information Theory (ISIT'03). Pacifico Yokohama, Kanagawa, Japan, p. 316.

Cominetti, R., Melo, E., Sorin, S., 2007. A simple model of learning and adaptive dynamics in traffic network games. Working paper.

Cooper, R., 1998. Coordination Games. Cambridge University Press.

Correa, J.R., Schulz, A.S., Stier-Moses, N.E., 2004. Selfish routing in capacitated networks. Math. Oper. Res. 29 (4): 961–976.

Cournot, A., 1838. Recherches Sur Les Principes Mathématiques de la la Théorie des Richesses. (Re-edited by Mac Millan in 1987).

Cover, T.M., 1975. Advances in Communications Systems. Some Advances in Broadcast Channels, pp. 229–260.

Cover, T.M., Thomas, J.A., 2006. Elements of Information Theory, second ed. Wiley-Interscience, New York.

Damme, E.V., 2002. Strategic Equilibrium, vol. 3. Elsevier Science Publishers.

Darwin, D., 1871. The Descent of Man, and Selection in Relation to Sex, first ed. John Murray, London.

DaSilva, L.A., 2000. Pricing for QoS-enabled networks: a survey. IEEE Commun. Surv. 3 (2), 2–8.

Debreu, G., 1952. A social equilibrium existence theorem. Proc. Natl. Acad. Sci. USA. 38 (10), 886–893.

Debreu, G., 1954. Decision Process, Representation of a Preference Ordering by a Numerical Function. John Wiley, New York, pp. 159–166.

Diaconis, P., Saloff-Coste, L., 1996. Logarithmic Sobolev inequalities for finite Markov chains. Ann. Appl. Probab. 6 (3), 695–750.

Dixit, A., Nalebuff, B., 1993. Thinking Strategically: The Competitive Edge in Business, Politics, and Everyday Life. Paperback, pp. 228–231 (Chapter 9).

Djeumou, B., Belmega, E.V., Lasaulce, S., 2009. Interference relay channels – part i: transmission rates. arXiv:0904.2585.

Driesen, B., 2010. Loss and Risk Aversion in Games and Decisions, PhD thesis. University of Maastricht.

Dumont, J., Hachem, W., Lasaulce, S., Loubaton, P., Najim, J., 2010. On the capacity achieving covariance matrix of rician MIMO channels: an asymptotic approach. IEEE Trans. Info. Theory 56 (3), 1048–1069.

Dumont, J., Loubaton, P., Lasaulce, S., 2006. On the capacity achieving transmit covariance matrices of MIMO correlated rician channels: a large system approach. In: IEEE Proc. of Globecom Technical Conference.

Dutta, P.K., 1991. A folk theorem for stochastic games. University of Rochester – Center for Economic Research (RCER).

Dutta, P.K., 1995. A folk theorem for stochastic games. J. Econ. Theory, 66 (1), 1–32.

Edgeworth, F.Y., 1881. An Essay on the Application of Mathematics to the Moral Sciences, vol. 150. Kegan Paul & Co., p. viii.

ElBatt, T., Ephremides, A., 2004. Joint scheduling and power control for wireless ad hoc networks. IEEE Trans. Wireless Commun. 3 (1), 74–85.

Engwerda, J.C., 2005. LQ Dynamic Optimization and Differential Games. Wiley.

Etkin, R., Parekh, A., Tse, D., 2007. Spectrum sharing for unlicensed bands. IEEE J. Sel. Areas Commun. Special issue on adaptive, Spectrum Agile and Cognitive Wireless Networks, 25 (3), 517–528.

Fan, K., 1952. Fixed point and minima theorems in locally convex topological linear spaces. In: Proceedings of the National Academy of Sciences, vol. 38, pp. 121–126.

FCC, 2002. Report of the spectrum efficiency working group. Technical report, Federal Commun. Commission.

Feinberg, E., Schwartz, A., 2002. Handbook of Markov decision processes: methods and applications. In: International Series of Operations Research and Management Science, vol. 40. Kluwer Academic Publishers.

Felegyhazi, M., Hubaux, J.-P., 2006. Game theory in wireless networks: a tutorial. Technical report, EPFL.

Fette, B.A., 2006. Cognitive Radio Technology. Elsevier.

Filar, J., Vrieze, K., 1997. Competitive Markov Decision Processes. Springer.

Fink, A.M., 1964. Equilibrium in a stochastic n-person game. J. Sci. Hiroshima Univ. 28, 89–93.

Fishburn, P., 1970. Utility Theory for Decision Making. John Wiley and Sons, New York. Available online [approved for public release].

Fleming, W.H., 1969. Optimal continuous-parameter stochastic control. SIAM Review 11, 470–509. MR41#9633.

Flesch, J., Schoenmakers, G., Vrieze, K., 2008. Stochastic games on a product state space. Math. Oper. Res. 33, 403–420.

Foschini, G.J., Milijanic, Z., 1993. A simple distributed autonomous power control algorithm and its convergence. IEEE Trans. Veh. Technol. 42 (4), 641–646.

Foster, D., Vohra, R.V., 1997. Calibrated learning and correlated equilibrium. Games Econ. Behav. 21, 40–55.

Friedman, J.W., 1971. A non-cooperative equilibrium for supergames. Rev. Econ. Stud. 38 (1), 1–12.

Fudenberg, D., Kreps, D., 1988. A Theory of Learning, Experimentation, and Equilibrium in Games. Mimeo, Working paper.

Fudenberg, D., Levine, D.K., 1998. The Theory of Learning in Games. The MIT Press.

Fudenberg, D., Takahashi, S., 2008. Heterogeneous Beliefs and Local Information in Stochastic Fictitious Play. Levines Working Paper Archive.

Fudenberg, D., Tirole, J., 1991. Game Theory. The MIT Press.

Fudenberg, D., Yamamoto, Y., 2009. The folk theorem for irreducible stochastic games with imperfect public monitoring. 1–18.

Gaitsgory, V., Nitzan, S., 1994. A folk theorem for dynamic games. J. Math. Econ. 23 (2), 167–178.

Geanakoplos, J., 1994. Common knowledge. In: Aumann, R.J., Hart, S. (Eds.), Handbook of Game Theory with Economic Applications, vol. 2. Elsevier.

Geanakoplos, J., Pearce, D., Stacchetti, E., 1989. Psychological games and sequential rationality. Games Econ. Behav. 1 (1), 60–79.

Gilboa, I., Matsui, A., 1991. Social stability and equilibrium. Econometrica, 59, 859–867.

Gillette, D., 1957. Stochastic games with zero stop probabilities. In: Dresher, M., et al. (Eds.), Contributions to the Theory of Games. Princeton University Press, Princeton, NJ, pp. 179–188.

Glicksberg, I.L., 1952. A further generalization of the kakutani fixed point theorem with application to Nash equilibrium points. In: Proceedings of the American Mathematical Society, vol. 3.

Goodman, D.J., Mandayam, N.B., 2000. Power control for wireless data. IEEE Pers. Commun. 7, 48–54.

Gossner, O., Tomala, T., 2009. Repeated games with complete information. In: Meyers, R. (Ed.), Encyclopedia of Complexity and Systems Science. Springer, New York.

Gupta, A., Jukic, B., Parameswaran, M., Stahl, D.O., Whinston, A.B., 1997. Streamlining the digital economy: how to avert a tragedy of the commons. IEEE Internet Comput. 1 (6), 38–46.

Gupta, P., Kumar, P.R., 1997. A system and traffic dependent adaptive routing algorithm for ad hoc networks. In: IEEE Conference on Decision and Control (CDC). San Diego, pp. 2375–2380.

Halpern, J., 2008. Beyond Nash equilibrium: solution concepts for the 21st century. In: ACM Proceedings of PODC'08. Toronto, Canada.

Hamilton, W.D., 1967. Extraordinary sex ratios. Science, 156 (774), 477–488.

Hamilton, J., Slutzky, S., 1990. Endogenous timing in duopoly games: stackelberg or cournot equilibria. Games Econ. Behav. 2 (1), 29–46.

Harsanyi, J.C., 1967. Games with incomplete information played by "Bayesian" players, i–iii. Part i. the basic model. Manage. Sci. 14 (3), 159–182.

Harsanyi, J.C., Selten, R., 2003. A General Theory of Equilibrium Selection in Games. MIT Press.

Hart, S., 1992. Games in Extensive and Strategic Forms, vol. 1. Elsevier Science Publishers.

Hart, S., Mas-Colell, A., 2000. A simple adaptive procedure leading to correlated equilibrium. Econometrica, 68, 1127–1150.

Hart, S., Mas-Colell, A., 2003. Uncoupled dynamics do not lead to Nash equilibrium. Am. Econ. Rev. 93: 1830–1836.

Haurie, A., Krawczyk, J.B., 2002. An Introduction to Dynamic Games. Internet Textbook.

Haykin, S., 2005. Cognitive radio: brain-empowered wireless communications. IEEE J. Sel. Areas Commun. 23, 201–220.

He, G., Debbah, M., Altman, E., 2010a. A Bayesian game-theoretic approach for distributed resource allocation in fading multiple access channels. EURASIP J. Wireless Commun. Netw. Article ID 391684, doi:10.1155/2010/391684.

He, G., Lasaulce, S., Hayel, Y., Debbah, M., 2010b. A multi-level hierarchical game-theoretical approach for cognitive networks. In: IEEE Proceedings of Crowncom.

Hillas, J., Kohlberg, E., 2002. Foundation of Strategic Equilibrium, vol. 3. Elsevier Science Publishers.

Hörner, J., Sugaya, T., Takahashi, S., Vieille, N., 2009. Recursive Methods in Discounted Stochasitc Games: An Algorithm for $\delta \rightarrow 0$ and a Folk Theorem. Mimeo, pp. 1–40.

Hofbauer, J., Sigmund, K., 1998. Evolutionary Games and Population Dynamics. Cambridge University Press.

Hofbauer, J., Sigmund, K., 2003. Evolutionary game dynamics. Bull. Am. Math. Soc. 40, 479–519.

Hofbauer, J., Sorin, S., Viossat, Y., 2009. Time average replicator and best-rely dynamics. Math. Oper. Res. 34 (2), 263–269.

Hopf, H., 1929. Ueber die Algebraische Anzahl von Fixpunkten. Math. Zeitung, 29.

Hopkins, E., Posch, M., 2005. Attainability of boundary points under reinforcement learning. Games Econ. Behav. 53, 110–125.

Inaltekin, H., Wicker, S.B., 2006. The analysis of a game theoretic MAC protocol for wireless networks. In: Proceedings IEEE SECON'06. Reston, VA.

Inaltekin, H., Wicker, S.B., 2008. The analysis of Nash equilibria of the one-shot random-access game for wireless networks and the behavior of selfish nodes. IEEE/ACM Trans. Netw. 16 (5), 1094–1107.

Isaacs, R., 1965. Differential Games. Wiley, New York.

Izquierdo, L.R., Izquierdo, S.S., Gotts, N.M., Polhill, J.G., 2007. Transient and asymptotic dynamics of reinforcement learning in games. Games Econ. Behav. 61, 259–276.

Jacquet, P., 2004. Geometry of information propagation in massively dense ad hoc networks. In: Proceedings of the ACM International Symposium on Mobile Ad Hoc Networking and Computing (Mobihoc). New York, pp. 157–162.

Jiang, J.-H., 1963. Essential components of the set of fixed points in the multivalued mappings and its application to the theory of games. Sci. Sin. 12, 951–964.

Kakutani, S., 1941. A generalization of Brouwer's fixed point theorem. Duke Math. J. 8 (3), 457–459.

Kaneko, M., Kline, J., 1995. Behavior strategies, mixed strategies and perfect recall. Int. J. Game Theory, 4 (2), 127–145.

Keshav, S., 1997. An Engineering Approach to Computer Networking. Addison-Wesley, Reading, MA, pp. 215–217.

Kiefer, J., Wolfowitz, J., 1952. Stochastic estimation of the maximum of a regression function. Ann. Math. Stat. 23 (3), 462–466.

Kleinrock, L., Tobagi, F.A., 1975. Packet switching in radio channels: Part II the hidden terminal problem in carrier sense multiple access and busy tone solution. IEEE Trans. Commun. 23, 1417–1433.

Knoedel, W., 1969. Graphentheoretische Methoden und ihre Anwendungen. Springer-Verlag.

Kohlberg, E., Mertens, J.-F., 1986. On the strategic stability of equilibria. Econometrica, 54 (5), 1003–1037.

Kolata, G., 1990. What if They Closed 42nd Street and Nobody Noticed? New York Times.

Koutsoupias, E., Papadimitriou, C., 1999. Worst-case equilibria. In: 16th Annual Symposium on Theoretical Aspects of Computer Science.

Kuhn, H., 1950. Extensive games. In: Proceedings of the National Academy of Sciences, vol. 38, pp. 570–576.

Kuhn, H., 1953. Contributions to the Theory of Games, Extensive Games and the Problem of Information. Princeton University Press, pp. 193–216.

Kumar, S., Raghavan, S.V., Deng, J., 2006. Medium access control protocols for ad hoc wireless networks: a survey. Ad Hoc Netw. (Elsevier Journal), 4, 326–358.

Kushner, H.J., Clark, D.S., 1978. Stochastic Approximation Methods for Constrained and Unconstrained Systems. Springer, New York.

Kushner, H.J., Yin, G.G., 1997. Stochastic Approximation Algorithms and Applications. Springer-Verlag, New York.

Lai, L., El Gamal, H., 2008. The water-filling game in fading multiple-access channels. IEEE Trans. Inf. Theory 54 (5), 2110–2122.

Laraki, R., Zamir, S., 2003. Théorie des jeux et analysis économique. Lecture notes. École Polytechnique.

Larsson, E.G., Jorswieck, E., 2008. Competition versus cooperation on the MISO interference channel. IEEE J. Sel. Areas Commun. Special issue on Game Theory Commun. syst. 26 (7), 1059–1069.

Lasaulce, S., Debbah, M., Altman, E., 2009a. Methodologies for analyzing equilibria in wireless games. IEEE Signal Process. Mag. Special issue on Game Theory for Signal Processing, 26 (5), 41–52.

Lasaulce, S., Hayel, Y., El Azouzi, R., Debbah, M., 2009b. Introducing hierarchy in energy games. IEEE Trans. Wireless Commun. 8 (7), 3833–3843.

Lasaulce, S., Suarez, A., Debbah, M., Cottatellucci, L., 2007. Power allocation game for fading MIMO multiple access channels with antenna correlation. In: ICST/ACM proceedings of the International Conference on Game Theory in Communication Networks (Gamecomm).

Lefschetz, S., 1929. Intersections and transformations of complexes and manifolds. Trans. Am. Math. Soc. 28, 1–49.

Lehrer, E., 1989. Lower equilibrium payoffs in two-player repeated games with non-observable actions. Int. J. Game Theory, 18 (1), 57–89.

Lehrer, E., 1990. Nash equilibrium of n-players repeated games with semi-standard information. Int. J. Game Theory, 19, 191–217.

Lehrer, E., 1992. Two-player repeated games with non-observable actions and observable payoffs. Math. Oper. Res. 17, 200–224.

Leitmann, G., Mon, G., 1967. On a class of differential games. In: Coll. Adv. Prob. Meth. Space Fit. Optim. University of Liege.

Levine, D.K., 1998. Modeling altruism and spitefulness in experiments. Rev. Econ. Dyn. 1 (3), 593–622.

Li, W., Chao, X., 2004. Modeling and performance evaluation of a cellular mobile network. IEEE/ACM Trans. Netw. 12 (1), 131–145.

Li, W., Fang, Y., Henry, R.R., 2002. Actual call connection time characterization for wireless mobile networks under a general channel allocation scheme. IEEE Trans. Wireless Commun. 1 (4), 682–691.

Ljung, L., 1977. Analysis of recursive stochastic algorithms. IEEE Trans. Automat. Control 22 (4), 551–575.

Long, C., Qian, Z., Bo, L., Huilong, Y., Xinping, G., 2007. Non-cooperative power control for wireless ad hoc networks with repeated games. IEEE JSAC 25 (6), 1101–1112.

Luce, R.D., Raiffa, H., 1957. Games and Decisions: Introduction and Critical Survey. Wiley.

Luo, Z.Q., Pang, J.S., 2006. Analysis of iterative waterfilling algorithm for multiuser power control in digital subscriber lines. EURASIP J. Appl. Signal Process. pp. 1–10.

Maillé, P., Stier-Moses, N.E., 2009. Eliciting coordination with rebates. Trans. Sci. 43 (4), 473–492.

Marden, J.R., Young, H.P., Arslan, G., Shamma, J.S., 2009. Payoff-based dynamics for multi-player weakly acyclic games. SIAM J. Control Opt. 48 (1), 373–396.

Maric, I., Dabora, R., Goldsmith, A., 2008. On the capacity of the interference channel with a relay. In: IEEE International Symposium on Information Theory. Toronto, Canada, pp. 554–558.

Maskin, E., Fudenberg, D., 1986. The folk theorem in repeated games with discounting or with incomplete information. Econometrica, 54 (3), 533–554.

Mazumdar, R., Mason, L., Douligeris, C., 1991. Fairness in network optimal flow control: optimality of product forms. IEEE Trans. Commun. 39 (5), 775–782.

Mériaux, F., Lasaulce, S., Kieffer, M., 2011a. More about base station location games. In: ACM Proceedings of the 5th International Conference on Performance Evaluation Methodologies and Tools (VALUETOOLS). Cachan (Paris).

Mériaux, F., Le Treust, M., Lasaulce, S., Kieffer, M., 2011b. Energy-efficient power control strategies for stochastic games. In: IEEE Proceedings of the 17th International Conference on Digital Signal Processing (DSP). Corfu, Greece.

Mertens, J.-F., Parthasarathy, T., 1991. Non zero sum stochastic games. In: Raghavan, T.E.S. et al. (Eds.), Stochastic Games and Related Topics. Kluwer Academic Publishers, pp. 145–148.

Mertens, J.-F., Sorin, S., Zamir, S., 1994. Repeated Games. Univ. Catholique de Louvain, Center for Operations Research & Econometrics.

Mertens, J.-F., Zamir, S., 1985. Foundation of Bayesian analysis for games with incomplete information. Int. J. Game Theory, 14, 1–29.

Mertikopoulos, P., Belmega, E.V., Moustakas, A., Lasaulce, S., 2011. Dynamic power allocation games in parallel multiple access channels. In: ACM Proceedings of the 5th International Conference on Performance Evaluation Methodologies and Tools (VALUETOOLS). Cachan (Paris), France.

Meshkati, F., Chiang, M., Poor, H.V., Schwartz, S.C., 2006. A game-theoretic approach to energy-efficient power control in multi-carrier CDMA systems. IEEE J. Sel. Areas Commun. 24 (6), 1115–1129.

Milchtaich, I., 1996a. Congestion games with player-specific payoff functions. Games Econ. Behav. 13 (1), 111–124.

Milchtaich, I., 1996b. Congestion models of competition. Am. Nat. 147 (5), 760–783.

Milchtaich, I., 1998. Crowding games are sequentially solvable. Int. J. Game Theory (IJGT), 27 (4), 501–509.

Milgrom, P., Roberts, J., 1990. Rationalizability, learning, and equilibrium in games with strategic complementarities. Econometrica, 58, 1255–1277.

Miyasawa, K., 1961. On the convergence of learning processes in a 2x2 non-zero-person game. Princeton University, Research Memorandum No. 33.

Mo, J., Walrand, J., 1998. Fair end-to-end window-based congestion control. In: SPIE International Symposium on Voice, Video and Data Communications.

Mochaourab, R., Jorswieck, E., 2009. Resource allocation in protected and shared bands: uniqueness and efficiency of Nash equilibria. In: Fourth International Conference on Performance Evaluation Methodologies and Tools (VALUETOOLS). Pisa, Italy.

Monderer, D., 1996. Potential games. Games Econ. Behav. 14, 124–143.

Monderer, D., 2007. Multipotential games. In: Proceedings of the 20th International Joint Conference on Artificial Intelligence (IJCAI), pp. 1422–1427.

Monderer, D., Shapley, L.S., 1996. Potential games. Econometrica, 14, 124–143.

Mosteller, F., Bush, R., 1955. Stochastic Models for Learning, Wiley.

Myerson, R., 1978. Refinements of the Nash equilibrium concept. Int. J. Game Theory, 7 (2), 73–80.

Myerson, R.B., 2008. Mechanism Design. The New Palgrave Dictionary of Economics Online.

Nagel, R., 1995. Unraveling in guessing games: an experimental study. Am. Econ. Rev. 85 (5), 1331–1326.

Nash, J.F., 1950. Equilibrium points in n-points games. Proc. Natl. Acad. Sci. USA 36 (1), 48–49.

Nash, J., 1951. Non-cooperative games. Ann. Math. 54 (2), 286–295.

Neel, J., Reed, J., Gilles, R., 2002. The role of game theory in the analysis of software radio networks. In: SDR Forum Technical Conference, November 2002.

Neyman, A., 1997. Correlated equilibrium and potential games. Int. J. Game Theory 26, 223–227.

Neyman, A., Sorin, S., 2010. Repeated games with public uncertain duration process. Int. J. Game Theory 39, 29–52.

North, D.C., 1981. Structure and Change in Economic History. Norton, New York.

Olama, M., Djouadi, S., Charalambous, C., 2006. Stochastic power control for time-varying long-term fading wireless networks. EURASIP J. Appl. Signal Process. (JASP) 2006, 1–13.

Osborne, M., Rubinstein, A., 1994. A Course in Game Theory. The MIT Press.

Ozarow, L.H., Shamai (Shitz), S., Wyner, A.D., 1994. Information theoretic considerations for cellular mobile radio. 43 (10), 359–378.

Papadimitriou, C.H., 2001. Algorithms, games, and the internet. In: ACM Proceedings on the 33rd Annual Symposium on Theory of Computing.

Pemantle, R., 1990. Nonconvergence to unstable points in urn models and stochastic approximations. Ann. Prob 18, 698–712.

Perlaza, S.M., Belmega, E.V., Lasaulce, S., Debbah, M., 2009a. On the base station selection and base station sharing in self-configuring networks. In: 3rd ICST/ACM International Workshop on Game Theory in Communication Networks.

Perlaza, S.M., Belmega, E.V., Lasaulce, S., Debbah, M., 2009b. On the base station selection and base station sharing in self-configuring networks. In: ACM/ICST Proceedings of the 4th International Conference on Performance Evaluation Methodologies and Tools (VALUETOOLS). Pisa, Italy.

Perlaza, S.M., Tembine, H., Lasaulce, S., 2010a. How can ignorant but patient cognitive terminals learn their strategy and utility? In: IEEE Proceeding of the 11th International Workshop on Signal Processing Advances for Wireless Communications (SPAWC).

Perlaza, S.M., Tembine, H., Lasaulce, S., Debbah, M., 2010b. Satisfaction equilibrium: a general framework for QoS provisioning in self-configuring networks. In: The IEEE Global Communications Conference (GLOBECOM). Miami, FL.

Pontryagin, L.S., 1967. Linear differential games. I. Soviet Mathematics. Doklady, 8, 769–771.

Poojary, N., Krishnamurthy, S.V., Dao, S., 2001. Medium access control in a network of ad hoc mobile nodes with heterogeneous power capabilities. In: IEEE Proceedings of the International Conference on Communication (ICC). pp. 872–877.

Rabin, M., 1993. Endogenous preferences in games. Am. Econ. Rev. 83, 1281–1302.

Ramesh, S., Rosenberg, C., Kumar, A., 1996. Revenue maximization in ATM networks using the CLP capability and buffer priority management. IEEE/ACM Trans. Netw. 4 (6), 941–950.

Renault, J., Tomala, T., 1998. Repeated proximity games. Int. J. Game Theory 27, 539–559.

Reny, P.J., 1999. On the existence of pure and mixed strategy Nash equilibria in discontinuous games. Econometrica, 67.

Robbins, H., Monro, S., 1951. A stochastic approximation method. Ann. Math. Stat. 22 (3), 400–407.

Robinson, J., 1951. An iterative method of solving a game. Ann. Math. 54, 296–301.

Rodriguez, V., 2003. An analytical foundation for resource management in wireless communication. In: IEEE Proceedings of Globecom. pp. 898–902.

Rose, L., Medina Perlaza, S., Lasaulce, S., Debbah, M., 2011. Learning equilibria with partial information in wireless networks. IEEE Commun. Mag. (Special Issue on Game Theory for Wireless Networks).

Rosen, J., 1965. Existence and uniqueness of equilibrium points for concave n-person games. Econometrica, 33, 520–534.

Rosenberg, D., Solan, E., Vieille, N., 2007. Social learning in one-arm bandit problems. Econometrica 75 (6), 1591–1611.

Ross, S., Chaib-draa, B., 2006. Satisfaction equilibrium: achieving cooperation in incomplete information games. In: 19th Canadian Conference on Artificial Intelligence.

Roughgarden, T., 2005. Selfish Routing and the Price of Anarchy. MIT Press.

Roughgarden, T., Tardos, E., 2002. How bad is selfish routing? J. ACM (JACM) 49, 236–259.

Roughgarden, T., Tardos, E., 2004. Bounding the inefficiency of equilibria in nonatomic congestion games. Games Econ. Behav. 47 (2), 389–403.

Saad, W., Han, Z., Debbah, M., Hjørungnes, A., 2008. Network formation games for distributed uplink tree construction in IEEE 802.16j networks. In: IEEE Global Telecommunications Conference (GLOBECOM). New Orleans, LA, November 2008.

Sahin, O., Erkip, E., 2007a. Achievable rates for the gaussian interference relay channel. In: Proceedings of the IEEE Global Communications Conference (GLOBECOM'07). Washington DC, pp. 786–787.

Sahin, O., Erkip, E., 2007b. On achievable rates for interference relay channel with interference cancellation. In: Proceedings of the IEEE Annual Asilomar Conference on Signals, Systems and Computers (invited paper). Pacific Grove, CA, pp. 805–809.

Sandholm, W.H., 2001. Potential games with continuous player sets. J. Econ. Theory 97, 81–108.

Sandholm, W.H., 2010. Population Games and Evolutionary Dynamics. MIT Press.

Sandholm, W. H., 2011. Population Games and Evolutionary Dynamics. Economic Learning and Social Evolution. MIT Press.

Saraydar, C.U., Mandayam, N.B., Goodman, D.J., 2002. Efficient power control via pricing in wireless data networks. IEEE Trans. Commun. 50 (2), 291–303.

Sastry, P.S., Phansalkar, V.V., Thatchar, M.A.L., 1994a. Decentralized learning of Nash equilibria in multi-person stochastic games with incomplete information. IEEE Trans. Syst. Man Cybern. 24, 769–777.

Sastry, P.S., Phansalkar, V.V., Thathachar, M.A.L., 1994b. Decentralized learning of Nash equilibria in multi-person stochastic games with incomplete information. IEEE Trans. Syst. Man Cybern. 24 (5), 769–777.

Savage, L.J., 1954. Foundations of Statistics, first ed. John Wiley, New York.

Scheinerman, E.R., 1995. Invitation to Dynamical Systems. Prentice Hall College Div.

Starr, A.W., Ho, Y.C., 1969. Nonzero-sum differential games. J. Opt. Theor. Appl. 3, 184–206.

Schuster, S., Kreft, J.-U., Schroeter, A., Pfeiffer, T., 2008. Use of game-theoretical methods in biochemistry and biophysics. J. Biol. Phys. 34 (1–2), 1–17.

Scutari, G., Barbarossa, S., Palomar, D.P., 2006. Potential games: A framework for vector power control problems with coupled constraints. In: International Conference on Acoustics, Speech and Signal Processing (ICASSP), May 2006.

Scutari, G., Palomar, D.P., Barbarossa, S., 2008a. Competitive design of multiuser MIMO systems based on game theory: A unified view. IEEE J. Sel. Areas Commun. Special Issue on game Theory, Vol. 26, 1089–1103.

Scutari, G., Palomar, D.P., Barbarossa, S., 2008b. Optimal linear precoding strategies for wideband non-cooperative systems based on game theory – Part I: Nash equilibria. IEEE Trans. Signal Process. 56, 1230–1249.

Scutari, G., Palomar, D.P., Barbarossa, S., 2008c. Optimal linear precoding strategies for wideband non-cooperative systems based on game theory – Part II: Algorithms. IEEE Trans. Signal Process. 56, 1250–1267.

Scutari, G., Palomar, D.P., Barbarossa, S., 2009. The MIMO iterative waterfilling algorithm. IEEE Trans. Signal Process. 57 (5), 1917–1935.

Selten, R., 1965. Spieltheoretische behandlung eines oligopolmodells mit nachfragetragheit. Zeitschrift fur die gesamte Staatswissenschaft. 12, 201–324.

Selten, R., 1975. Reexamination of the perfectness concept for equilibrium points in extensive games. Int. J. Game Theory, 4 (1), 25–55.

Sendonaris, A., Erkip, E., Aazhang, B., 2003. User cooperation diversity – Part II: implementation aspects and performance analysis. IEEE Trans. Commun. 51 (11), 1939–1948.

Shah, V., Mandayam, N.B., Goodman, D.J., 1998. Power control for wireless data based on utility and pricing. In: IEEE Proceedings of the 9th International Symposium on Indoor and Mobile Radio Commun. (PIMRC), vol. 3, pp. 1427–1432.

Shapley, L.S., 1953a. Contributions to the Theory of Games II, A Value for n-Person Games. In: Annals of Mathematics Studies, vol. 28. Princeton University Press, pp. 307–317.

Shapley, L., 1953b. Stochastic games. Proc. Natl. Acad. Sci. USA 39 (10), 1095–1100.

Shapley, L.S., 1964. Some topics in two-person games. In: Dresher, M., Shapley, L.S., Tucker, A.W. (Eds.), Advances in Game Theory. Princeton University Press, Princeton, pp. 1–28.

Shen, H.-X., Basar, T., 2004. Incentive-based pricing for network games with complete and incomplete information. In: Proceedings of the 11th International Symposium on Dynamic Games and Applications, December 2004.

Shen, H.-X., Basar, T., 2006. Hierarchical network games with various types of public and private information. In: IEEE Proceedings of the 45th Conference on Decision and Control. pp. 2825–2830.

Silva, A., Bernhard, P., Altman, E., 2008. Numerical solutions of continuum equilibria for routing in dense ad hoc networks. In: Workshop on Interdisciplinary Systems Approach in Performance Evaluation and Design of Computer & Communication Systems (InterPerf). Athens, Greece, October 2008.

Simon, H.A., 1972. Theories of bounded rationality. In: McGuire, C.B., Radner, R. (Eds.), Decision and Organization. North-Holland Publishing Company, pp. 408–423.

Smart, D.R., 1974. Fixed Point Theorems. Cambridge University Press, Cambridge.

Smith, M.J., 1984. The stability of a dynamic model of traffic assignment – an application of a method of Lyapunov. Transp. Sci. 18, 245–252.

Smith, M., Price, G.R., 1973. The logic of animal conflict. Nature 246, 15–18.

Special issue. 2008. Game theory in communication systems. IEEE J. Sel. Areas Commun. 26 (7).

Sobel, M., 1971. Noncooperative stochastic games. Ann. Math. Stat. 42, 1930–1935.

Sobel, M., 1973. Continuous stochastic games. J. Appl. Prob. 10, 597–604.

Solan, E., 1998. Discounted stochastic games. Math. Oper. Res. 23, 1010–1021.

Sorin, S., 1992. Repeated games with complete information. In: Handbook of Game Theory with Economic Applications, vol. 1. Elsevier Science Publishers.

Sorin, S., 2002. A First Course on Zero-Sum Repeated Games. In: Mathématiques et Applications, vol. 37. Springer.

Srivastava, V., Neel, J., MacKenzie, A.B., Menon, R., DaSilva, L.A., Hicks, J.E., et al., 2005. Using game theory to analyze wireless ad hoc networks. IEEE Commun. Surv. Tutorials 7 (4), 46–56.

Stackelberg, V.H., 1934. Marketform und Gleichgewicht.

Stanford Encyclopedia of Philiosophy. 2007. Prisoner's dilemma.

Starr, Ho, 1969.

Takahashi, M., 1962. Stochastic games with infinitely many strategies. J. Sci. Hiroshima Univ. Ser. A-I 26, 123–134.

Tarski, A., 1955. A lattice-theoretic fixed point theorem and its applications. Pac. J. Math. 5, 285–309.

Taylor, P., Jonker, L., 1978. Evolutionary stable strategies and game dynamics. Math. Biosci. 40, 145–156.

Telatar, E., 1999. Capacity of multi-antenna Gaussian channels. Eur. Trans. Telecommun. ETT 10 (6), 585–596.

Tembine, H., 2009a. Population games in large-scale networks: time delays, mean field and applications. In: LAP Lambert Academic Publishing, forthcoming.

Tembine, H., 2009b. Population games with networking applications. In: Ph.D. Dissertation, University of Avignon.

Tembine, H., Altman, E., El-Azouzi, R. Hayel, Y., 2008a. Evolutionary games with random number of interacting players with application to access control. In: Proceedings of WiOpt.

Tembine, H., Altman, E., El-Azouzi, R., Hayel, Y., 2009a. Evolutionary games in wireless networks. In: IEEE Transactions on Systems, Man, and Cybernetics, Part B. Special Issue on Game Theory.

Tembine, H., Altman, E., El-Azouzi, R., Sandholm, W.H., 2008b. Evolutionary game dynamics with migration for hybrid power control in wireless communications. In: 47th SIAM/IEEE CDC.

Tembine, H., Lasaulce, S., Jungers, M., 2010. Joint power control-allocation for green cognitive wireless networks using mean field theory. In: IEEE Proceedings of the 5th International Conference on Cogntitive Radio Oriented Wireless Networks and Communications (CROWNCOM), Cannes, France.

Tembine, H., Le Boudec, J.Y., El-Azouzi, R., Altman, E., 2009b. Mean field asymptotics of Markov decision evolutionary games and teams. In: Proceedings of the IEEE Gamenets.

Teschl, G., 2010. Ordinary differential equations and dynamical systems. Available at http://www.mat.univie.ac.at/gerald/ftp/book-ode/.

Thathachar, M.A.L., Sastry, P.S., 1991. Learning automata in stochastic games with incomplete information. In: Madan, R.N., Viswanadham, N., Kashyap, R.L. (Eds.), Systems and Signal Processing. New Delhi, pp. 417–434.

Thathachar, M.A.L., Sastry, P.S., 2003. Networks of Learning Automata: Techniques for Online Stochastic Optimization. Springer.

Thuijsman, F., Raghavan, T., 1997. Perfect information stochastic games and related classes. Int. J. Game Theory, 26 (3), 403–408.

Tian, G., 2009. The existence of equilibria in games with arbitrary strategy spaces and payoffs: a full characterization. In: Levine, D.K. (Ed.), Levine's Working Paper Archive.

Tomala, T., 1998. Pure equilibria of repeated games with public observation. Int. J. Game Theory 27 (1), 93–109.

Topkis, D.M., 1979. Equilibrium points in non-zero sum n-person submodular games. SIAM J. Cont. Optim. 17 (6), 773–787.

Topkis, D.M., 1998. Supermodularity and Complementarity. Princeton University Press.

Touati, C., E. Altman, E., Galtier, J., 2006. Generalized Nash bargaining solution for bandwidth allocation. Comp. Nets. (Elsevier), 50, 3242–3263.

Toumpis, S., 2006. Optimal design and operation of massively dense wireless networks: or how to solve 21st century problems using 19th century mathematics. In: Workshop on Interdisciplinary Systems Approach in Performance Evaluation and Design of Computer & Communication Systems (InterPerf).

Treust, M.L., Lasaulce, S., 2010a. A repeated game formulation of energy-efficient decentralized power control. IEEE Trans. Wireless Commun. 9 (9), 2860–2869.

Treust, M.L., Lasaulce, S., 2010b. Approche Bayésienne pour le contrôle de puissance décentralisé efficace énergétiquement. In: Proceedings of ROADEF (Société Franaise de Recherche Opérationnelle et d'Aide à la Décision.).

Treust, M.L., Lasaulce, S., Debbah, M., 2010a. Implicit cooperation in distributed energy-efficient networks. In: IEEE Proceedings of 4th International Symposium on Communication, Control, and Signal 012 Processing.

Treust, M.L., Tembine, H., Lasaulce, S., Debbah, M., 2010b. Coverage games in small cells networks. In: IEEE Proceedings of the Future Network and Mobile Summit (FUNEMS). Florence, Italy.

Tulino, A., Verdú, S., 2004. Random matrix theory and wireless communications. Foundations and Trends in Communications and Information Theory. Now Publishers Inc., The Essence of Knowledge.

Tulino, A., Verdú, S., 2005. Impact of antenna correlation on the capacity of multiantenna channels. IEEE Trans. Inf. Theory 51 (7), 2491–2509.

Vidal, J., 2006. Heart and Soul of the City. The Guardian.

Vieille, N., 2000a. Equilibrium in 2-person stochastic games I: a reduction. Isr. J. Math. 119, 55–91.

Vieille, N., 2000b. Equilibrium in 2-person stochastic games II: the case of recursive games. Isr. J. Math. 119, 93–126.

Vieille, N., 2002. Stochastic games: recent results. In: Aumann, R.J., Hart, S. (Eds.), Handbook of Game Theory with Economic Applications, vol. 3, pp. 1833–1850.

Ville, J., 1938. Sur la théorie générale des jeux où intervient l'habileté des joueurs. Borel's Applications de la théorie des probabilités aux jeux de hasard. Section 6, pp. 105–113.

von Neumann, J., Morgenstern, O., 1944. Theory of Games and Economic Behavior. Princeton University Press, Princeton.

von Stackelberg, H., 1952. The Theory of the Market Economy. Oxford University Press, Oxford, England.

Voorneveld, M., 2000. Best-response potential games. Econ. Lett. 66 (3), 289–295.

Waldegrave, J., 1713. Letter from James Waldegrave to Pierre-Remond de Montmort.

Wardrop, J., 1952. Some theoretical aspects of road traffic research. Proc. Inst. Civ. Eng. Part II, 1, 325–378.

Weibull, J., 1997. Evolutionary Game Theory. The MIT Press.

Wichardt, P.C., 2008. Existence of Nash equilibria in finite extensive form games with imperfect recall: a counterexample. Games Econ. Behav. 63 (1), 366–369.

Willard, G.J., 1978. On the equilibrium of heterogeneous substances. Conn. Acad. Sci. 1875–1878.

Willems, F.M.J., 1983. The discrete memoryless multiple access channel with partially cooperating encoders. IEEE Trans. Inf. Theory IT-29 (3), 441–445.

Wu, H.-Yi., Miao, X.-N., Zhou, X.-W., 2010. A cooperative differential game model based on transmission rate in wireless networks. Oper. Res. Lett. 38, 292–295.

Wu, Y., Wang, B., Liu, K.J.R., Clancy, T.C., 2009. Repeated open spectrum sharing game with cheat-proof strategies. IEEE Trans. Wireless Commun. 8 (4), 1922–1933.

Wyner, A.D., 1974. Recent results in Shannon theory. IEEE Trans. Inf. Theory 20 (1), 2–10.

Xing, Y., Chandramouli, R., 2008. Stochastic learning solution for distributed discrete power control games in wireless data networks. IEEE/ACM Trans. Netw. 16 (4), 932–944.

Yao, D., 1995. S-modular games with queueing applications. Queueing Syst. 21, 449–475.

Yates, R.D., 1995. A framework for uplink power control in cellular radio systems. IEEE J. Sel. Areas Commun. 13 (7), 1341–1347.

Yee, G., Korba, L., Song, R., Chen, Y.-C., 2006. Towards designing secure online games. In: IEEE Proceedings of the 20th International Conference on Advanced Information Networking and Applications (AINA).

Yeung, D.W.K., Petrosyan, L.A., 2010. Cooperative Stochastic Differential Games. Springer Series in Operations Research and Financial Engineering.

Young, H.P., 1994. Cost allocation. In: Aumann, R.J., Hart, S. (Eds), Handbook of Game Theory with Economic Applications, vol. 2. Elsevier, pp. 1194–1235 (Chapter 34).

Young, H.P., 2004. Strategic Learning and Its Limits. Oxford University Press.

Young, H.P., 2009. Learning by trial and error. Games Econ. Behav. Elsevier, 65 (2), 626–643.

Yu, W., Ginis, G., Cioffi, J.M., 2002. Distributed multiuser power control for digital subscriber lines. IEEE J. Sel. Areas Commun. 20 (5), 1105–1115.

Yu, W., Rhee, W., Boyd, S., Cioffi, J.M., 2004. Iterative water-filling for Gaussian vector multiple-access channels. IEEE Trans. Inf. Theory 50 (1), 145–152.

Zamir, S., 2009. Encyclopedia of complexity and systems science. In: Bayesian Games: Games with Incomplete Information. Springer, New York, pp. 426–441.

Zermelo, E., 1913. Ueber eine Anwendung der Mengenlehre auf die Theorie des Schachspiels. In: Proceedings of the 5th Congress Mathematicians. Cambridge University Press, pp. 501–504.

Zhu, K., Niyato, D., Wang, P., 2010a. Optimal bandwidth allocation with dynamic service selection in heterogeneous wireless networks. In: Proceedings of the 53rd IEEE Global Communications Conference. Miami, Florida.

Zhu, Q., Tembine, H., Basar, T., 2009a. A constrained evolutionary gaussian multiple access channel game. In: The IEEE proceedings of GameNets.

Zhu, Q., Tembine, H., Basar, T., 2009b. Evolutionary game for hybrid additive white gaussian noise multiple access control. In: IEEE Proceedings of Globecom.

Zhu, Q., Tembine, H., Basar, T., 2010b. Heterogeneous learning in zero-sum stochastic games with incomplete information. In: Proceedings 49th IEEE Conference on Decision and Control. Atlanta, GA.

Zhu, Q., Tembine, H., Basar, T., 2010c. Network security configuration: a nonzero-sum stochastic game approach. In: IEEE Proceedings of American Control Conference.

Zimmermann, H., 1980. OSI reference model. IEEE Trans. Commun. COMM-28 (4), 425.

Index

Printed and bound by CPI Group (UK) Ltd, Croydon, CR0 4YY

03/10/2024

01040313-0010